Modern Risk Quantification in Complex Projects

Modern Risk Quantification in Complex Projects

Non-linear Monte Carlo and System Dynamics Methodologies

YURI G. RAYDUGIN

OXFORD
UNIVERSITY PRESS

OXFORD
UNIVERSITY PRESS

Great Clarendon Street, Oxford, OX2 6DP,
United Kingdom

Oxford University Press is a department of the University of Oxford.
It furthers the University's objective of excellence in research, scholarship,
and education by publishing worldwide. Oxford is a registered trade mark of
Oxford University Press in the UK and in certain other countries

First Edition published in 2020

Impression: 1

Published in the United States of America by Oxford University Press
198 Madison Avenue, New York, NY 10016, United States of America

British Library Cataloguing in Publication Data
Data available

Library of Congress Control Number: 2020940963

ISBN 978–0–19–884433–4

Printed and bound by
CPI Group (UK) Ltd, Croydon, CR0 4YY

To us

Foreword

This book by Dr. Yuri Raydugin addresses a serious concern of, and challenge to, those of us in the profession of project risk analysis and management who use Monte Carlo simulation (MCS) to analyze the schedule and cost risk of large, complex industrial projects.[1] From the author: "The purpose of this book is to re-discover the power of the Monte Carlo modeling methodology in its non-linear version and unleash it upon complex projects."[2]

One thread of his argument, as Dr. Raydugin notes, is that simulation results for project schedule and cost risk of overruns have been rather timid and lacking in reality when they are compared to actual results of such projects over the years. Risks specified in Monte Carlo simulation of schedule and cost models are typically considered in isolation, or if not, it is not known how to express their interactions that influence the project in non-linear ways. In short, Monte Carlo simulation exercises usually underestimate the probability and magnitude of project schedule and cost overruns because they overlook the risk interaction factors that drive this book.

Dr. Raydugin puts the cause of these timid results to the linear nature of the standard Monte Carlo approach to handling risks individually. First, he argues that many risk interactions should be understood as leading to non-linear responses. He posits that risks can interact with each other in identifiable ways that make their joint impact greater than the sum of their individual impacts on schedule and cost.

Therefore, if we implement non-linear reactions to risks the results will be more realistic than typical simulation-based risk analysis exercises. The methods that Dr. Raydugin proposes will not be welcomed by many project mangers, owners or contractors, not because they produce incorrect results but because their results may be so pessimistic as to kill projects before the final investment decision (FID). There is other research cited in his references (Flyvbjerg (2011)) that typical projects tend to overrun their risk analyses results sufficiently to promote great displeasure of their stakeholders.

Many of the projects we have "risked" over the years have been large ones. In each of these we try to represent clearly and honestly the degree of risk of schedule and cost overruns using advanced risk data collection and modeling techniques based on Monte Carlo simulation of schedule and/or cost models.

[1] This Foreword addresses the quantitative risk analysis based on Monte Carlo simulation. The author also deals with Systems Dynamics methodology. It provides valuable insight on non-linear interactions in project systems, could be used as a standalone risk quantification method but is not reviewed here.

[2] Many quotations, sentences, phrases or paraphrased have been taken from the book and are included in this Foreword usually without quotation marks. Dr. Raydugin's words often cannot be improved upon.

In 2006 we worked with Pertmaster to develop the Risk Drivers method (they called it the Risk Factors Module) of representing the impact of risks using multiplicative factors to represent the impact of any risk on activity durations and costs. With Risk Factors, if a risk occurs on any iteration, its impact multiplies the activities' durations in the schedule rather than adding days to them. If multiple risks occurred in the same iteration their impacts multiplied each other.

Early on, around 2009, we realized that we were assigning multiple highly (e.g., from 70% to 100%) likely threat-type risks to many of the same activities. During simulation, on many iterations several risks occurred, and their values impacts were multiplied together to produce non-linear consolidated multipliers. This effect delivered activity durations that seemed to be unrealistically long. This multiplication of numbers greater than 1.0 was serendipitous because we had no underpinning of a theory that caused the risks other than a general belief that impacts of multiple risks may behave in a multiplicative non-linear way. In fact, our response to challenges concerning those activity durations that were too long was to ask Monte Carlo software providers to give us the option of placing the risks in parallel or in series based on the type of risk. The net result provided us a way to avoid the message that Monte Carlo simulation was trying to tell us.

A second thread that is captured in Dr. Raydugin's book is the discussion of project systemic conditions. In recent years there have been articles, presentations and books about the need to implement systems thinking. The project itself as well as delivering the project (see his references to Merrow (2011), for example) are both best thought of as systems that must interact jointly to be successful.

Systemic risks or project system issues are conditions that exist and affect the success or lack thereof of the project outcomes including schedule, cost and quality. Such systems include the strength of the owner's system of managing successfully the delivery of the project through a project team that is large and experienced enough to handle the project. Other systemic conditions include the design and execution of a project before having thoroughly thought it out or its engineering challenges resolved. These also include a lack of understanding the challenges of performing the project in its socio-economic-political environment.

These systemic conditions are broad, pervasive and unlikely to be overcome in the space/time of any one project. These project delivery failings can magnify the impact of project-specific risks above those risks if they occur in well-resourced and managed project.

The third thread that Dr. Raydugin addresses is the increased size and complexity of the projects we analyze. Such projects are often called "mega-projects" or "complex projects." Project complexity is sometimes referred to as large projects with many interdependencies in the presence of ambiguity (see the International Center for Complex Project Management). Another word for ambiguity, of course, is uncertainty and risk. These forces often make large projects more prone than simpler projects to failing. The sources of ambiguity are many and persistent. With many interdependencies these risks tend to spread their influence throughout the project structure in unpredicted and unprepared-for ways. Megaprojects are sometimes considered fragile because they are exposed to many failure events that have knock-on influences beyond their direct consequences.

Dr. Raydugin brings a novel but coherent approach to the analysis of complex projects that are potentially challenged by risks to schedule and to cost. His basic approach is to look at performing projects as the interplay between two systems. His view of systems is summarized as two main coexisting and interdependent subsystems.

- PSS or Project Structure Subsystem, describing the structural complexity of what should be delivered, the complexity of what the project is and how it must be planned for, executed and commissioned before turn-over.
- PDS or Project Delivery Subsystem, the structure of organizing, planning and fielding resources and business systems to deliver the project.

The concept of a mismatch between PSS – PDS is fundamental to understanding the risk of any complex project. If these two subsystems are not balanced, with delivery tools adequate and available to deliver the components and the entirety of the complex project structure necessitated by the objective, the project execution will suffer in ways that can be represented by their characteristics and expressed by revised bowtie diagrams of causes, risks and effects and by mathematical equations reflecting the structure of the revised bowtie.

That these two subsystems are highlighted in this book is not to the exclusion of other systems that impinge on project success. For example, Dr. Raydugin mentions a third project subsystem could be introduced – external environment subsystem (EES). It is to be believed that mismatch of any involved system could be examined and included in the explanation of nonlinear behavior of project cost and schedule. In fact, Risk Breakdown Structures (RBS) shown in the PMBOK® Guide of the Project Management Institute typically include technical, external, organizational and even project management as sources of risk to the project.

Any misalignment or imbalance between the two subsystems, PDS and PSS, can be understood as a source of dynamic behavior of an entire project system. Dr. Raydugin focuses on three general types of risks and represents their effect using the revised bowtie diagrams.

- Dynamic risk-induced project system issues due to occurrence of standalone non-interacting project risks that occur and interact with the risk-induced dynamic PSS – PDS mismatches.
- One type of the revised bowtie diagram describes the internal risk amplification when a given standalone risk is amplified in presence of chronic project system PSS – PDS mismatch. Hurricanes or sandstorms that took place previously in the project location area could be treated as regular risk causes. Another example could be a need for adopting a new technology by a project. But, in the presence

of a PSS – PDS mismatch the effect of the regular risk impact on schedule and cost is raised to a power greater than 1.0.

- Dynamic risk-interaction-induced project system issues due to knock-on interactions with a risk that occurs in a later project phase. This normally happens when quality risks occur in early project stages, for instance errors in engineering, manifest themselves with a consequent delay during commissioning or project start-up. This situation gives rise to a late rework cycle, which the author demonstrates in chapters dedicated to System Dynamic modeling.
- The impact on the quality objective inflicted by the first risk causes an additionally induced dynamic project system issue for the subsequent risk: it serves as a dynamic systematic cause with a multiplicative factor greater than 1.0 on the impact parameters for the down-the-line risk. Flawed engineering found during commissioning in the presence of a weak commissioning team would have a non-linear effect toward the end of the project.
- Dynamic risk-interaction-induced project issues associated with the dynamic PSS – PDS mismatches if two risks occur together, due to compounding. The resulting impacts on schedule and cost objectives are expected to be higher than sums of individual impacts. In practice, more than two project risks may get compounded depending on the dynamic risk pattern supported by a project system.

The three-point estimates of impact (low, most likely, high days added if the risk occurs) that are found in the risk registers address the one risk independently and do not recognize any impact on impact of a PSS – PDS system mismatch. In Dr. Raydugin's book, the typical risk register assumes a linear project environment rather than recognizing a possible non-linear project situation. The impact multipliers are usually sourced from the project's risk register before they are enhanced by non-linear concerns, and they are augmented by non-linear considerations.

Each of the three-point impact values from the risk register is altered by multiplicative factors that reflect the intensity of the interaction on the risk's consequences caused by the mismatch of PSS and PDS in ways that Dr. Raydugin examines. In general, if the system mismatch influence on the value of a risk's impact on schedule and cost is significant, each of the three-point range estimates of impact grows multiplicatively with some exponent, in other words, non-linearly.

If the impact of the PSS – PDF mismatch is missing or faint, the multiplicative factor is zero (no interaction of the mismatch with the risk) or small and the risk problem reduces to the linear case.

Applying non-linear factors reflecting the project and its delivery systems' mismatch adjusts those three-point estimates by a factor greater than 1.0 before Monte Carlo simulation. Because the augmentation function is non-linear, the multiplicative effect on the three points of the impact distribution is ordered by low < most likely < high. In other words, depending on the degree of non-linearity the resulting relative impacts on coefficients before simulation for the high case might be significantly greater than that for the moderate case. This effect increases the skew of the distribution of impact factors from which the simulation software chooses for each iteration.

The equations that are developed to change the linear risk register three-point estimates for activity durations and for cost line items into parameters reflecting internal amplification, knock-on, and compounding non-linearity parameters are developed by Dr. Raydugin depending on the type of interaction represented. The equations that he proposes derive from the revised risk-interaction-specific bowtie diagrams he develops and presents. Their derivation from analytical bowtie structures to mathematical expressions is left to the reader to discover in the book.

Dr. Raydugin specifies that the non-linear parameters in the model should be calibrated to establish values of non-linearity parameters corresponding to 10% and 25% above P-70 duration and cost results in the linear case model.[3] It is suggested that overruns beyond the desired level of 10% or less are acceptable, those between 10% and 25% are to be addressed aggressively, and those over 25% represent failed projects.

Several runs of the modified risk register are undertaken to develop the non-linear parameters that would produce the 10% and 25% result. Each run has one value of non-linearity parameters for all schedule risk interactions. This allows the analyst to pin down values of non-linearity parameters that give rise to exactly 10% and 25% of duration and cost overruns, correspondingly. Results of 10% overrun or less are associated with a mature project delivery subsystem (PDS), those between 10% and 25% are viewed as an ordinary maturity delivery subsystem and those that overrun of more than 25% of the desire level of confidence are viewed as the result of an immature PDS.

Finally, there is a review by project participants and risk practitioners of the assessed risk interactions for a project. The review looks at each risk, keeping in mind calibration results that show which non-linearity parameter ranges correspond to a mature, ordinary, and immature project system characterized by weak, medium, and strong static and/ or dynamic PSS – PDS mismatches, correspondingly. Non-linearity parameters are assessed for each risk. These assessments are done based on observations and opinions of project risk specialists and available empirical data from previous projects.

The introduction of expert judgment is implemented by the approach that structures the discussion in specific ways, thus conditioning the data gathering on a few items per risk. In the non-linear case the collected information is used to define non-linearity parameters and non-linear multipliers. This approach provides a well-defined structure for the collection of risk information. The described structure allows the risk analyst to focus expert judgements on the PDS – PSS mismatch that will reduce bias. Individuals are required to assess very specific and defined risk exposure features.

The non-linear impact parameters from two examples of the risks applied to the book's case study are shown below. This table compares, for two example risks, the parameters used in a linear model that is based on the risk register and the parameters that consider intra-risk and cross-risk interactions (a non-linear model). The

[3] The organization can choose any desired level of confidence, but the P-70 is used for illustration.

non-linear impact parameters are larger than the risk register's minimum, most likely and maximum values.

Example: two risks linear and non-linear impact parameters

			Schedule impact parameters (days)			
Risk examples	**Type of model**	**Probability**	**Min**	**Most likely**	**Max**	**Non-linear adjustments made**
R02	Linear	20 %	10	20	40	None
R02	Non-linear	20 %	14	40	134	Internal amplification and compounding with R01
R04	Linear	20 %	30	50	70	None
R04	Non-linear	20 %	53	106	273	Internal amplification, knock-on from R03 and compounding with R05

One benefit of this approach that relies heavily on the mismatch of the project structure subsystem (PSS) and the project delivery subsystem (PDS) is the searchlight that it puts on the need to take care of the issues of the project delivery subsystem. The method requires comparing that subsystem with the project's inherent complexity. Often the project owner does not give the delivery subsystems/ staffing chart a cold-eyed review compared to the project complexity, thinking that they can get away with fewer indirect staff and save money. The consequences of that approach are that the staff is spread thin and becomes ineffective and that known issues will not be staffed consistently with the challenges and consequences of failure.

The discipline of considering the relationship of risks in amplification, knock-on and compounding dimensions focuses the mind on the task of revising the linear risk register impacts. Two examples can illustrate this point:

- Having a weak or inexperienced team will lead to complex project interdependencies' not being anticipated. In the face of complex projects, the team needs to be large and experienced enough to meet the challenges of the whatever complexity is reflected in the PSS. The team needs to be both flexible and adaptive, since new issues arise, and the team needs to identify and respond appropriately. If this condition exists, ordinary linear risks become non-linear in their impact and the project can get out-of-hand quickly without any chance to recover.
- Starting execution before all technical problems are resolved or progressing to the field with incomplete or incorrect ready-for-construction (RFC) drawings causes quality challenges that may be found too late when it is expensive in time and money to fix. This risk can be more avoided by increased QC/QA systems and manpower.

Risk addressing the project team's strength and maturity facing the project's complexity will pay great dividends since it affects the intensity of many risks' impacts. According to Dr. Raydugin, a complex project should be an adaptive and flexible system to address project risks effectively through avoiding over-planning and, as a result, to ensure the project's success.

In summary, Dr. Raydugin is providing us with both a challenge and a developed approach.

The fundamental and welcome challenge is to focus our attention on the strategic issues that can cause project delivery systems to be inadequate in handling the project's complexity. This 10,000-foot perspective is often forgotten in gathering data for a risk register or augmenting those data even with confidential interviews.

This strategic perspective will focus attention on fundamental issues of the project delivery subsystem when compared to the project's complexity that may be embarrassing to the organization or difficult to address. In practice many people "don't want to hear" about these. Also, these system issues are the most difficult to deal with since they may involve fundamental changes to the organization, current staff and its cultural attitude toward project risk. Without identifying this mismatch many risk practitioners focus on lower-level less-important standalone risks in their comfort zone.

The approach in this book is based on a well-documented, mature thought process that is focused on the ways in which systems' mismatch can magnify the effects of identified risks that drive the Monte Carlo simulation approach through amplification, knock-on and compounding relationships.

It is important to state that once the three-point estimates of impacts are adjusted to reflect the non-linear influences of the risks on the project, then the Monte Carlo simulation method is the standard one, using existing software, that are familiar to most quantitative risk analysis practitioners.

Dr. Raydugin has identified and reduced to understandable equations the three effects' affecting the influence on the impact of risks. Once the equations are developed the risk data collection can be focused on the most important issues and in a productive way. This way leads to parameters that can be understood as derived from direct explained equations. He has developed a framework and followed it to its beneficial conclusions.

The application of the equations leads to identifying the non-linear factors that traditional Monte Carlo applications miss. Sometimes analysts attempt to correlate risks, but that just means that they are associated, probably because they are symptoms of a common root cause risk. Any of these traditional methods fails the test that is at the foundation of this important book, that there is a systematic and disciplined model that can be used to aid in developing more realistic parameters to be used in simulation.

This is a transformational book that all project risk analysis practitioners should ponder, understand, and implement when looking at the risk of complex projects.

David T. Hulett, Ph.D., FAACE
Hulett & Associates, LLC
Los Angeles, California 26 December 2019

Preface

A term *zemblanity* is coined by William Boyd (1998) in his novel *Armadillo* that means 'unpleasant unsurprise' or 'unhappy, unlucky and unexpected discoveries occurring by design'. Apparently, this word stems from Novaya Zemlya[1] or 'a New Land'—an archipelago of extremely cold and barren islands in arctic Russia. Anybody who ends up out there would supposedly be in trouble 'by design'.[2]

A virtual land of project management seems to be quite comparable to the Novaya Zemlya environment. Too many projects of the past were completed with significant delays, astonishing overspending, or severe ongoing operational issues for many months after their formal completion (Raydugin (2013)). Many of them were 'unpleasant unsurprises' being predictably expected. A selected list of some of them provided by Flyvbjerg (2017) looks appalling. For example, it includes Sydney Opera House project, Australia, infamous for its 1,400% cost overrun, or Boston Big Dig Artery/Tunnel project, USA, which resulted in 220% overspending, among almost three dozen projects. This list that keeps infrastructure projects mostly could be easily enriched by comparable data for oil and gas, mining, IT, defence, airspace, pharmaceutical, consumer goods, and other projects. Flyvbjerg (2011) even puts forward an iron law of megaprojects: 'over budget, over time, under benefits, over and over again'. Many of these projects could be legitimately referred to as 'zemblanity projects' as their failure seemed to be predictable 'unpleasant unsurprise' happening 'over and over again'. It feels like some characteristics of these projects and a way they were delivered have had severe fundamental intrinsic issues—project system issues.

This opinion is well-aligned with observations by Merrow (2011). In addition to profound analysis of major project failure drivers based on regression analyses used in descriptive statistics, Merrow (2011) points out that cost contingency development methods are notoriously and optimistically inadequate in many cases: actual project cost outcomes are much higher than approved costs.

A simple litmus test should be proposed to verify if a particular risk quantification method is good enough. Namely, any adequate risk quantification method should be capable of predicting the outcomes of notorious projects like Sydney Opera House or Boston Big Dig Artery/Tunnel project cited above with a decent accuracy. Similarly, outcomes of projects with abnormally high delays should be predictable too. Otherwise the method should be either modified or played down.

[1] Strictly speaking, the right term should be 'zemlyanity' deriving from 'Zemlya' which became transformed to 'zemblanity' in the course of translation to English. The first reference to this term is done by Vladimir Nabokov (1989) in his 1962 novel *Pale Fire*.

[2] An additional piece of information on the authentic severity of this place is that 224 nuclear bomb tests were conducted up there in the second part of the twentieth century.

The notoriously low accuracy of risk quantification methods and models seems to have three distinctive facets. They might be put together under a project placebo effect umbrella: they do produce some sorts of results that are optimistically inadequate systematically.

The first facet reflects on disconnect of applied risk quantification models with contexts of upcoming projects. As every model has its distance to project reality, it is important to be aware of risk factors that are relevant and included in a risk quantification model. Sometimes it is even more important to pay heed to the factors that are left out. It might turn out that a given risk quantification method is simply irrelevant or inapplicable to a given project either at all or in a particular phase of project development. In any case the produced results could look credible, acceptable, and even encouraging. Their utilization in decision-making could be compared with a positive impact of a placebo effect in medicine which has clearly psychological bias roots. However, this is a somewhat limited comparison as pharmaceutical placebos do not inflict any harm except missing opportunities for a real treatment, whereas project placebos do.

The second facet shines with various realizations of psychological and organizational bias. Any realization of bias can distort risk data collection, model building, and interpretation of results (Raydugin, 2013). A most notorious type of bias—a hidden agenda—is rooted in a desire to make projects look more attractive to stakeholders and decision-makers and to push those to sanctioning. This often results in playing down some critical risks and underestimating of required schedule and cost contingencies on purpose. Flyvbjerg (2017) discusses the Hiding Hand principle linked to a behavioural economics mechanism of deception and self-deception. Essentially, the self-deception means that if decision-makers knew or really wanted to know all relevant risks and real durations and costs from the start, the majority of projects would be never approved. The Hiding Hand is just a mechanism that makes decision-makers less risk-avert in a hope that project risks would fall over time and project benefits should grow miraculously.[3] Is this not another proven form of the project placebo effect?

One can easily discover direct and organic causal links between the project placebo effect, self-deception, and the iron law of mega projects. These links explain quite amazing endurance of some inadequate risk quantification methods. Also, it could not be fully ruled out that some stakeholders and decision-makers might willingly be deceived having their own hidden agendas.

Astonishingly, some books from the author's high-school curriculum on Russian literature ring a bell in this regard. For instance, the prominent poet Alexander Pushkin who is sometimes called the Russian Shakespeare has several romantic lines that seem to allude well to project management. However, this has a simple

[3] As Viktor Chernomyrdin, a former CEO of Gazprom and a prime minister of Russia used to say, 'we tried our best, but it turned out as always'. Apparently, representatives of many other organizations involved in project management could endorse this brilliant aphorism.

explanation as both romances and project management can easily succumb to various although pretty much the same realizations of psychological bias. On one occasion Pushkin (2017) writes in his 1830 poem *The Hero* that 'a deception that elevates us is dearer than a host of low truths'. In his 1826 poem *Confession* he accentuates, that 'it's so easy to deceive me . . . I long to be deceived myself'.

The former citation seems to point to some realizations of a hidden agenda and possibly to strategic misrepresentation as knowing low truths of actual project risk exposure as well as project schedule and cost outcomes beforehand might be slightly undesirable and even distasteful! Whereas the latter seems to connote for a more innocent but still quite harmful self-deception of stakeholders in project decision-making.

It would be worthwhile to get back to the observation by Merrow (2011) and establish a fact that underestimated project contingencies regularly produced by both empirical and probabilistic methods do play into the Hiding Hand and encourage placebo-type self-deception in decision-making. A link between some inadequate risk quantification methods and the Hiding Hand is tracked in this book on several occasions.

A recent example illustrating this link points to a hydro-power generation mega-project in Newfoundland and Labrador, Canada, that the author of this book is intimately aware of. This project succumbed to two types of bias: optimism bias and strategic misrepresentation. Project management and decision-makers elected to establish a cost contingency considered for a final investment decision (FID) in 2012 at a critically low level of around 10% to keep the project economically attractive enough to major stakeholders. The actual project risk exposure was completely ignored. The actual project cost almost doubled since then followed by endless public enquiries and hearings (see e.g. Telegram (2017) or National Post (2018)). The so-called '10% syndrome' in contingency development introduced by Baccarini (2005) is discussed on several occasions too.

The topic on various realizations of psychological and organizational bias is in the background of all discussions in this book. Bias being a systematic error can easily distort inputs and outputs of any risk quantification method as well as interpretation of the results. In many respects, project risk quantification can be distinguished as a branch of political mathematics by analogy with political economics. Various realizations of bias related to risk quantification are revealed openly and honestly in this book despite some topics being rather touchy, although not doing this would permit a much larger room for harmful realizations of bias in project quantification.

Finally, the third important low-accuracy facet is glowing with possible technical shortcomings in risk quantification methods. This covers fundamental deficiencies of risk quantification methods themselves, poor quality of input data, and banal model's errors and omissions. Simply speaking, some statistical models used for risk quantification are not statistically robust. Some empirical methods are not quite empirical using subjective expert judgements. Some Monte Carlo methods miss out very top components of project risk exposure and cannot be accurate by design. Usually these issues are overlooked unintentionally in most cases and may be explained by special types of subconscious bias—cognitive and perceptional—or by a lack of modelling skills and in some cases by a lack of formal mathematical education. In most cases, this again results in a much lower than required schedule or cost contingencies. This unintentional self-deception in a hope to miraculously get adequate risk quantification results is almost irrational although it is well qualified as a project placebo effect too.

Most empirical risk quantification methodologies actually are not risk quantification methodologies at all. They pass up risk identification, addressing, and assessments. This makes claims on their fit for risk quantification over-ambitious: it is difficult to quantify risks without identifying as well as addressing them. Being based on projections of outcomes of previously completed projects onto upcoming projects they are capable of predicting duration or cost outcomes at best that include risks in a blend with a multitude of other factors. These empirical methods can be conditionally called project outcome forecasting methods. Despite this dichotomy between the risk quantification and the outcome forecasting methods, the latter are still reviewed in this book being tagged as unconventional risk quantification methods to reveal their features. The former methods are referred to as conventional risk quantification methods.

This book has two goals. One, to provide an explanation why various risk quantification and outcome forecasting methods are inadequate so often. Two, to put forward a modified probabilistic Monte Carlo methodology that could adequately quantify project risks and accurately forecast cost and schedule contingencies.

Any project is considered a system in this book that comprises two interacting subsystems: project structure subsystem (PSS) and project delivery subsystem (PDS). The former reflects on a project as a deliverable (structural complexity), the latter represents an organization that is to deliver the project (delivery complexity). Ideally PDS should match PSS. Any PSS–PDS mismatch reflects of various static and dynamic project system issues inside the subsystems and in-between them and defines dynamical properties of the project system including non-linear interactions among project risks. The project system approach that is used when modifying Monte Carlo methodology is directly supported by system dynamics modelling. Several system dynamic models that mirror corresponding Monte Carlo models are developed in parallel to better understand non-linear risk interactions in project systems.

When risk interactions are strong enough and cannot be ignored, the traditional linear Monte Carlo methodology cannot be accurate. According to the widely-used project complexity classification, the linear projects can be tagged as simple. Three other types of projects are in the primary focus of this book: complicated, complex, and chaotic. All three non-linear types are referred to as complex for simplicity purposes.

As shown in this book this requires a consistent factoring of various non-linear risk interactions as well as static and dynamic project system issues standing behind them into risk quantification models. These factors are largely overlooked by the traditional Monte Carlo risk quantification methods. The traditional Monte Carlo methodology treats complex projects as simple leaving dynamic features of complex systems out. The dynamic risk patterns stemming from affinities of interacting risks should be consistently factored into the Monte Carlo models. The modified Monte Carlo methodology that does take into account risk interactions is referred to as a non-linear Monte Carlo methodology in this book as opposed to the traditional linear Monte Carlo methodology.

Notably, besides unfriendly Novaya Zemlya, Boyd (1998) also mentions a much friendlier land of Serendip, which is a former name of Ceylon and Sri Lanka. The word *serendipity* originated as a result. In the context of this book, serendipity should mean a happy case of a simple project system that does not possess any issues and does not support any risk interactions. As a result, a traditional linear Monte Carlo methodology could still be successfully utilized that counts standalone non-interacting risks. Unfortunately, the majority of undertaken projects belong to the different category—the category of complex zemblanity-type projects. The newly-developed non-linear Monte Carlo methodology is a helpful navigation tool to survive in the land of zemblanity.

Contents

List of Figures

List of Tables

List of Abbreviations

AACE	Association for Advancement of Cost Engineering
ANN	artificial neural networks
AR	activity ranging
AWP	advanced work packaging
BT	bathtub model
CER	cost estimating relationship
CII	Construction Industry Institute
CPM	critical path method
CRA	cost risk analysis
CSU	commissioning and startup
DCF	discounted cash flow
DJIA	Dow Johns Industrial Average
EES	external environment subsystem
EIA	environmental impact assessment
EPC	engineering, procurement, construction
FAT	factory acceptance tests
FEL	front-end loading
FID	final investment decisions
GAO	US Government Accountability Office
GOSP	gas-oil separation plant
GRNN	generalized regression neural networks
HAZOP	hazard and operability studies
IFB	issued-for-bid
IFC	issued-for-construction
IRBEB	integrated risk-based and economic-based
IRR	internal rate of return
ISBL	inside-battery limit
ISO	International Organization for Standardization
IT	Information Technology
JCL	joint confidence level
LIR	line-item ranging
L-SCRA	linear schedule and cost risk analysis
MLFN	multi-layer feedforward networks
ML	most likely
MVP	most valuable player
NASA	National Aeronautics and Space Administration
NGO	non-governmental organizations
OSBL	outside-battery limit
P&ID	piping and instrumentation diagram
PDRI	project definition rating index
PDS	project delivery subsystem

PERT	programme evaluation and review technique
PFD	process flow diagram
PMC	project management consultancy
PMI	Project Management Institute
PNN	probabilistic neural networks
PO	purchase order
PRM	project risk management
PSS	project structure subsystem
QA/QC	quality assurance, quality control
RAM	risk assessment matrix
RBS	risk breakdown structure
ROI	return on investments
SAT	site acceptance tests
SCRA	schedule and cost risk analysis
SD	system dynamics
SRA	schedule risk analysis
TRA	technology readiness assessments
TRL	technology readiness levels
VAR	value assurance review
WBS	work breakdown structure

List of Models

All workable Monte Carlo and system dynamics models discussed in this book are available on the book's companion website at www.oup.co.uk/companion/Raydugin.

For Chapter 5:

- The linear Monte Carlo schedule and cost risk analysis model L-SCRA

For Chapter 6:

- The system dynamics bathtub model BT1
- The system dynamics bathtub model BT2
- The system dynamics bathtub model BT3
- The system dynamics bathtub model BT4
- The system dynamics bathtub model BT5

For Chapter 7:

- The system dynamics model SD1
- The system dynamics model SD2
- The system dynamics model SD3

For Chapter 15:

- The non-linear Monte Carlo schedule and cost risk analysis model N-SCRA-A
- The non-linear Monte Carlo schedule and cost risk analysis model N-SCRA-B

For Chapter 16:

- The system dynamics model SD4
- The system dynamics model SD5
- The system dynamics model SD4 (with compounding)

For Chapter 17:

- The non-linear Monte Carlo schedule and cost risk analysis model N-SCRA-C

All Monte Carlo models are built using Safran Risk v.7.3. A free trial version of the tool can be retrieved from https://www.safran.com/risk-free-trial to immediately run all included Monte Carlo models.

All system dynamics models are built using Vensim PLE Plus v. 7.3.5 by Ventana Systems Inc. A free version of Vensim PLE could be uploaded from https://vensim.com/free-download/ to immediately run all included system dynamics models.

Introduction

This book is designed as a fundamental upgrade of previously written texts on Monte Carlo risk quantification.[1] A modified non-linear Monte Carlo methodology is developed to adequately forecast project outcomes in complex projects where the traditional Monte Carlo methodology fails.

There are multiple complaints that traditional Monte Carlo risk quantification methods are not accurate in many cases. This failure is associated with a major risk exposure component of complex projects—non-linear risk interactions—that is missed out in the traditional risk quantification models. Simply, the traditional Monte Carlo methodology treats complex projects as simple. Hence, it can be accurate only in the case of simple projects. Non-linear risk interactions and underlying static and dynamic project system issues are in the major focus of this book to carry out the upgrade. Two main types of non-linear risk interactions are considered in this book—intra-risk interactions (amplification of standalone risks) and inter-risk interactions (knock-on and compounding interactions).

For example, the knock-on interactions take place when a risk occurred in the earlier project phase, impacts down-the-line risk due to its delayed discovery. This normally happens when earlier occurred quality risks manifest themselves with a delay, say, during a project startup and give rise to a late rework cycle: project startup phase is infamous for throwing 'banquets of consequences[2]' referring to quality issues overlooked in previous phases.

The multiple failures to predict project outcomes contributed to the iron law of mega projects that are 'over budget, over time, under benefits, over and over again' (Flyvbjerg (2011)). However, cases when project outcomes could be predicted quite precisely do take place every now and then. One such happy example is reviewed by Raydugin (2017a): a project team successfully completed a carbon capture and storage project in Alberta, Canada, a few weeks prior to an approved completion date. Corresponding schedule contingency was developed by the author of this book through applying the traditional linear Monte Carlo methodology.[3] It should be made clear however that that project can be characterized by a highly mature project system (a strong project team, well-developed and implemented procedures, a clearly developed project scope, full support by all main stakeholders, etc.). For this reason, that project could be considered simple, although the vast majority of projects seem to follow the iron law.

[1] These texts include the author's previous book (see Raydugin (2013)).
[2] This is a reference to a 2015 book *A Banquet of Consequences* by Elizabeth George (see George (2015)).
[3] A so-called P50 schedule contingency (confidence level 50 per cent) was approved during the final investment decision assuming a stretched completion date target with one in two chances to complete the project on time or earlier. An actual completion date turned out to be about P42.

Modern Risk Quantification in Complex Projects: Non-linear Monte Carlo and System Dynamics Methodologies. Yuri G. Raydugin, Oxford University Press (2020). © Yuri G. Raydugin.
DOI: 10.1093/oso/9780198844334.001.0001

A topic on project complexity has become very popular recently. Multiple publications are available on the topic that describe in detail various aspects of project complexity. However, a project practitioner might simply ask how the project complexity concept may help manage his or her project. The non-linear Monte Carlo methodology developed in this book is a first part of the answer. It should allow properly managed project risks and develop accurate project contingencies.

The second part of the answer is based on understanding of the postulate that projects as complex systems must be highly adaptive to be successful. This logically should imply a high degree of flexibility in project planning and executing. A rigid over-planning and total control that is often observed in practice is discussed as one of the most severe chronic project system issues. Practical recommendations on making complex projects adaptive systems are outlined in this book including flexible short-term planning to suppress work constraints and ensure work availability.

This book has four target auditoria that define its general approach and structure.

First, it is written for project professionals in any industry and in both private and public sectors.[4] In the context of this book this auditorium could be broken down to two major subgroups. The first bigger subgroup comprises project specialists that are not involved in Monte Carlo risk modelling directly although may deal with risk management in general and take part in risk management workshops including risk quantification workshops occasionally. Project managers, engineers, estimators, schedulers, operations specialists, decision-makers, corporate planners, and any other project specialists are present in this subgroup. The second smaller subgroup of project specialists consists of Monte Carlo modellers who actually build and run Monte Carlo risk models.

Second, this book is written for instructors and students of various certification, undergraduate, graduate, and postgraduate programs that include project management and project risk management in their curricula. As in the case of project professionals, the student's group can be divided to the same two subgroups depending on goals and scope of their studies.

Third, representatives of academia who are involved in project management studies including risk management studies could be interested in this book. Due to the author's background, the book could be viewed as a means to bridge the gap between academia and project risk management practitioners. In particular, academia could be interested in deeper integration of non-linear Monte Carlo methodology with system dynamics. Another intriguing area would be integration and alignment of non-linear Monte Carlo methodology with ANN method. The ANN topic is touched in this book although it is not explored in detail. This could be done by proponents of regression-based methods who should consider ANN a natural upgrade to low-precision regression-based parametric methods.

[4] The author of this book has project risk management experience that almost entirely belongs to the oil and gas industry in North America, Asia, and the Middle East. A combined budget of projects the author took part in as a risk specialist is about $150B. Although general risk quantification principles are universal, representatives of the other industries should adopt these principles keeping in mind the unique context of their projects which may drastically differ from the usual context of typical oil and gas projects.

Finally, this book could be of interest to specialists involved in project disputes and claims resolution including corporate lawyers, members of tribunals handling claims and disputes cases, judges, etc. Notably, system dynamics applications to project management stem from some litigation cases. Additional insights provided by integration of non-linear Monte Carlo methodology with system dynamics approach could be of practical litigation (aka monetary) value to representatives of both project owner and EPC contractor organizations.

The purpose of this book is to re-discover the power of the Monte Carlo modelling methodology in its non-linear version and unleash it upon complex projects. To make the book accessible to both the modellers and non-modellers, the presented information is organized in three layers.

First, the main text is devoted to risk quantification ideas and principles that do not require Monte Carlo modelling experience or experience in mathematics and statistics.

Second, several appendices support development of all Monte Carlo and system dynamics models developed in this book. These appendices might be helpful to better understand main concepts and modelling methods including input data development although without undertaking actual modelling.

Third, all discussed Monte Carlo and system dynamics models are available as supplementary materials on the book companion website at www.oup.co.uk/companion/Raydugin. These models could be retrieved and run using required Monte Carlo[5] and system dynamics[6] software packages. This should further accelerate a steep learning curve though making actual modelling. It should also facilitate the educational process as instructors could use provided off-the-shelf models in their classes or would be able to amend them if required.

An in-depth discussion of general risk management principles and traditional linear Monte Carlo methodology could be found in the author's previous book (Raydugin (2013)). Various important risk management and risk quantification topics (e.g. cost escalation and currency exchange rate risks) that are not supposed to utilize Monte Carlo quantification methodology are also discussed there.

Although the material in this book is organized in consecutive chapters there are elements of controlled chaos in presenting main concepts. First, some aspects are repeated on several occasions on purpose. Second, on several occasions the cart is put before the horse on purpose to anticipate upcoming discussion in later chapters: references to upcoming chapters are done regularly to anticipate upcoming topics and establish a link with them. For example, the forward-looking references help to link well-known features of the traditional linear Monte Carlo modelling with ones pertaining to its anticipated non-linear version. This semi-structured approach should be written off on the complexity of intertwined and convoluted topics discussed in this

[5] Safran Risk v.7.3 has been used for SCRA modelling. A free trial version of the tool can be retrieved from https://www.safran.com/risk-free-trial to immediately run all included Monte Carlo models.

[6] A software package Vensim PLE Plus v. 7.3.5 by Ventana Systems Inc. is used in this book. A free version of Vensim PLE as well as required user manuals could be uploaded from https://vensim.com/free-download/ to immediately run all included system dynamics models.

book but not on the author's writing style, in most cases. This allows taking a look at the major concepts from various angles and in different contexts and provides a sense of familiarity with them in later chapters.

A final point is about political aspects of applicability of the newly developed non-linear Monte Carlo methodology to projects. To be honest, it directly confronts the deception and self-deception practice associated with the Hiding Hand principle and project placebo effect overviewed in the Preface. The proponents of the Hiding Hand would not like the non-linear Monte Carlo methodology. Actually, schedule and cost outcomes similar to notorious ones for Sydney Opera House, or Boston Big Dig Artery/Tunnel project[7] could be forecast (i.e. revealed before sanctioning). This gives a reason to call the non-linear Monte Carlo risk quantification methodology the Revealing Hand. At the same time, the Revealing Hand[8] would win out only if an entire project risk management approach is overhauled at corporate (in case of private projects) and government (in case of public projects) levels. This could be achievable through adopting adequate risk quantification methods and practising of truly independent third-party and cold-eye reviews to generate a reliable, accurate, and truthful information on possible project outcomes. Decision-makers should not be able to ignore important information that is really critical. This seems to be the only way to avoid willing or unwilling deception and self-deception. In other words, this is the only way to avoid 'unpleasant unsurprises' or 'unhappy, unlucky and unexpected discoveries occurring by design'. At least adequate risk quantification results would not be swept under the carpet. A promising example of this in the government environment is described by Volden and Samset (2017). In the private sector, there should be a full responsibility of boards of directors to ensure the unbiased decision-making process. This is about the political part of risk quantification as a branch of political mathematics.

In the light of this discussion, one additional and highly controversial point on political aspects of risk quantification casts doubt. Two notoriously unsuccessful capital projects–outliers are cited in this book on several occasions: Sydney Opera House and Boston Big Dig Artery/Tunnel projects. Yes, they are characterized by immense overspending—1,400 per cent and 220 per cent correspondingly. Yes, if the decision-makers knew about actual outcomes beforehand, it is highly likely these projects would not be approved. Yes, one of them is notorious for being known as the most disastrous construction project in the history, the other as the most expensive US highway project. Yes, both burned piles of Australian and American taxpayers' money. However, putting aside the rightful indignation, two questions should be asked and answered. First, would Sydney be a better place if Sydney Opera House had not been constructed at all? Second, would Boston be a better place if the Big Dig were never approved?

[7] These two projects are cited in this book on several occasions.

[8] The Hiding Hand is played by handing out placebos to approve unhealthy projects but it completely misses opportunities to cure (or otherwise) them before they become dead in execution; the Revealing Hand should certainly connote handing out bitter pills keeping in mind that there is no goal to save all unhealthy projects at sanctioning.

Questions of this kind should be thought through carefully as the newly-developed non-linear Monte Carlo methodology has a capacity to have a say in the argument of the Hiding and the Revealing Hands.

To recap this discussion, risk quantification has two angles to look from. The first is about mathematical robustness of adopted risk quantification methods. The second is about ways decision-makers use or mis-use risk quantification results. Both angles are covered in this book although the primary focus is on the former. Complaints could be heard from project risk practitioners that decision-makers ignore or mis-use their risk quantification results. First of all, this is why risk quantification should be called political mathematics in the first place. Some decision-makers ignore robust risk quantification results because they can. As discussed above, if there is no mandatory governance mechanism to develop, review, and use risk quantification results, the unpleasant unsurprises are inevitable.

Second, there are known cases when the results were rejected simply because adopted risk quantification methodologies—both parametric and Monte Carlo—were not robust enough. Sometimes it sounds like the complaints in the second case are more vocal. Apparently, risk quantification if used in decision-making should be a two-way street.

A few opening chapters of this book are devoted to cases when quantification methods are not adequate. Their deficiencies are uncovered and ways to address shortcomings are proposed. At the end of the day, the selection of adequate risk quantification methodologies is part of the overall decision-making too.

The roadmap of the book is shown by Figure I.1. Although the numeration of chapters and appendices in figure I.1 is kept the same as in the table of content, chapters' titles are abridged.

The book starts with the overview of major risk management methods at large and risk quantification methods in particular written in a form of semi-structured essays (Chapters 1–4). The main purpose of Chapters 1–4 is to provide a helicopter view of the risk quantification landscape as a starting point of non-linear Monte Carlo methodology development. Features of both conventional and unconventional methodologies are overviewed including their strong points and shortcomings. A project Zemblanity business case devoted to risk quantification using the traditional linear Monte Carlo methodology is introduced in Chapter 5. Actually, Chapter 5 can be considered a crash course on traditional integrated Monte Carlo schedule and cost risk analysis (SCRA). The corresponding model is referred to as the linear SCRA model or the model L-SCRA.

Chapters 8–12 lay bricks one at a time to develop the non-linear Monte Carlo method. Chapter 8 is devoted to overview of various aspects of complexity in general and project complexity in particular. Chapter 8 provides an exhaustive literature review on non-linear interactions in complex project systems. Chapter 9 outlines a structure of a project system that should consist of two subsystems—PSS and PDS. The detailed description of these two subsystems is provided in Chapters 10 and 11. Chapter 12 merges concepts related to the project subsystems and their complexity to bring

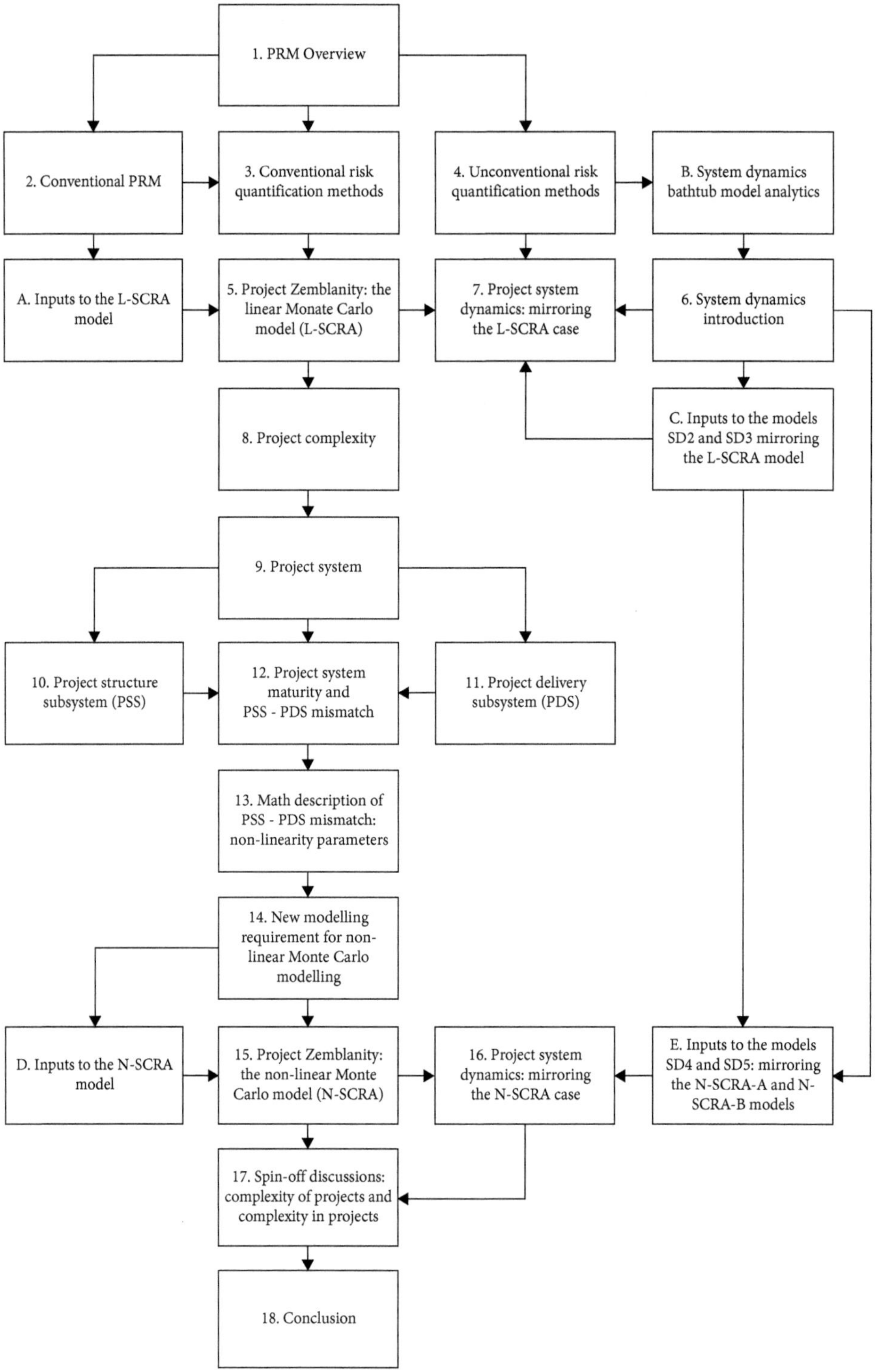

Figure I.1 The roadmap of the book

forward a workable qualitative method to evaluate a maturity of a project system. Chapter 13 is devoted to development of mathematical foundation of the non-linear Monte Carlo methodology. Chapter 14 summarizes discussions and findings of Chapters 8–13 to put forward a data input specification for non-linear N-SCRA modelling. Chapter 15 presents the second project Zemblanity case where the initial linear L-SCRA model is modified based on requirements of Chapters 13 and 14 to yield the non-linear N-SCRA models.

It should be pointed out that in parallel with development of N-SCRA methodology, project system dynamic modelling is undertaken. Chapter 6 brings forward a basic system dynamics methodology regardless of any project applications. Five basic so-called bathtub models BT1–BT5 are demonstrated having an increasing level of sophistication. Analytical old-fashion last-century solutions to differential equations standing behind these five models are derived and compared with solutions obtained from system dynamics modelling. Chapter 6 also introduces all required system dynamics concepts and terminology that are used in Chapter 7 for actual project system dynamics modelling. The modelling in Chapter 7 mimics the Monte Carlo model L-SCRA in comparing results. Similarly, Chapter 16 mirrors the non-linear N-SCRA model of Chapter 15. The link between the models is established through recalculation of parameters of the N-SCRA model to required parameters of the system dynamics model. Chapter 17 is devoted to a sideline discussion on project complexity. Several fundamental project complexity concepts are discussed and explained. For instance, an additional Monte Carlo modelling is undertaken to investigate conditions when probabilistic distributions may have two peaks. A direct analogy with physics is drawn referring to a behaviour of a quantum system in a two-well potential. Conditions when there should be a single project failure mode attached to the two-peak distributions pushing a project to chaos are formulated. Another topic is devoted to discovery of the origin of general uncertainties in linear systems and internal risk amplifications in non-linear systems. Finally, the topics on projects as adaptive systems, on reverse-engineering approach to develop project contingencies, and on the role of stretched targets are explored.

Chapter 18 concludes the book providing a summary of main results and recommendations.

Appendices A–E are devoted to support Monte Carlo and system dynamics models discussed in this book including development of the required inputs.

Each chapter of the book is concluded by a section with the chapter's markup and takeaway except Chapter 18 that itself is the markup and takeaway for the entire book.

The author has a vested interest in receiving readers' feedbacks and possibly learn about particular practical risk management challenges of their projects. Any feedbacks, comments, or suggestions could be sent to yuri.raydugin@risk-service.com.

Dhahran, Saudi Arabia
27 October 2019

PART ONE

RISK QUANTIFICATION IN SIMPLE PROJECTS

1
High-level Overview of Project Risk Management (PRM)

1.1 Introduction

Implications of risk management for business undertakings has been recognized for centuries. For instance, in 1688 La Vega (1957) confesses that 'the business is such that it is always in dark, it is always risky, and it is always frightening'. A number of risk management methodologies applicable to various business areas has been developed since then. This chapter provides a high-level overview of project risk management (PRM) methods.

For the purpose of this overview, all PRM methodologies are divided conditionally to two groups: conventional and unconventional. PRM methodologies are referred to as conventional if they are based on traditional PRM steps to identify, assess, address, and monitor risks. These include deterministic (scoring) and probabilistic (Monte Carlo) techniques. Chapter 2 is devoted to the detail overview of conventional PRM methodologies. Chapter 3 zeroes in on their risk assessment steps only.

The second group (see Chapter 4) deals mostly with types of risk assessment. All the other traditional steps are usually not involved or have a secondary importance. These methodologies are devoted to forecasting of expected project duration or cost outcomes including risks among many other factors. As such they cannot be qualified as risk quantification methods being outcome forecasting methods: it is impossible to quantify risks without their identification as well as addressing. Nevertheless, methodologies in this group are tagged as unconventional risk quantification methods. This group includes regression-based parametric methods as well as two emerging methodologies: system dynamics and artificial neural networks (ANN). However, if the other traditional steps of PRM process are included to an unconventional PRM methodology, it could be reconsidered as conventional. This transformation is done for system dynamics modelling in this book as identified and addressed risks are used as inputs to some project system dynamics models (see Chapters 7 and 16). Chapter 5 is devoted to the traditional Monte Carlo integrated schedule and cost risk analysis (SCRA) referred to as the linear Monte Carlo SCRA methodology and corresponding models are labelled as L-SCRA.

Chapters 1–5 can be used as a snapshot of PRM methodologies as they existed prior to writing of this book.

Modern Risk Quantification in Complex Projects: Non-linear Monte Carlo and System Dynamics Methodologies. Yuri G. Raydugin, Oxford University Press (2020).
DOI: 10.1093/oso/9780198844334.001.0001

Starting with Chapter 6, bricks for the non-linear Monte Carlo methodology are being laid to take into account previously overlooked non-linear phenomena in project systems—risk interactions and dynamic risk patterns. Corresponding non-linear Monte Carlo models are labelled as being N-SCRA.

Hence, this book can be treated as a logical upgrade of books on traditional (linear) Monte Carlo methods including the book by the author (Raydugin (2013)) which provides an exhaustive discussion on deterministic (scoring) and traditional (linear) Monte Carlo methods and their integration as discussed in Chapters 2, 3, and 5.

1.2 Two Fundamental Goals of Risk Management

PRM has two fundamental goals. First and foremost, managing of risks should be understood as addressing risks. This constitutes the first PRM goal. Risk addressing is a most practical and important risk management step whereas all other risk management steps are supposed to serve and support this fundamental activity. Second, presuming that risks are addressed in a proper way, the other challenge is to quantify risks after addressing and to develop required risk contingencies for residual risks that cannot or should not be further addressed. This constitutes the second PRM goal related to decision-making.

Depending on the nature and context of particular risk management activities in various industries, contingencies could relate to various quantifiable objectives. For example, in the case of insurance or investment industries contingencies usually relate to required monetary reserves to cover potential losses trying to match value at risk; in case of IT, marketing, warfare, or space exploration projects, risk contingencies and reserves have completely different purpose and meaning; in PRM these usually relate to project durations and budgets. Project risk quantification and development of the adequate contingencies (the second PRM goal) are in the primary focus of this book, although this goal cannot be achieved without proper risk addressing (the first PRM goal). In other words, conventional PRM methodologies pursue both PRM goals, whereas unconventional try to achieve only the second PRM goal without paying due attention to the first one.

1.3 Inanimate, Animate, and Mixed Systems and Risks

Achieving of both PRM goals is simple and viable in some cases and rather difficult in others. This depends on types of project systems and risks pertaining to them that are supposed to be managed and quantified. The following three sorts of systems and associated risks may be told apart for the purpose of this discussion.

- Inanimate,
- Animate, and
- Mixed.

The first sort of systems are not influenced by human nature and behaviour at all. Phenomena outside of PRM from flip-of-a-coin to behaviour of subatomic particles could be described and modelled statistically. In project systems poor weather or natural disaster-related risks could be treated as being inanimate. This example is rather trivial though. What if a project organization works like clockwork? It may have a world-class project development and execution procedures that are consistently implemented, all project functions are manned by top industry professionals, all internal and external stakeholders are amazingly aligned, contracts awarded only to best industry contractors and suppliers, project cost estimates and schedules as well as related project contingencies are developed based on best industry practices in an unbiased way, etc. Clockwork project organizations could be treated as inanimate in a good sense for the purpose of this discussion. Some regular project risks can still exist although the outcome for a clockwork project would be quite certain—a project should be a success which is highly likely. Projects like this do exist although not terribly often.

The second sort is largely defined by human nature and could be referred to as animate. A best non-project example points to capital markets where decisions of multiple investors—institutional and individual—define overall market trends. Some people say that capital markets are driven by supply and demand, others by fear and greed. Investors do have their goals, ambitions, opinions, expectations, fears, biases as well as psychological, cultural, and competitive traits. No doubts that information about economic trends and fundamentals is taken into account too. However, these trends are largely shaped by human beings again including government regulations (central bank's interest rates, taxation, tariffs, etc.) or international politics (armed conflicts, trade wars, pandemics, protectionism, alliances, etc.). Inanimate risks like natural disasters may still play a role. For example, there could be some impacts of poor weather or natural disasters on consumer-goods sales, although these don't play a cardinal role in overall market trends. Despite a certain success of mathematical modelling and behavioural sciences, animate systems like capital markets are highly volatile and unpredictable.[1] In project environment very early project phases devoted to development of business cases that involve stakeholders, markets, competition, project economics, government regulations, etc are a good example of animate systems. Naturally, the early project phases devoted to capital investments resemble capital markets quite closely. Their forward-looking modelling of project outcomes is quite futile and is usually reduced to backward-looking parametric modelling based on data of previously completed similar projects. Depending on relevance of projects included to the databases, regression analysis results could be relevant to upcoming projects in early phases to a certain degree.

[1] For example, a value at risk (VaR) methodology that utilizes the Monte Carlo modelling technique is believed to predict possible financial losses only until a set of assumptions used for VaR modelling is still valid, which is usually no more than a few days.

The third type of systems and risks is mixed and positioned somewhere in-between inanimate and animate systems. A non-project example could point to centralized planned economy or highly regulated capital markets. They have traits of both animate and inanimate systems. In project environment this sort of system corresponds to projects in later project phases unless corresponding projects are ideal and work like clockwork.

In early project phases prior to final investment decisions (FID) a structured front-end loading (FEL) methodology is traditionally used to suppress animate aspects of project development. Certainly, it is an implicit and futile call for making projects work smoothly. However, project practitioners keep grumbling over low factual efficiency of the FEL methodology which apparently requires some overhaul (see e.g. Westney and Dahl (2019)).

As discussed in this book on several occasions, most advanced Monte Carlo integrated schedule and cost risk analyses (SCRA) methods can be fully adequate in cases of clockwork-type projects yielding quite accurate duration and cost forecasts. These projects also referred to as serendipity-type projects in introduction. However, a vast majority of projects cannot belong to the clockwork-type category. As they usually belong to zemblanity-type category, even the most advanced risk quantification methods fail on many occasions and do require a substantial overhaul.

This book is devoted to improvement of the traditional Monte Carlo quantification methods in order to make them adequate not only in case of inanimate or clockwork-type projects but in case of regular mixed inanimate-animate projects too. The spread of the new modified methodology on to the purely animated project systems is over-ambitious and it is outside of the scope of the book.

1.4 PRM Context—PRM Method Mismatch

The two PRM goals—risk addressing and contingency development—imply existence of two different PRM contexts that should be kept in mind when selecting methodologies to achieve these two goals.

1.4.1 A Role of PRM Planning

Most of the PRM methodologies are supposed to pursue both PRM goals. Although some of them are good at addressing risks only, the others are a fit for risk quantification purpose in some project phases, and some belong to neither category. A selection of adequate PRM methodologies is not an easy task as it depends on deep understanding of a PRM context of a given project in a given phase. A notion of context differentiators can be introduced to accentuate unique features of a given project.

It is a very rare occasion when a given off-the-shelf PRM methodology or a tool can be immediately used as-is in any project and in any phase. If this happens, this usually means ignoring some relevant and important elements of PRM context. To avoid such a risk to risk management, a proper planning is required at the very beginning of the PRM process. Chapter 2 further explores the important role of planning in paying proper heed to PRM context which should be established as a first and mandatory step in the PRM process.

1.4.2 Project Phases and Selection of PRM Methodologies

Although PRM context of a given project has unique features related to its scope, size, assumptions, constraints, stakeholders, locations, technologies, execution strategy, etc, there are high-level features that are common to most projects depending on project development phases. Whatever the other elements of project context might be, projects that are in the same phases have comparable features in terms of current challenges, methods, and deliverables.

The same gated project workflow with some variations is used (or should be used) in capital and other projects. It is broken down to several project phases with decision gates and value assurance reviews in between. Several phases can be combined in stages. Phases prior to project sanctioning or final investment decision (FID) constitute front-end loading (FEL) that is followed by an execute stage. The workflow ends with operations. Phases prior to FID aim at project value identification, and phases following FID—zero in on project value realization. This accentuates a role of value assurance review and decision gates: to ensure that the value is identified and delivered every phase of the workflow.

As a rule, FEL contains four phases: FEL0 (initiation), FEL1 (business case), FEL2 (scope definition), and FEL3 (project definition). FEL0 is devoted to identification of a business need and corresponding business opportunity. FEL1 handles all business aspects of an upcoming project and establishes business goals. FEL2 identifies all possible paths to meet the business goal, selects a best possible alternative, and defines its scope. However, it is not unusual that FEL2 (scope definition) is further broken down to two distinctive subphases: FEL2A (alternative's selection) and FEL2B (scoping) as FEL2 is devoted to development of scope for a selected alternative. Selection of alternatives has an extremely important handover role between a business case and scope development. Its role is often underestimated. By this reason it is better off keeping FEL2 divided to two subphases (Raydugin (2017b)). The purpose of the FEL2A phase is to select an alternative with lower risk exposure although with acceptable economics: this is a way to avoid risks in the first place instead of successfully identifying and addressing them in later phases.

The last FEL phase—FEL3 (project definition)—is devoted to development of all major project execution plans followed by a final investment decision (FID).

The execute stage usually includes detail engineering (E), procurement (P), and construction (C) phases and often called an EPC stage by this reason. A standalone phase related to commissioning and startup (CSU) can be distinguished as a bridge between construction and operations.

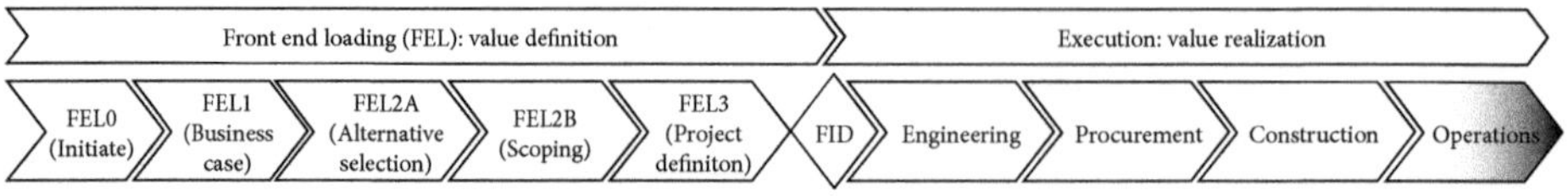

Figure 1.1 A typical gated project workflow

It would be worthwhile to point out that a true decision to proceed with a project is done by the end of FEL2 not FEL3 phase. That is when a project business case and scope are well-understood, frozen, and endorsed by all main stakeholders. The final investment decision (FID) following FEL3 should be more like a rubber-stamp action although subject to successful completion of the more technical FEL3 phase to define the project. Despite this, only one decision gate is shown by Figure 1.1 that relates to FID. Any project phase prior to FID should be concluded by a value assurance review (VAR) and a decision gate.

Different industries, not to mention particular organizations, could have somewhat different sets of project phases and decision gates. Even a project workflow followed by a project owner and its PMC or EPC contractor may differ for the same project due to different contexts of owner's and contractor's activities.

A simplified version of a generic gated project workflow used in this book is introduced in Figure 1.1.

A quick look at the gated project workflow points to the fact that represented phases should possess very different PRM context. Dominating business risks in early phases get on the back burner in the course of project development. Technical risks start taking over in FEL2 and represent a bulk of all risks in execution phases.

Due to this single reason, selection of a right PRM methodology to meet the both PRM goals directly depends on a phase a project is in. Some methods could be indispensable in earlier project phases, although should be avoided in the later phases having better alternatives. Chapters 2 and 3 discuss this topic in detail.

1.4.3 Bias and PRM Zealotry

This section handles some really touchy aspects of PRM describing various real-life realizations of psychological bias in its most picturesque forms. Although, as bias plays a pivotal and highly negative role in risk quantification, it is not possible to drop this section and ignore the topic. It should help to clearly understand root causes of some usual PRM blunders.

In most notorious cases the PRM context—PRM method mismatch leading to misapplication of risk methodologies, may take a form of PRM zealotry when one and the only one methodology is proclaimed the only true one. In some instances, this phenomenon could be tagged as a high-horse zealotry.

The PRM zealotry has five bias root causes that are overviewed below. They are further discussed in Chapter 2 in the context of risk quantification method's selection.

First, PRM context could be sincerely misunderstood and nature of applied risk methodologies innocently misapprehended. This could be called a channel vision zealotry stemming from so-called cognitive bias. This case may include activists that can be referred to as enthusiastic amateurs. For instance, one might come across an individual who is eager to learn new mathematical concepts trying to enthusiastically and immediately implement them in PRM or project controls. Yesterday it could be just a discovery of a simple regression analysis, today it would be mesmerizing generative adversarial networks, and tomorrow it might be the overwhelming artificial intelligence at large.[2] Or there could be an example of an individual who has a robust coal mining background who unexpectedly becomes a data mining specialist, etc. These cases should be fully tolerable and supported unless there are attempts to use some poorly understood methods aggressively in project decision-making.

Second, there could be an inertia to use a particular outdated methodology based on a particular previous experience or long-time preferences. This phenomenon may be tagged as anchoring zealotry which could be explained by perceptual bias. One example that is observed too often relates to some outdated Monte Carlo risk quantification methodologies. As a matter of fact, prior to 2004 Monte Carlo risk quantification tools did not have standalone risk register modules at all. All risks were treated as general uncertainties which constituted activity ranging (AR) for schedule risks and line-item ranging (LIR) for cost risks Monte Carlo methodologies. As discussed in Chapter 3 (Merrow (2011)), these two methodologies have notoriously low accuracy. Despite risk modelling tools obtained the risk register modules in 2004 (see e.g. Patterson (2018)) and certainly improved accuracy since then, some specialists still use these two outdated ranging methods. A most alarming fact is that some new-generation risk managers still consider these methods quite modern and top-of-the-line.

Third, there could be a conscious propensity to promote a particular methodology[3] based on commercial interests. This could include access to proprietary project databases which could be marketed profitably along with consulting services. This phenomenon could be called hidden agenda zealotry stemming from conscious bias that is fuelled by commercial interests.

Fourth, there could be a reasoning that decision-makers understand only particular and usually oversimplified or outdated methodologies which are promoted to please customers and ensure repetitive business. This could be labelled a servile zealotry and may be explained by a combination of customer's cognitive bias and seller's hidden agendas. The AR and LIR Monte Carlo methodologies are a frequent subject of this type of zealotry.

Fifth, developers and marketers of risk management software packages often promote their tools as the only relevant ones to any type of PRM. It should be fair to call

[2] The topic on artificial intelligence is extremely exciting. Although it would be fair to state that before it is fully applicable to project risk management, any major realizations of natural stupidity should be eradicated first.

[3] PRM is often pursued and driven by consulting organizations. As consulting is a very competitive business, this inevitably leads to competition of methods which is sometimes based on the true method's merits.

this a marketing zealotry. This usually takes place when hidden but explainable agenda of marketers meets customers' cognitive or perceptual bias.

It should be pointed out that in most notorious cases the zealotry (any kind) may take a form of *bahramdipity*. According to Sommers (2001), bahramdipity means the suppression of a scientific discovery by a more powerful individual or a self-serving promotion of an often-unreliable discovery by a powerful individual (*bahram*). Both instances of bahramdipity involve revenge on those who have different opinions. It might sound like quite exotic phenomenon to many. Although bahramdipity is not something terribly extraordinary: the less robust a methodology is, the more zealotry it radiates to survive. One of the observed bahramdipity realizations might even be described by an interesting conscious-bias motto that 'if you are not confusing people enough, you are not important enough'. Bahramdipity as a limiting zealotry case should be kept in mind when discussing and selecting PRM methodologies at large and risk quantification methodologies in particular.

In practice these five zealotry root causes could intertwine in a spectacular way leading to both risk mismanagement and decision-making blunders[4] due to failures to meet both fundamental PRM goals. The discussion on PRM zealotry provided here is fully applicable to risk assessment (see Chapter 3). Comparable if not the same traits are observed when risk quantification methods are selected.

1.4.4 Conventional and Unconventional PRM Methods

Proper risk quantification is not possible without risk identification and risk addressing. When risk addressing is excluded (i.e. ignored), the undertaken risk quantification would correspond to a situation before (i.e. without) risk addressing. Along with a failure of contingency development this means that an important project control to manage risks is ignored. This means a failure of the first PRM goal and almost certain failure to meet the second PRM goal.

Even if only some risks are missed out it would not be possible to properly quantify and address them. This leads to higher room for unknown unknowns. In practice steps should always be taken to reduce a room for unknown unknowns down to an acceptable size (Raydugin (2014)). It is preferable that only risk methodologies that align with this intent are selected. Although methodologies that completely ignore risk identification and addressing do exist. They are referred to as unconventional and overviewed in Chapter 4.

It seems appropriate to conclude this section on PRM context—PRM method mismatch by a citation of a story on 'cargo cult science' by a prominent physicist and Nobel Prize winner Richard Feynman (1997). Despite this story is alluding to scientific

[4] Due to high sensitivity of this topic, no references to particular papers, books or discussions at conferences that point to particular instances of PRM zealotry are provided.

theories, it still provides valuable insights on selecting PRM methodologies in accordance with PRM context.

During World War II local tribes on some South Seas islands saw military cargo airplanes landing with a variety of attractive consumer goods. After the War was over, they still wanted the same things to continue happening. So these people (aka 'cargo cult people') arranged things that somewhat resembled 'runways', put fires along the 'runways', made a wooden 'air traffic controller' hut for a man (aka 'a traffic controller') to sit in with two wooden pieces on his head like 'headphones' and bamboo bars sticking out like 'antennas'. And they waited for the iron birds to land...They did everything right in that it looked to them exactly the same as before. Unfortunately, they were missing something very essential and it did not work: no airplanes landed. Feynman (1997) uses this story to call for looking into 'theories that don't work and science that isn't science'...Using some risk quantification methods would be comparable with believing that an airplane is just an iron bird or an automobile is simply a mechanical horse or camel.

1.4.5 Sampling Theory Basics

This section could be a better fit for Chapter 2 where overview of quantification methods is undertaken in general or for Chapters 3 and 4 where more detailed review of Monte Carlo, regression-based, and other methods is carried out. The regression-based and Monte Carlo methods are essentially statistical methods. So, it is worthwhile to review some basics of statistics including sampling theory before a detail review of these methods is undertaken later in the book.

1.4.5.1 Two kinds of statistics

There are two primary kinds of statistics in a statistical analysis as a discipline—descriptive and inferential (see Doane et al. (2010) or Spiegel and Stephens (1999)). The former handles collection, organization, presentation, and summary of data as well as establishing statistically significant relations among variables. The latter deals with estimating unknown parameters, drawing conclusions, and making decisions. Regardless of a selected method—parametric or Monte Carlo—the inferential statistics utilizes sampling theory that is a study of relations between 'a population' of interest and 'samples' drawn from the 'population'.

1.4.5.2 A general approach towards sampling

A population in statistics is understood as a collection of similar items of interest, with statistical parameters of a population being unknown. A population can consist of tangible or physical objects (e.g. all vehicles in England) or some intangible characteristics (e.g. production years of vehicles in England). A sample is defined as a subset of the investigated population which statistical parameters (or 'sample statistic') are known. If several samples are drawn from a population, sample statistic distributions also called estimators or predictors could be derived, for example, a distribution of sample's means or standard deviations, etc. to approximate parameters of the investigated population. Apparently, each sample provides a one-point value estimate used in a predictor. How samples are defined and drawn is called a sampling experiment.

A cornerstone of the sampling theory is a requirement that samples should be representative of a population (unbiased samples). If samples are truly random (i.e. each item of a population can become a part of a sample) the samples are considered representative and unbiased: corresponding estimators/predictors can provide accurate information about the population's parameters of interest.

If samples are not quite random (biased samples), they are not able to produce accurate predictions due to substantial statistical bias or systematic errors. There are several distinguished types of non-random biased sample types (see Doane et al. (2010)), for example, a judgement sampling method that relies on the expertise of the sampler to choose items that are representative of the population. A convenience sampling method relies on handy samples that are easily available and often used indiscriminately. They are easy to obtain and use and easy to produce biased results.

All elements of population are supposed to exist when samples are drawn. Normally any sample is smaller than the entire population. A limiting case when a single sample is used that is equal in size to the entire population of interest is called *census*.

The inference is undertaken to define statistics of an existing population, not future elements of population or the population with newly added elements. For example, a number of samples could be drawn to learn the share of cars on the roads in England that were produced from 2017 to 2019. Or a question can be asked about a share of Porsche in all cars on the roads in England, etc. If samples are collected only in Kensington, Chelsea, and West Brompton (apparently a quite biased sampling) the results will differ significantly from a case when samples are drawn randomly in multiple locations in England. Whatever the Porsche results are in both sampling experiments, no futuristic projections can be asked about the Porsche's share in England in 2022 or 2024 from this particular sampling experiment.

When the number of samples drawn from the population is small (below 30), a so-called small sample theory should be applied due to extremely low accuracy of predictions.

1.4.5.3 Sampling in PRM

A cardinal difference of sampling and inference in PRM in comparison with the usual case outlined in section 1.4.5.2 is that the population of interest—an expected actual duration or cost outcome of an upcoming project—do not exist when the inference is undertaken. The project has yet to be completed.

Undoubtedly actual duration or cost outcomes will be one-point numbers. These numbers can be evaluated using a specific confidence interval. (A particular confidence interval should contain the expected actual outcome with the established probability or confidence.) To undertake such evaluation a statistical distribution for the project duration or cost is required, which defines the population of interest in PRM inference regardless of a selected statistical inferential method.

1.4.5.4 Parametric sampling and inference

Available databases of previously completed projects are used for inference purposes when a parametric method is adopted. The details of the parametric methodology are discussed in section 1.6 and Chapter 4. Without going into detail, it should be pointed out that a key weakness of parametric method sampling is that in many cases used samples are biased. They have traits of both convenience and judgement sampling. Moreover, statistics of the available database as the used sample is forcefully imposed on upcoming projects through ignoring context differentiators.

If the used sample is small produced results will be especially inaccurate (see section 1.4.5.6).

In essence the existing sample that can be biased provides an ersatz-distribution for possible outcomes of an upcoming project. A possible issue is that an upcoming project is not obliged to belong to the sample or look like one of the sample elements. It might be way outside as in the case of previously cited Sydney Opera House or Boston Big Dig Artery/Tunnel projects. What databases may predict outcomes of these two projects—1,400 per cent and 220 per cent overspending, respectively—when using the parametric inferential method?

1.4.5.5 Monte Carlo sampling and inference

In case of the Monte Carlo method a sampling experiment is designed differently. Namely, one sample per Monte Carlo iteration is generated: each sample is based on a unique combination of input parameters directly relevant to an upcoming project—a base schedule or estimate, general uncertainties, and relevant uncertain events. The details of the Monte Carlo methodology are further overviewed in section 1.5 and Chapter 3.

If a Monte Carlo model is run using, say, 5,000 iterations, 5,000 samples are produced where each sample (aka iteration) represents one fully relevant 'project'. No small sample theory would be required in case of Monte Carlo modelling. Together all 5,000 estimates produce a statistic distribution (the estimator/ predictor).

By design, an expected project outcome should belong to one of the samples. In theory this is a right statement if the Monte Carlo sampling is not biased. In contrast with the parametric biased sampling the Monte Carlo biased sampling stems from missed components of project risk exposure. In this case all 5,000 iterations may reflect on a risk exposure that does not have much to do with a real project risk exposure. By this reason an actual project outcome might be way outside any confidence intervals of the derived Monte Carlo distribution. Chapter 3 provides an additional discussion on the Monte Carlo biased sampling and a systematic error when various important components of project risk exposure are missed out—

overlooked risks (all or some) including schedule-driven cost risks, ignored risk interactions, etc.

1.4.5.6 Implications of the small sample theory

Figure 1.2 demonstrates how the number of samples influences inferential results. Four Monte Carlo charts represent approximations of a simple triangular distribution that has (min., most likely, max.) parameters (0, 10, 40) when 10, 20, 50, and 5,000 Monte Carlo iterations are used. This distribution has a mean value of 16.7. The first three charts are called for to mimic typical databases of completed projects used in parametric modelling that contain 10, 20, and 50 fully relevant previously completed projects.[5] The last chart represents results for 5,000 iterations which is typical for Monte Carlo models. As Figure 1.2 shows, distributions for 10, 20, and 50 iterations yield mean values 18.3, 15.8, and 15.4, correspondingly. Only the 5000-iteration

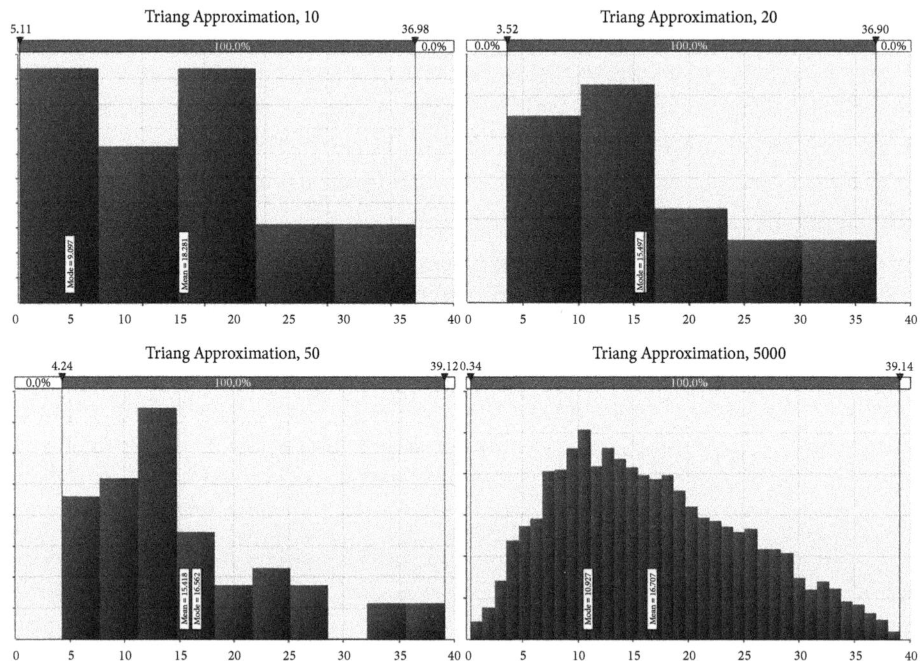

Figure 1.2 Approximation of a triangular (0, 10, 40) distribution using 10, 20, 50, and 5,000 Monte Caro iterations

[5] According to Merrow (2011), the widely-cited database used by the IPA Institute contained 318 capital projects broken down to eight industrial sectors as of in 2011. The biggest group contained 130 oil and gas projects and the smallest group contained only seven pipeline projects. First, the oil and gas industry has dozens of facility types on-shore and off-shore. By this reason, the cited number of 130 oil and gas projects does not look terribly high when a task is to infer results for a particular upcoming project. Second, using just seven pipeline projects constructed in different countries all over the world, in different terrains and climate conditions to infer an outcome of an upcoming pipeline project would not be a convincing undertaking either. In deference to Merrow (2011), no attempts to carry out inferential predictions are included to his book.

distribution provides the right mean value. Visually, approximations based on smaller numbers are rough and not quite representative of the triangular distribution.

Unfortunately, databases containing a critically low number of fully relevant completed projects may be often used in parametric inferential analysis. Actually, the databases may contain a critically low number of relevant projects along with quite a high number of irrelevant ones, which does not contradict the definition of the convenience sampling.

1.4.6 A List of methods Used In or In Place of Inferential Statistics

The Monte Carlo methodology is in the primary focus of this book, although some other risk quantification methods should be mentioned and overviewed. Some methods belong to inferential statistics, some are used in place of it. Far from being exhaustive, their list includes as follows.

- Expert judgement: an empirical method to utilize educated but often biased opinions of specialists.
- Predetermined guidelines: an empirical method to prescribe the same cost or schedule contingency value as a per cent of base durations or estimates across entire project portfolio (see for example the article by Baccarini (2005) on 10 per cent syndrome).
- Parametric modelling: an empirical multi-regression analysis methodology to infer results for an upcoming project through using data of previously completed projects.
- Scoring method: a simplified risk quantification method based on applications of risk assessment matrices (RAM) and review of risk severity scores.
- Simulation analysis: a term normally used as a placeholder for various versions of Monte Carlo modelling including
  - range estimating: Monte Carlo modelling based on considering general schedule (activity ranging (AR)) and cost (line-item ranging (LIR)) uncertainties only,
  - expected value: Monte Carlo modelling based on considering either schedule or cost uncertain events (or both) usually along with general uncertainties.
- Hybrid modelling: an eclectic method of combining parametric and Monte Carlo models.
- System dynamics: a simulation method that allows modelling and quantifying non-linear interactions in a project system.
- Artificial neural networks (ANN): an empirical inferential method of statistical data modelling with complex (non-linear) input-output relations, which could be considered an upgrade of parametric models, to search patterns in arrays of data.

The above-mentioned methods are reviewed at higher level in sections 1.5 and 1.6 first, and second in Chapters 3 and 4 more specifically.

1.5 A High-level overview of Conventional PRM Methodologies

1.5.1. Two Important Conventional PRM Methodologies

All traditional PRM methods do have risk addressing and quantification steps that are organically embedded into the overall PRM process (see Chapter 2 for further detail). Risk management methodologies based on typical risk management process steps—plan, identify, assess, address, implement, monitor, accept, close—can and should be referred to as conventional. They are supported by various industry risk management standards and guidelines developed by the International Organization for Standardization (ISO) (see ISO (2018)), the Project Management Institute (PMI) (see PMI (2009)), Construction Industry Institute (CII) (see CII (2013)), etc and multiple regional and country risk management organizations. The conventional PRM methodologies discussed here are not fully aligned although they have comparable PRM process steps in the cited standards. Using a helicopter view on all these methodologies for the purpose of this discussion, they could be considered conditionally identical.

The conventional risk management methodologies can be further divided into two subgroups: deterministic and probabilistic. As a matter of fact, a cardinal difference between deterministic and probabilistic conventional methodologies relates mostly to risk assessment steps (see Chapter 2).

1.5.2 Advantages and Limitations of Conventional PRM Methodologies

Deterministic and probabilistic assessment methods of conventional PRM methodologies have both limitations and strong traits.

Deterministic (scoring) methods are good for addressing of risks with impacts on any project objectives: both quantifiable ('hard') and non-quantifiable ('soft').[6] Scores could be attached to both probabilities and impacts on cost, schedule, scope/quality, cash flow, safety, reputation, environment, etc objectives. The role of the scoring assessment is to compare risk severity before and after addressing and come up with risk ranking. From this viewpoint deterministic methods are more powerful and comprehensive than probabilistic methods in terms of the first PRM goal: the latter can handle only quantifiable (hard) objectives such as cost, schedule, and sometimes

[6] Non-quantifiable ('soft') objectives, as discussed in Chapter 2, can be considered quantifiable but only in terms of scoring. Forced applications of Monte Carlo method to non-quantifiable objectives requires inventing of artificial quantification parameters (e.g. impacts on reputation) that can be used as inputs to Monte Carlo models. This is an interesting academic exercise that does not have any practical significance.

scope/quality. However, they are virtually useless for adequate contingency development (the second PRM goal). That is where Monte Carlo methodology should step in and take over. Conversely, probabilistic methodology is virtually useless in assessments of risks with impacts on soft objectives.

Probabilistic modelling may have another fundamental shortcoming related to missing components of risk exposure. This is about some risk kinds that might be overlooked and missed out in Monte Carlo models. For example (see Chapter 3), Monte Carlo ranging methods actually do not take into account risk events or do this in a terribly inconsistent way. In turn, some more advanced Monte Carlo cost risk analysis models do take into account cost risk events although are not able to take schedule-driven costs into account consistently. The fully integrated Monte Carlo SCRA models that do take into account schedule-driven costs usually ignore non-linear phenomena in project systems such as risk interactions. Actually, the last point is absolutely common for the time being which defines this book's primary purpose: to take into account non-linear risk interactions. Some valuable insights on maturation levels of various Monte Carlo methodologies are provided by Hulett (2019), although non-linear phenomena in project systems are not discussed.

In practice, the scoring and Monte Carlo methods should work hand in hand: the former serves to reach the first PRM goal; the latter is designed to reach the second PRM goal through using the scoring method results as inputs especially risk addressing.

Moving from ranging Monte Carlo modelling to linear Monte Carlo SCRA, to the emerging non-linear Monte Carlo SCRA could be considered a natural Monte Carlo methodology progression. A next frontier after this could be a full integration of the non-linear Monte Carlo SCRA methodology with system dynamics and, possibly, ANN modelling, which is briefly discussed in Chapter 4.

1.6 A High-level Overview of Regression-based Methodologies

1.6.1 Flaws of Averages

Key unconventional methodologies are based on regression analysis and can be referred to as regression-based methods. They are used both in descriptive and inferential statistics. A simple bivariate regression analysis allows the identification of significant interdependencies between an independent and dependent regression parameter which is widely used in descriptive statistics. Multiple regression analyses involving several independent regression parameters (predictors) are used in inferential statistics. These regression-based inferential methods are referred to as parametric methods.

Parametric methods are widely used in cost estimating since the middle of the last century. Simply, existing cost estimates could be scaled up or down for projects with similar scopes but different sizes. This does not imply presence of any risks or uncertainties except material take-off uncertainties. When branching out to risk management, the regression-based estimating methods do without the remainder of the typical PRM steps: they try to project historic data from previously completed projects that do include contributions of occurred risks on to upcoming projects. These methods seem to have several intrinsic limitations.

1.6.1.1 A first limitation: predictions are limited by the used sample

As already mentioned in section 1.4.5, the first limitation of regression-based methodologies is that they cannot predict outcomes of project-outliers comparable with Sydney Opera House or Boston Big Dig Artery/Tunnel projects. These or comparable projects are too unique and unpredictable to become meaningfully handled by regression-based inferential methods. An excellent book by Savage (2009) further explores general pitfalls and flaws of averaging in situations of uncertainty.

1.6.1.2 A second limitation: a proclivity to use convenience and judgement sampling

A second limitation is that they have a proclivity to consume so-called convenience and judgement samples as inputs. These are some publicly and mostly commercially available regression-based models and datasets that include information on some completed projects of the past used to infer on schedule and cost contingencies of upcoming projects (convenience sampling). Available data samples always cover a limited number of projects which are characterized by a variety of context differentiators: variations in size, locations, complexities, technologies, stakeholders, features (strengths or weaknesses) of project teams, used project procedures and tools, execution strategies including various execution contract types, risk addressing (no or failed risk addressing vs. persistent and successful risk addressing), etc. Rules on why particular projects made it to get included in a database and others did not are usually not clear (judgement sampling). Usually, only very few projects in any given sample are actually comparable with a given upcoming project.

A best-ever illustration on convenience and judgement sampling is an investigation on up-armament of military aircraft during World War II. The US military investigated areas where bombers were shot up during their missions to Nazi Europe. Upon returning home airplanes were investigated to identify locations of damaged areas. It turned out that most of the bullet holes were found on wingtips, fuselages, and tail planes. The military proposed to up-armour these areas. However, a brilliant statistician Abraham Wald proposed to up-armour undamaged parts of planes including engines (see e.g. Savage (2009) or Ellenberg (2014)). As a statistician he understood only too well a role of right sampling to draw meaningful inferences: the simple reason planes were coming back with fewer or no bullet holes to the engines was that planes hit in their engines were rarely coming back!

What is the relevance of Wald's example to project regression analysis? The relevance is direct. Putting aside the variety of project context differentiators discussed above, one of presumed features of databases used for regression analyses could be that projects that were cancelled before completion or failed too spectacular did not make it to the available databases. There could be other limitations such as commercial agreements to include or not to include some particular projects to databases. Simply speaking it is not always clear which criteria (judgements) have been used to build the project databases and why some projects did not get included.

Another example of convenience and judgement sampling refers to a situation when a sample used to infer information about a population does not even belong to the population or just partially overlaps with the population. Drug testing experiments on animals (nonclinical laboratory studies) prior to the testing of a drug's applicability on people (clinical laboratory studies) are an example of this. However, statistical information about drug testing collected during nonclinical studies is never used for final inference being often only partially relevant.

In addition, it is worthwhile mentioning a remarkable one-hundred-year growth of the Dow Johns Industrial Average (DJIA) Index of thirty top publicly traded US companies as a typical example of convenience and judgement sampling. The fact that companies that fail are regularly replaced by more successful growing companies is less well-known. For the time being, there is no General Electric, Sears, AT&T, Eastman Kodak, General Motors, Bank of America, etc that failed miserably among the current thirty DJIA companies. This is another example of convenience sampling (DJIA provides a quick and easy snapshot of the US economy) and judgement sampling (by obvious reasons).

The situation with convenience and judgement samples could be rather desperate if a project is somewhat unique. Putting aside an apparent example of applications of new technologies, a routine oil and gas industry example would relate to a de-bottlenecking of existing processing facilities to improve their overall efficiency and utilization. Normally projects like this are highly tailor-made, extremely intertwined with existing facilities, and, hence rarely allow any meaningful comparison with projects included to available convenience samples. A more extreme example could be about the upcoming NASA projects to travel to Mars or just to the Moon. What sort of convenience and judgement sampling could be offered in this case?

Simply speaking the higher degree of sample's bias and irrelevance, the lower the accuracy.

1.6.1.3 A third limitation: ignored historic aspect of data collection

A third limitation is that a historic aspect of project data collection is often ignored. Project databases used for regression-based inferential methodologies often contain projects completed in the middle of the last century. New technologies and execution methods emerged since then being applicable to many types of capital projects. For example, oil and gas projects are done very differently now due to extensive use of new drilling methods, lean design, design standardization, modularization, prefabrication, new materials (e.g. non-metal pipes), step-by-step project phasing, etc. A recent oil price meltdown was a major driver behind this (see, for example, Rassenfoss (2017)). On the top of this, locations of oil and gas projects differ largely from ones undertaken just few decades ago as conventional or easy oil and gas plays are less and less available. This moves projects to remote locations including Arctic and offshore areas more and more often. For instance, using a regression analysis based on previous oil and gas projects completed in Texas in the mid-1950s to predict an outcome of an upcoming project in the Russian Arctic would not be a terribly clever idea.

1.6.1.4 A fourth limitation: misalignment with mathematics fundamentals

Parametric modelling reveals three alarming mathematical aspects: judgement sampling (sample conditioning), collinearity of predictors, and ignored non-linearity. They are considered one by one below.

First, as already discussed in section 1.4, the parametric sample's distribution is forcefully imposed on the upcoming project. This is a major underlying and quite questionable assumption of any parametric model. By this reason proponents of parametric inferential methods keep repeating a mantra that 'your project does not differ from previous ones'. This is equivalent to a request: 'you must ignore any specific context differentiators of your project so that the parametric method could be relevant instead'. As discussed in Chapter 4, a conditioning of statistical characteristics of the used sample should be undertaken to make the sample better relevant to the upcoming project. This has all the traits of the judgement sampling (see section 1.4).

Second, multi-regression analysis models may have a collinearity of (or correlations among) predictors (see e.g. Doane et al. (2010)). Predictors should be independent to ensure reliability of regression analysis models and credibility of their inferential outcomes. In practice predictors are often strongly correlated and tend to 'move in unison' due to the correlations. This drastically reduces accuracy of the inferential outcome.

Third, a linear regression analysis has been extensively used for decades. However, in many cases non-linear regression analyses should be used to obtain meaningful or more precise results. This is not done too often due to explainable inertia.

All these topics are further discussed in Chapter 4.

1.6.2 Low Accuracy of Parametric Inferential Methods and a Room for Their Applicability

As mentioned in section 1.6.1. there is an often and quite onerous urge for conditioning of regression analysis inputs to ensure a perception of an apple-to-apple comparison with upcoming projects. When parametric modellers start talking apples, highly likely samples used for inference are convenience or biased ones. They try to make them even more biased by applying judgement sampling in a form of conditioning of parametric inputs. Such conditioning of inputs to parametric models usually takes a form of various subjectively developed scoring or weighting coefficients and allowances (see e.g. Hollmann (2016)). The assignment of the subjective scoring coefficients is usually done in the name of empirical experience or data. Even by this single reason, parametric methods get the accuracy similar to pretty biased and subjective expert judgement methods. But this is not the only reason for their low accuracy. An interesting and revealing situation might happen if another set of samples—another database of completed and somewhat relevant projects—is used for inference: the results of the second parametric modelling may substantially differ from the results based on the first sample.

All in all, parametric methods do not try to predict an outcome of an upcoming project. Apparently, it is rather an attempt to predict and explain the past, not the future, which is exactly a domain of the descriptive statistics. Instead, they just try to forcefully impose averaged although tweaked outcomes of projects included to a database on the upcoming project. By design, such database (the sample) is biased being a child of convenience and judgement samplings. However, the situation with parametric modelling is not completely desperate as might be inferred from the above discussion.

The statement on applicability of parametric modelling in early project phases where parametric modelling produces relatively brilliant results should be pointed out. In FEL1 (business case) and early FEL2 (scope development) the parametric methods should be considered seriously as the only meaningful alternative to develop overall project durations and costs with contingencies. The reason for this is that Monte Carlo methods cannot be specific enough: project risk registers have yet to be developed as project scopes are not yet defined.

A low accuracy of project estimates required in the early project phases naturally commensurates with an explainable low intrinsic accuracy of parametric methods. To be precise, the parametric modelling does not predict an outcome of an upcoming project. It predicts with low accuracy an average outcome of a given class of projects. Many features (the context differentiators) are not known yet in FEL1 and FEL2A before a preferable project alternative is selected in FEL2A. That is why most of these context differentiators can be successfully ignored and the parametric modelling is a good fit for the purpose. The Monte Carlo modelling does not work yet for the same reason: the context differentiators required to identify, address, and quantify risks have yet to be defined. Hence, this positions parametric contingency development methods as a method of choice but only in very early project phases FEL1 and FEL2A.

In practice, the Monte Carlo methods are ready to take over in FEL2B as soon as a preferable project alternative is selected in FEL2A. Sometimes, if time and resources permit, Monte Carlo methods might be used prior to selection of alternatives in FEL2A to support alternative selection decisions.

Results of parametric modelling can still be used in later project phases although for information purposes mostly. Namely, empirical data might be used as sanity checks to compare with Monte Carlo modelling results but certainly should not be used for decision-making due to low accuracy.

The majority of industry professionals do recognize limitations of regression-based inferential methods and their applicability only to earlier project phases. As a case in point, the Association for Advancement of Cost Engineering (AACE) recognizes this (see e.g. AACE (2011)) pointing out that other methods should be used for developing of contingencies in later phases when projects become defined. This section just listed reasons for this.

1.6.3 Remarkable Achievements of Regression-based Methods in Descriptive Statistics

Despite the discovered regression-based method's shortcomings in inferential statistics, they did bring up a significant PRM value that belongs to descriptive statistics. Two valuable instances of this should be pointed out.

First is a discovery of statistically significant interdependencies between project schedule and cost outcomes and several predictors such as (Merrow (2011)):

- levels of business case definition including stakeholder alignment,
- levels of project definition,
- levels of project team integration and skills,
- types of EPC contract,
- etc.

Some of these findings are formalized as various project readiness-type indices (e.g. front-end loading index by the IPA Institute (Merrow (2011) or project definition rating index (PDRI) by CII (1996)). These findings are extremely important for Monte Carlo modelling too as possible additional causes of risks—project system issues—that lead to non-linear interactions in a project system (see Chapter 9).

The second is about description of accuracy classes of base estimates. Regression analyses of previously completed projects in various phases gave rise to low and high ranges (bands) of accuracy (Merrow (2011)) that are formalized in AACE practices (see e.g. an AACE recommended practices 87R-14 (AACE (2015a)) and 10S-90 (AACE (2012))). They are widely used in the estimating of capital projects.

The issue with the accuracy classes is that they have broad low and high bands around base estimates that make them not quite valuable and precise for inference. For instance, at funding authorization, which correspond to the end of FEL3 or FID (Figure 1.1), the lower range is (-10%; -20%) and the high range is (+10%; +30%). Apparently, various projects with different contexts included to the regression analysis formed these broad ranges. Unfortunately, many companies use a simple cost contingency allocation method: project cost contingencies are always equal to 10 per cent of base estimate by the end of FEL3. This is a lowest possible contingency size that complies with the AACE class 3. This practice is called predetermined guidelines (see AACE (2008)) and perfectly ignores context differentiators at large and actual project risk exposure in particular. For instance, Baccarini (2005) refers to contingency development methods like this 'the 10 per cent syndrome'. Moreover, this apparently plays into the Hiding Hand (Flyvbjerg (2016)) mentioned in the introduction. Despite

AACE procedures are pointing to natural pitfalls and limitations of predetermined guidelines (AACE (2008)), the genie is already out of the bottle and the damage is done: predetermined guideline method is a great Hiding Hand supporter as it allows risks to be ignored and keeps contingencies down to show projects off as attractive to stakeholders. This probably explains the remarkable endurance of the predetermined guidelines method.

True accuracy ranges for a given project should not be declared using AACE classes but derived from a consistent Monte Carlo modelling: the derived (not declared) accuracy ranges are called primary accuracy ranges (Raydugin (2013)).

The discussion provided in this section clearly positions regression-based methods as extremely important and informative descriptive statistics methodologies, although their power in inferential statistics in a form of parametric methodology is seemingly restricted by several natural limitations pointed out above.

1.7 High-level Introduction of Emerging Unconventional Risk Quantification Methods: System Dynamics and ANN

Two emerging risk management methodologies—system dynamics and ANN—are worth being introduced despite them still being considered recondite by many project practitioners. Both can be qualified as unconventional PRM methodologies as they zero in on risk quantification. However, both methods can be re-qualified if information about identified risks and their addressing is factored into the models. This is demonstrated for system dynamics in Chapters 7 and 16 where identified project risks after addressing are used as inputs to project system dynamics models.

System dynamics is a powerful method to uncover and recognize non-linear interactions in systems including project systems (Sterman (2000)). Two case studies are included in this book (see Chapters 7 and 16) to demonstrate the power and value of system dynamics. System dynamics provides valuable insights on non-linear interactions in project systems, pushes for taking them into account in Monte Carlo modelling and, hence, supports a promotion of non-linear Monte Carlo methodology to quantify the non-linear risk interactions. A comparison of system dynamics and Monte Carlo SCRA modelling is done to establish compatibility of the two modelling methods (see Chapters 5 vs. 7 and 15 vs. 16).

Similar to a Russian doll, ANN can be considered a part of machine learning, which in turn is a part of artificial intelligence. Both machine learning and artificial intelligence topics are fully left out in this book as irrelevant. Only a high-level introduction of ANN is carried out in Chapter 4 to demonstrate its possible relevance and applicability to PRM. Actually, ANN might be seen as a logical upgrade to regression-based parametric methods to improve risk quantification accuracy. At the same time, neither parametric nor ANN modelling are in the focus of this book as they are not actually risk quantification methods belonging to the unconventional category. In addition, they both have practical limitations mostly related to the quality of used samples. This allowed the suppressing of a temptation to include an ANN business case in the book.

The word 'emerging' in the title of this section is slightly misleading as both system dynamics and ANN methods have been well-known to academicians for decades (see e.g. Forrester (1961, 1968), Ferrari (1978), Braunschweig and Huot (1984), Coyle (1996), Chao and Skibniewski (1994), Murtaza and Fisher (1994) or Chen and Hartman (2000)). However, due to their perceived sophistication and complexity they have yet to make it to the PRM mainstream. In any case, an attempt to marry systems dynamics to PRM is one of the goals of this book. Both methods—system dynamics and ANN—are further discussed in Chapter 4.

1.8 Chapter's Markup and Takeaway

- The first fundamental goal of PRM is risk addressing. The second fundamental goal of PRM is risk quantification for contingency development.
- Systems can be described as inanimate, aminate, and mixed. Most project systems can be treated as mixed. The better a project is organized and run, the more inanimate it could be considered. Unsuccessful projects are usually more animate due to various realizations of psychological and organizational bias.
- PRM planning identifies PRM context of an upcoming project to ensure the context/risk quantification method match.
- Bias and zealotry can impede selection of adequate risk and risk quantification methods. Channel vision, anchoring, hidden agenda, servile, and marketing zealotry types should be kept in mind when selecting the methods.
- There are two types of statistics: descriptive and inferential. Risk quantification relates to inferential statistics.
- All risk quantification methods can be understood as either conventional (based on the traditional PRM process steps) or unconventional (focused on the risk assessment step of the PRM process only).
- Main conventional risk quantification methods are the scoring and the Monte Carlo methods. The main unconventional is the parametric method. Two emerging unconventional methods are artificial neural networks (ANN) and system dynamics.
- The scoring method is a perfect match to achieve the first PRM goal only. It can be used for the addressing of risks with impacts on any project objectives—both 'soft' and 'hard', although it has a critically low accuracy to be used for contingency development.
- The Monte Carlo method should be used for risk quantification. By design, its application is limited to handling risk impacts on quantifiable ('hard') objectives only.

- A key shortcoming of various versions of Monte Carlo methodologies is that main components of project risk exposure might be missed out resulting in quite low accuracy of predictions.
- The regression-based parametric inference methods have remarkably low accuracy due to multiple limitations related to biased sampling (convenience and judgement sampling) and method's design. Although they may be used in very early phases when the high accuracy is not expected and the Monte Carlo method does not work yet as the context differentiators have still to be defined.
- The parametric method can be used for forecasting the average outcome of a project class in early project phases, not of a particular project.
- The regression-based methods are successfully used in descriptive statistics. They allow uncover important interdependencies among regression parameters.
- The regression-based method is used for justification of AACE accuracy classes, which unfortunately resulted in proliferation of contingency development predetermined guidelines ('a 10% syndrome').
- ANN can be treated as a long-overdue upgrade to parametric modelling in a hope of increasing accuracy of predictions.
- System dynamics modelling allows uncover and quantify non-linear interactions in project systems including non-linear risk interactions.

2

Conventional Project Risk Management (PRM) Methodologies

To continue the discussion from Chapter 1, this chapter overviews the conventional PRM methodologies based on the traditional PRM process in more detail. It includes steps related to risk planning, identification, assessment, addressing, and monitoring, although the detailed review of risk assessment steps of the PRM process is put off until Chapter 3. There is no chapter in this book on unconventional PRM methodologies as those usually have the PRM process step devoted to risk quantification only: the entire PRM process is actually reduced to risk quantification. Chapter 4 overviews unconventional risk quantification methods.

Reviewing and telling apart comparable conventional risk management methodologies supported by various standardization and certification organizations mentioned in section 1.5.1 are outside the scope of this book. If the 10,000-foot view is taken, these methodologies have cosmetics differences. Instead, a high-level overview of a conventional PRM methodology is evaluated in this chapter based on a book by Raydugin (2013).

Conventional PRM methods overviewed in this chapter treat all risks as standalone risks. No risk interactions are contemplated. As a result, the Monte Carlo methodology reviewed in this chapter is the traditional Monte Carlo methodology. In addition, some PRM tools discussed in this chapter (e.g. a bowtie diagram) are developed for standalone risks.

2.1 A Definition of Project Risk

The following definition of project risk is used in this book (Raydugin (2013, 2017a)).

Risks are possible impacts on project objectives.

An identical definition could be used interchangeably that reads 'risk are possible deviations from project objectives'. This definition does not contradict other more verbose definitions of risk (see e.g. ISO (2018), PMI (2009), etc) and usually is very well received by project practitioners for its precision and conciseness. Representatives of other industries may replace the word 'project' in the definition for words reflecting on contexts of their activities (e.g. IT, operations, market, investment, insurance, military, political, etc.).

Modern Risk Quantification in Complex Projects: Non-linear Monte Carlo and System Dynamics Methodologies. Yuri G. Raydugin, Oxford University Press (2020). © Yuri G. Raydugin.
DOI: 10.1093/oso/9780198844334.001.0001

The proposed definition implies following five realizations of uncertainty pertaining to risks of any project.

2.1.1 Occurrence and Probability Uncertainty

Probabilistic nature of risks has two angles. First, a risk event might happen or might not having a probability of occurrence $P < 1$. This is an occurrence uncertainty. Second, the occurrence uncertainty has two scenarios. A first generic scenario is realized when probability P cannot be defined as a one-point number (probability uncertainty) and should be accessed as a probability uncertainty range $(P_{min}; P_{max})$. A second less common scenario takes place when probability P can be clearly defined as a one-point number (probability certainty) based on historic data, modelling or mathematical calculations. For instance, in a flip-of-a-coin experiment, tails and heads' probabilities are equal $P_t = P_h = 1/2$ and are not characterized by probability ranges. In some cases, $P = 1$ which corresponds to a full occurrence and probability certainty. This presumes two situations. First, it can be a future event that should happen for sure but has not happened yet. It can be tagged as a certain future event. Second, it could be a given or a fact including a previously occurred uncertain event which is called an issue.

2.1.2 Uncertainty of Impact

An exact impact I of a risk on a selected project objective is usually unknown beforehand and can be assessed as an uncertainty range $(I_{min}; I_{max})$. This is a situation of an uncertain or unknown impact. In a limiting case a future impact might be known and could be defined by a one-point number $I = I_{min} = I_{max}$ (a certain or known impact).

Combining all possible scenarios for probabilities (section 2.1.1) and impacts (section 2.1.2), four main risk objects that should be remembered in project risk management can be derived as follows.

- An uncertain event: occurrence uncertainty $(P < 1)$ *vs.* impact uncertainty $(I_{min}; I_{max})$, where a probability could be either a range $(P_{min}; P_{max})$ or a one-point number $(P = P_{min} = P_{max})$.
- A discrete uncertain event: occurrence uncertainty $(P < 1)$ *vs.* impact certainty $(I = I_{min} = I_{max})$, where a probability could be either a range $(P_{min}; P_{max})$ or a one-point number $(P = P_{min} = P_{max})$.
- A general uncertainty: occurrence and probability certainty $(P = P_{min} = P_{max} = 1)$ *vs.* impact uncertainty $(I_{min}; I_{max})$
- An issue: occurrence and a probability certainty $(P = P_{min} = P_{max} = 1)$ *vs.* impact certainty $(I = I_{min} = I_{max})$.

Probability uncertainty is secondary in this classification as soon as there is an occurrence uncertainty ($P < 1$). In many practical cases a probability range is introduced for discussion purposes in risk workshops whereas all Monte Carlo software tools actually intake one-point probability numbers in any event equal to an arithmetic average of P_{min} and P_{max}.

The cardinal difference between uncertain events and discrete uncertain events is a level of impact uncertainty. In case of a discrete uncertain event, if this occurs, its impact will be a one-point number. For example, there could be a risk that an environmental impact assessment (EIA) might recommend the installation of a particular set of equipment which was purchased previously for another project but not used. The cost risk impact, if the EIA risk does occur, will be a one-point number as the cost of the previously purchased equipment is known precisely.

There is a cardinal difference between general uncertainties that are givens or facts with unknown impacts and general uncertainties that are future events. The former are the subject to activity and line-item ranging Monte Carlo methodologies. The latter should be treated the same way as future risk events that are quite certain in terms of probability of happening. This topic is further explored in Chapter 13.

A term 'risk' is used interchangeably with a term 'uncertain event' in this book.

Apparently, these four main risk objects may become transformed to each other in the course of project development: they are limiting cases to each other. For instance, an occurred uncertain event is transformed to an issue as it loses all uncertainty characteristics becoming an actual. Uncertain events as well as discrete uncertain events are normally populating project risk registers and are not told apart. Issues and general uncertainties are often left out to be managed outside PRM. However general uncertainties are factored in to risk quantification models, notably, to Monte Carlo models along with uncertain events.

A role of issues is mostly overlooked in PRM. Although issues including occurred uncertain events play a pivotal role: they serve as causes of uncertain events in bow-tie diagrams (see section 2.2.3.2). As further discussed in Chapter 9, some types of issues that are referred to as static and dynamic project system issues can serve as systematic risk causes and play a paramount role in description of non-linear risk interactions.

2.1.3 Uncertainty of Favourability

Possible impacts on project objectives could be either favourable and upside (opportunities) or unfavourable and downside (threats). Uncertain events normally result in either upside or downside impacts (not both). On the contrary, general uncertainties are capable of giving rise to both downside and upside impacts on objectives simultaneously. In many cases favourability depends not only on what risk objects are discussed but also on how corresponding baselines are developed. Namely, aggressive schedule and cost baselines attract more negative impacts, whereas prudently developed ('inflated' or 'padded') baselines have a room for optimization and positive impacts.

2.1.4 Uncertainty of Manageability

As argued in section 1.2, a primary PRM goal is to address risks or, in other words, to suppress threats and enhance opportunities. Required risk addressing actions and controls should be developed and implemented to achieve this goal. As the outcome of some future addressing actions might not be fully controlled by a project team or the project team is not skilful enough, outcomes of risk addressing efforts may be uncertain: either design or implementation of risk addressing actions (or both) may fail. As a result, risk severity levels after addressing could be not as good as expected. As the second goal of PRM is to carry out risk quantification and come up with adequate contingencies after addressing (section 1.2), risk addressing failure will lead to failure of risk quantification. That is why project risk contingencies after addressing should have a standard condition attached: all approved addressing actions should be successfully implemented. This condition may be rather optimistic in many cases.[1] If there are reasons to believe that all addressing actions may completely fail, risk assessment before addressing should be used in decision-making.

Possible failure to implement risk-addressing actions and controls points to a risk manageability uncertainty.

2.1.5 Uncertainty of Identification

All speculations about uncertainties of occurrence, impact, favourability, and manageability are suitable only if all relevant risks were identified in the first place. For multiple reasons why a particular risk might not be identified to stay unknown see in Raydugin (2014):

- project novelty,
- maturity of industry,
- phase of project development, and
- psychological and organizational bias pertaining to a project team.

Obviously, a risk that has not been identified cannot be managed and assessed as it does not appear on the PRM radar screen. It stays unknown until it happens. This implies failure of both PRM goals—risk addressing and assessment.

A possible risk identification failure points to risk identification uncertainty.

[1] Some modern Monte Carlo software packages (e.g. Safran Risk) have a useful built-in functionality to factor a possibility of risk-addressing failure into modelling. This corresponds to a case 'after partial addressing' and should be considered quite realistic.

2.2 Three Components of a Conventional Risk Management System

A project risk management system is comparable with a symphonic orchestra that includes several instruments. It requires proper structure and fine tuning. Its parts should be well synchronized to ensure a professional performance or to avoid a cacophony at least.

An efficient project risk management system is supposed to manage successfully all PRM aspects including five realizations of uncertainty introduced in section 2.1. It should be structured to have the three following cornerstone components (Raydugin (2013)).

- PRM context vs. adopted methods: specific characteristics and risks of a project (context differentiators) as well as PRM requirements and methods adopted by a project organization.
- PRM process: steps to identify, assess, address, and monitor risks.
- PRM tools: instruments to support PRM process steps.

These three components are listed in accordance with their importance. The PRM context should shape the other two components, whereas the PRM process should be supported by required PRM tools. For example, a very common mistake is to use the same standardized PRM tools and the PRM process to any types of projects in any project phase without any adjustments. This is a typical PRM cart-before-the-horse situation.

Full definition of the adopted PRM system (all three components) should be represented in a PRM plan. This document can be created as early as in FEL1 phase and should be updated in the course of project development at the beginning of every new project phase. For example, the PRM context, the process, and tools in FEL2A when project alternatives are considered, significantly differ from the ones used in FEL1, FEL2B, or FEL3. A planning step as a very first step of the PRM process is devoted to developing and updating of the PRM plan.

2.2.1 PRM Context

First, a PRM context reflects on features of a project and its internal and external environment that all may be tagged as context differentiators. Anticipating a discussion in Chapters 9–11, a project system is made of two interacting subsystems: a project structure subsystem (PSS) and project delivery subsystem (PDS). The former comprises all features of a project as a deliverable: scope, size, components, locations, technologies, etc. The latter describes an organization that is to deliver the project: leadership, required technical skills and procedures, equipment, contractual obligations, tools, etc. Unique PSS and PDS features constitute context differentiators of a given project.

The project system (both subsystems) evolves in the course of project development. A PSS initially takes a form of business and economic studies, technology reviews, location surveys, memoranda of intentions, etc. As soon as a business case is converted to a selected project alternative in FEL2A, the PSS takes a form of specific technical and commercial studies, data sheets, drawings, reports, estimates, for the selected alternative. In FEL3 a full project definition should be completed and submitted for the FID approval. As soon as a project moves to execute, it obtains more and more tangible forms (detailed engineering drawings and datasheets, purchased installed equipment, roads, camps, concrete works, pipe racks, pipelines, erected facilities, etc.).

Similarly, the PDS evolves from a small team in FEL1 and FEL2 to a much larger integrated engineering team in FEL3. As soon as a project is sanctioned, an EPC contractor is selected and awarded a contract to execute the project. An EPC team is mobilized to conduct engineering, procurement, and construction activities.

A PRM plan should reflect on features of the PSS and the PDS in every given phase of the project development. Any inefficiencies, mismatches, discrepancies, or imbalances inside these subsystems and especially between them should be identified as these can become additional systematic causes of project risks. These issues are referred to as project system issues in this book. They should be documented and used for risk identification along with regular risk causes that do not pertain directly to the project system (weather, underground conditions, availability of contractors, regulations, permits, market situation, etc).

Second, PRM context should dictate which objectives are important to PRM. Typical project objectives of importance (Raydugin (2013)) are:

- Scope/quality: to deliver a facility that has production rate, reliability, availability, etc according to an approved specification;
- Cost: to deliver a facility according to an approved budget;
- Schedule: to deliver a production facility by an approved date;
- Safety: to comply with all applicable safety requirements and standards in order to avoid safety incidents;
- Environment: to comply with all applicable environmental regulations and standards to avoid impacts on environment;
- Reputation: to assure that no negative impact on the company's reputation and relations with main stakeholders occurs during project delivery and in operations.

Many organizations treat soft objectives as 'goal zero' ones meaning that there should be no negative impact on safety, environment, reputation, etc.

Additional objectives can be added to the list depending on PRM context (security, legal, employee retention, profitability, cash flow, business continuity, etc.). Depending on a type of a project some objectives could be amended or further specified. As a case in point, an IT security project may use 'a response time' objective in place of a generic quality objective; a housing development project may use a 'lease-ability' objective as a primary quality objective to ensure that developed housing units are of any interest to potential leasees, etc.

Third, PRM context covers organizational aspects including PRM and other procedures used for managing risks and developing project contingencies. These inevitably include scheduling and cost estimating procedures that provide input information for schedule and cost risk identification, addressing, and quantification.

The PRM plan should include a method statement on risk activities planned in a current project phase. This should specify types of considered risks and applicable methodologies. As a case in point, the PRM plan should specify if schedule-driven costs are part of the consideration, or if cost escalation and currency exchange risks should be included,[2] etc.

Finally, PRM context should stipulate roles and responsibilities of project team members in PRM. This aspect should be also reflected in the PRM plan.

PRM process to support PRM context is described in section 2.2.2.

2.2.2 PRM Process

A typical PRM process is represented by Figure 2.1. Table 2.1 summarizes purpose and features of PRM process steps (Raydugin (2013)). Table 2.1 points to expected status of risks and actions.

Risk assessment steps are discussed in detail in Chapter 3. Whichever risk assessment methods are adopted, risk assessments should be undertaken twice (a two-step assessment): before addressing ('current' or 'as-is' assessment) and after ('future' or 'to-be' assessment). Any previously implemented addressing actions or controls should be factored in to the before-addressing assessment although any future addressing should be excluded. On the contrary, an after-addressing assessment should factor in all proposed and approved future addressing actions and controls in addition to all previously implemented ones. After-addressing assessments presume that all proposed and approved addressing actions were already implemented despite in reality most of them would only be implemented in some distant future. This accentuates a supportive role of risk assessments to evaluate expected efficiency of proposed addressing actions referring to the first PRM goal.

The second PRM goal to develop contingencies after addressing becomes achievable when development of addressing is completed. It is done presuming that approved addressing actions will have been successfully implemented in due time.

According to Figure 2.1, the PRM process contains a feedback loop from monitoring step to risk identification step. The reason for this is that risk exposure evolves in the course of project development because:

[2] Cost escalation and currency exchange risks and their modelling are outside the scope of this book. See Raydugin (2013) to review this topic.

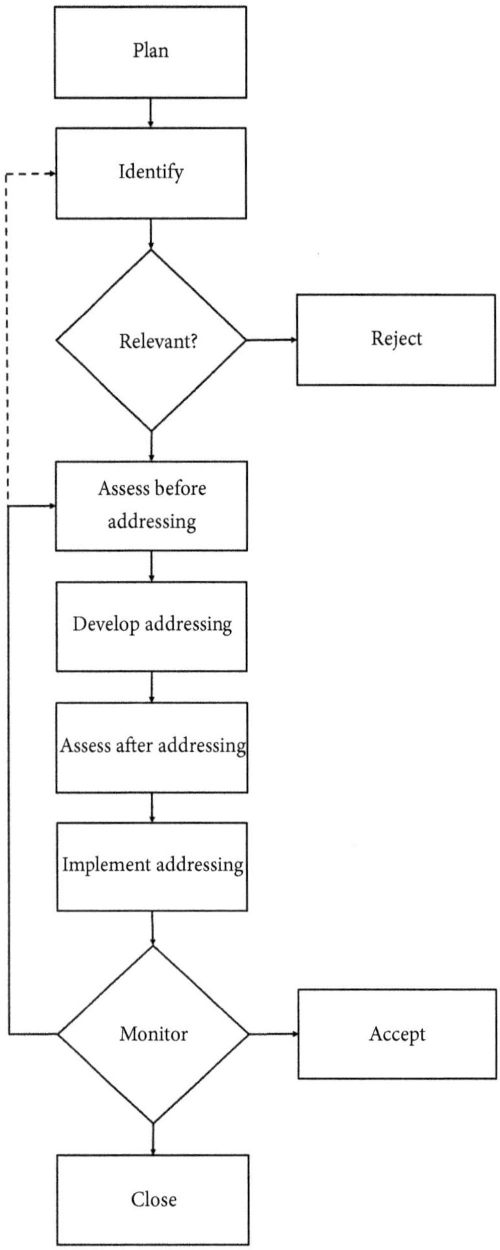

Figure 2.1 A PRM process

- new risks may emerge,
- some previously identified risks may have happened,
- windows of occurrences of some identified risks may be over,
- new information about probabilities and impacts of some risks could be collected,
- changes to a project may be carried out,
- secondary risks may be induced due to implementation of risk-addressing actions or realization of opportunities,
- etc.

Table 2.1 Description of PRM process steps

Step	Purpose	Outcome
Plan	To fully recognize PRM context and plan PRM activities	Fully developed PRM plan
Identify	To identify all causes, risk events, and risk impacts	All identified risks are included to a project risk register (risk status: proposed/inactive)
Relevant?	To verify if identified risks are relevant	Irrelevant risks are pointed out (risk status: proposed/inactive)
Reject	To screen out irrelevant risks	Irrelevant risks are left out, some risks are amended (risk status: rejected/inactive or approved/active)
Assess before addressing	To undertake current risk assessments 'as-is'	Risks are assessed by a selected assessment method (risk status: in progress/active)
Develop addressing	To put forward and approve a set of addressing actions and controls	Each risk has required addressing actions and controls (risk status: in progress/active; action status: approved/active)
Assess after addressing	To undertake risk assessments 'to-be' as if all addressing actions were implemented	Risks are assessed by a selected assessment method (risk status: in progress/active)
Implement addressing	To implement approved addressing actions and controls	All approved addressing actions and controls are implemented (risk status: in progress/ active; action status: completed/inactive)
Monitor	To monitor risk addressing implementation and changes in risk exposure	Progress of risk addressing implementation is tracked, changes in risk exposure documented
Close	To close risks that became irrelevant	All irrelevant risks are closed (risk status: closed/inactive; action status: closed/ inactive)
Accept	To draw a conclusion that no more addressing for a risk is required or possible	All fully addressed risks are accepted (risk status: accepted/active; action status: closed/ inactive)

By the reasons listed above, PRM process should repeat itself regularly in every project phase and for each particular active risk. Frequencies of PRM reviews of all risks or particular risks (e.g. most severe risks) as well as required PRM reporting should be established in a PRM plan.

To sum up, identify, relevant, and reject steps of PRM process aim at the identification uncertainty reduction (see section 2.1.5) with a goal of reducing a room for so-called unknown unknowns as well as at the favourability uncertainty reduction (section 2.1.3). Assess before and after addressing steps evaluate the occurrence and probability uncertainties (section 2.1.1) and the uncertainty of impact (section 2.1.2), although they also do support evaluation of the manageability uncertainty (section 2.1.4). This proves that the proposed PRM process is capable of managing all five realizations of project uncertainties.

2.2.3 PRM Tools

A number of tools are required to support PRM process every step of the way. All selected PRM tools should be described in the PRM plan.

Risk identification step utilizes two tools: a risk breakdown structure (RBS) and a bowtie diagram.

2.2.3.1 Risk breakdown structure

A risk breakdown structure (RBS) could be viewed as a cousin of a project work breakdown structure (WBS). An RBS contains categories that describe where particular risks might occur in a project (sources of risks). A typical RBS normally has two or three levels. Many companies keep the same RBS categories for entire project portfolios at the top RBS level making lower level categories tailor-made to reflect on project contexts. This seems to be a best approach as this allows slice and dice project risks and roll them up to enterprise risk management (ERM) categories. This is a way to keep an RBS flexible enough to reflect on the PRM contexts of projects and keep it consistent enough at the top ERM level.

Table 2.2 demonstrates a sample RBS. Normally the RBS categories at the PRM level are further specified. For example, all relevant engineering disciplines should be added under engineering category. If required, stakeholders could be listed, etc. Until a business case is settled down, all five RBS categories at the ERM level should be kept, although technical category is usually kept in an embryonic state in early phases. In later phases some ERM categories may die out beginning with the economic one.

Some risks can belong to more than one category. For example, permitting risks could belong to regulatory, stakeholder, or construction categories.

A situation when a risk seems to belong to several RBS categories simply points to several risk causes. This does not bring forward any major issues as soon as all relevant

Table 2.2 A sample risk breakdown structure (RBS)

RBS categories (ERM level)	RBS subcategories (PRM level)
Political	External stakeholders Regulatory Consortium partners
Organizational	Internal stakeholders Resources and skills Procedures and tools Change management
Commercial	Market conditions (vendors and contractors) Agreements and contracts
Economic	Return on investment Other benefits
Technical	Engineering Procurement Construction Commissioning and startup Operations

risks are eventually identified. A PRM plan should specify RBS categories and rules to apply them.

Sometimes RBS categories are called memory joggers by practitioners used for risk identification, and an RBS itself could be understood as a risk register without risks. The ultimate purpose of an RBS is to consistently scan the project's internal and external environment for risk sources to facilitate risk identification.

2.2.3.2 Bowtie diagram for risk identification

A bowtie diagram is a simplification of an engineering fault-tree analysis approach (Ebeling (1997)) that logically links causes of risks, risk events, and impacts on project objectives. Figure 2.2 introduces a typical bowtie diagram that got its name because of its shape.

Typical causes of risks relate to the following categories (Raydugin (2013)):

- novelty,
- lack of info,
- ambiguities,
- assumptions,
- inefficiencies,
- decisions,
- concerns,
- choices,
- changes,
- facts,
- givens,
- bias,
- occurred risks,
- etc.

As could easily be figured out, causes of risks that explain why risk might happen directly reflect on features of PRM context. It is worthwhile to point out that the bowtie diagram of Figure 2.2 is suitable for identification of standalone risks only. Chapter 9 demonstrates how the bowtie diagram should be modified in case of non-linear risk interactions in a project system.

Structure of the bowtie diagram (causes–risk even–impacts) points to a possibility of identifying risks using three-part naming: several causes may lead to a risk event that in turn will result in impacts on several objectives (Raydugin (2013)).

Figure 2.3 shows the bowtie diagram for a most simple case when two standalone risks are interacting being correlated. Causes C and D could be similar or the same for both risks yielding various degrees of correlations. It is important to remember that one-point values cannot be correlated, only distributions can be. Correlations are used to screen out iterations with unrealistic combinations of impacts (Raydugin (2013)).

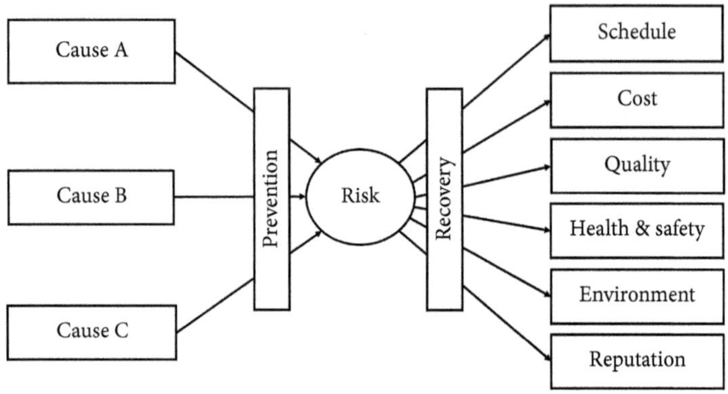

Figure 2.2 A sample bowtie diagram for a case of downside risks

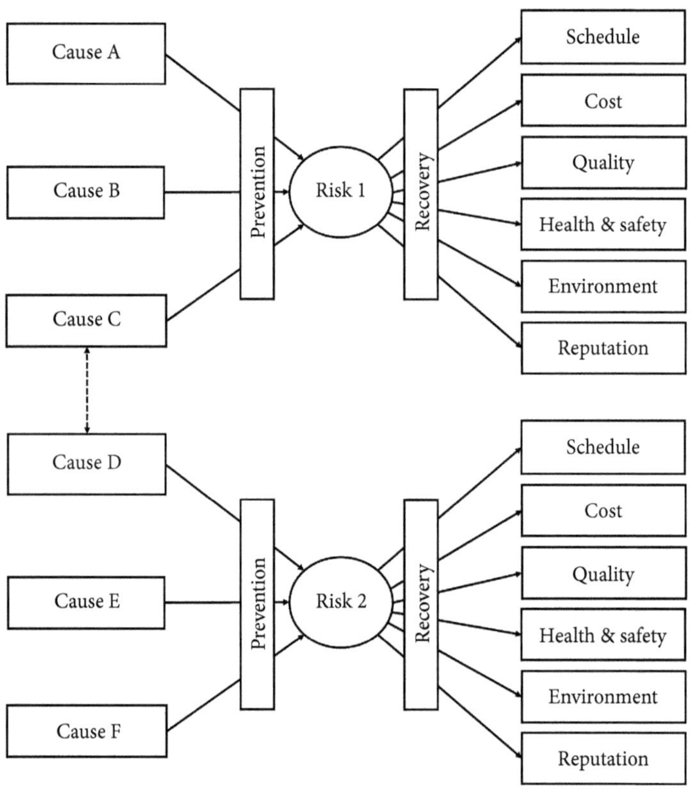

Figure 2.3 A bowtie diagram for two standalone correlated risks

They are taken into account in traditional linear Monte Carlo models as embryonic realizations of non-linearity.

The version of the bowtie diagram introduced in this section contains several project objectives. It is good for identification of risks with impacts on both 'hard' (quantifiable) and 'soft' (unquantifiable) objectives. In Chapter 9 only three quantifiable objectives are kept—schedule, cost, and quality. It is done in full alignment with the declared purpose of this book—risk quantification.

2.2.3.3 Bowtie diagram for risk addressing

The bowtie diagram is a powerful PRM tool not only for risk identification but for risk addressing too. According to its structure (Figure 2.2), there are two barriers that can be used to address risks. In case of downside risks or threats, the first barrier is about prevention when logical links between causes and a risk event are either weakened or completely blocked. This is an attempt to reduce the probability of risk occurrence. If all links are fully blocked, the probability of the risk event should be estimated as zero which corresponds to risk avoidance. The second barrier is about recovery when a risk does happen and becomes an issue although its impact on objectives is either reduced or fully eliminated.

Five risk-addressing strategies can be used in case of threats (Raydugin (2013)):

- avoid (preventive),
- mitigate—prevent (preventive),
- mitigate—recover (recovery),
- transfer (recovery), and
- accept (recovery).

As opposite to threats, opportunities are welcome in projects unless they lead to severe secondary risks themselves. In case of opportunities the five addressing strategies take the following form:[3]

- exploit (magnifying): to pull it to the scope,
- enhance—magnify (magnifying): to try to increase its probability,
- enhance—amplify (amplifying): to try to increase its positive impact,
- share (amplifying): to share it with project partners, and
- take (amplifying): to accept it when the other applicable strategies have been applied.

Strictly speaking, all risk-addressing strategies are fully applicable to uncertain events only that are characterized by both occurrence and impact uncertainty. In case of general uncertainties and issues only recovery strategies are applicable as $P = 1$.

2.2.3.4 Risk assessment tools

Chapter 3 describes details of risk assessment steps of the PRM process. Tools to conduct risk assessments in case of deterministic (scoring) and probabilistic (Monte Carlo) methodologies are introduced there. A risk assessment matrix (RAM) to assess risk severities before and after addressing is introduced in section 3.2. Probabilistic

[3] In case of opportunities the term 'barrier' could be replaced by a better term that is associated with opportunity's enhancement.

(Monte Carlo) tools are discussed in section 3.4. Selected and adopted risk assessment tools should be reflected in the PRM plan.

The previous sections describe the roles of the traditional bowtie diagram in risk identification and addressing. Besides these roles the bowtie diagram is kept in mind when probabilities of risk occurrences are evaluated. Namely, if causes A, B, and C are fully independent, a probability of the risk shown by Figure 2.2 may be assessed as $P = P_A + P_B + P_C$, where P_A, P_B, P_C are probabilities associated with these causes. By default, $P \leq 1$.

This approach implies that causes do not have any say in defining impacts on project objectives. This assumption is challenged in this book through a presumption that some causes can influence the impacts too. As argued in Chapter 9, these causes—systematic causes associated with project system issues—can and do influence risk consequences if risk interactions in a project system are taken into account.

To anticipate this discussion in Chapter 9, an example of a human immune system that is discussed in the introduction can again be called for. If two individuals who have very different statuses of their immune systems are exposed to a risk of contracting a flu or a corona virus, not only a probability to contract it is different in these two cases. The actual health impact, if a virus is contracted, can be very different too. The status of the immune system can be understood as a system issue and a systematic cause of the virus contracting risk. Its capacity to define both the probabilities and the impacts of the flu contraction should be pretty different in these two cases.

2.2.3.5 Project risk registers

Risk registers should be considered one of the most important PRM tools that are to keep entire risk information collected during risk workshops and reviews. Structure of a project risk register for a given project is dictated by PRM context including PRM requirements and the PRM process and should be fully defined by the PRM plan.

Risk register should contain core risk information about (Raydugin (2013)):

- selected project objectives,
- identified risks (three-part naming) and their owners,
- full information about risk probabilities and impacts before and after addressing,
- full description of addressing actions including their owners and implementation timing,
- status options for risks and addressing actions, and
- adopted RBS.

Traditionally project risk registers keep information about uncertain events only. Any other items such as issues, concerns, facts, assumptions should not be collected as standalone entries. However, they all should be attached to specific risks as causes in accordance with the bowtie framework (Figure 2.2). As discussed in Chapter 9, special attention should be paid to discovery of project system issues as systematic causes of project risks.

2.3 Role of Bias in PRM

Bias in PRM may be defined as a systematic error that may pertain to any PRM process step. Apparently, bias as a systematic error can be an ingredient of any project activity from engineering, estimating, and scheduling, to stakeholder management, and to leadership. Two sorts of bias are relevant to PRM as follows.

- Psychological bias that stems from psychological features of individuals taking part in PRM activities.
- Organizational bias that reflects on features of PRM procedures and methods used by organizations.

Psychological and organizational types of bias are highly interrelated and feed each other. Returning to section 1.3, various realizations of bias are associated with activities of human beings involved in project development. Hence, bias should pertain to project systems that are mixed-type systems by default.

2.3.1 Psychological Bias

Three generic sorts of psychological bias can be told apart as follows.

- Perceptual (subconscious): it is based on previous experience or gut feelings of an individual as related to particular types of risks or risk methods.
- Cognitive (subconscious) bias: this comes out of lack of understanding and knowledge of particular PRM concepts, methodologies, and tools.
- Motivational (conscious): this is an intentional sort when results of particular PRM activities are distorted on purpose.

In PRM lingo, perceptual bias is often tagged as 'anchoring'. Most malicious motivational bias may be described by terms such as 'hidden agenda', 'double-dipping', 'window dressing', 'red herring', 'optimism bias', 'group think', 'strategic misrepresentation', and, as discussed in the introduction, as 'a Hiding Hand'.[4] In some cases these three generic sorts of bias may be realized through some of the five PRM zealotry types discussed in Chapter 1.

2.3.2 Organizational bias

An organizational bias constitutes itself when PRM methodologies used by a company are not consistently implemented or they are not robust enough. The former

[4] Concerning cognitive bias, it would be impolite to cite here some jargon terms used to label it.

case includes a situation of the PRM context—PRM method mismatch discussed in section 1.4. Additional factors instigating organizational bias may relate to organizational culture and risk tolerance, qualification of PRM specialists, and resources allocated to PRM activities. Psychological and organizational sorts of bias are closely related. Moreover, the organizational bias could be described very often by the same jargon terms as psychological bias.

Any PRM process steps could be impacted by various realizations of bias. Some risks might not be identified staying unknown, other risks could be wrongly assessed, a set of incomplete or unattainable addressing actions could be proposed. Monitoring of risk could fail, some still active risks (for example, residual risks) could be closed, etc.

For example, as discussed in section 1.4, organizational bias associated with the PRM context—PRM method mismatch reflects on sample—population misfit in the case of parametric methods when biased samples are used. In case of Monte Carlo methodology organizational bias may be realized through missing important risk exposure factors that are excluded from Monte Carlo models.

Realizations of PRM systematic errors stemming from both psychological and organizational bias can be mathematically described through statistical bias discusses in Chapter 3. These realizations of organizational and psychological bias give rise to higher systematic errors and lower accuracies of risk quantification methods.

2.4 Chapter's markup and takeaway

- Risks are possible deviations from project objectives.
- The following five uncertainty realizations should be managed: occurrence, impact, favourability, manageability, and identification uncertainties.
- A PRM system has three components organized in a hierarchy: context, process, and tools. The cart should not be put before the horse as PRM tools should serve PRM context and the process, and the PRM process should serve context.
- A PRM plan should define all aspects of the PRM system adopted by a project.
- Besides planning, the steps to identify, assess, address, and monitor risks are key components of the PRM process.
- The risk breakdown structure (RBS) is a tool to pin down sources of risks.
- The bowtie diagram used for risk identification links causes of risks with a risk event and the risk event with impacted objectives.
- The bowtie diagram is also used to develop risk-addressing actions through breaking or weakening the links and using five risk-addressing strategies: avoid, mitigate-prevent, mitigate-recover, transfer, and accept.
- Risk assessment matrix (RAM) can be used for scoring assessments of risk probabilities and impacts on project objectives of both types: quantifiable ('hard') and unquantifiable ('soft').

- Two-step risk assessment should be undertaken: before and after addressing to support development of risk addressing (the first PRM goal).
- The Monte Carlo method should be used for quantification of impacts on 'hard' objectives and development of required risk contingencies (the second PRM goal).
- Bias as a systematic error in PRM steps including assessment steps might be either psychological (perceptual, cognitive, and motivational) or organizational, although both types are closely intertwined.

3
Overview of Conventional Risk Quantification Methods

This chapter provides a more detailed overview of the conventional risk quantification methods initiated in Chapter 2. These methods pertain to risk assessment steps that comprise the PRM process together with the other PRM steps (identification, addressing, monitoring, etc.). Most scoring and simulation methods may be treated as conventional as they cannot be consistently implemented without the other PRM process steps. Their power and shortcomings are discussed in this chapter. A cornerstone concept of statistical bias and systematic error is introduced to explain accuracy levels of various risk quantification techniques.

Unconventional risk quantification methods that do not pay much attention to risk identification, addressing, and monitoring are reviewed separately in Chapter 4.

3.1 A Challenge to Select Adequate Risk Quantification Methods

A list of relevant conventional and unconventional risk quantification methodologies is discussed in section 1.4.6. A true challenge is to select an adequate risk quantification method from the list that is a match to a given context of an upcoming project. This is a challenge to:

- evaluate the capability of each risk quantification method to produce accurate results,
- select those that are a best fit to a context of an upcoming project, or
- develop new methodologies if existing ones are inadequate.

Apparently, expert judgement, predetermined guidelines, parametric modelling, hybrid modelling, system dynamics and artificial neural networks modelling are unconventional risk quantification methods as they are focused mostly on risk quantification. Is this not amazing to undertake risk quantification without risk identification? In essence, all these unconventional methods try to assess overall project durations and costs that are supposed to include relevant risks whatever they are without their identification. Needless to say, an important project controls aspect—risk addressing (the first PRM goal)—is largely missed out too. Some of these methods are overviewed in Chapter 4. However, if an unconventional risk quantification method is

Modern Risk Quantification in Complex Projects: Non-linear Monte Carlo and System Dynamics Methodologies. Yuri G. Raydugin, Oxford University Press (2020). © Yuri G. Raydugin.
DOI: 10.1093/oso/9780198844334.001.0001

at least partially integrated in the overall PRM process and starts using at least one additional step it might be treated as conventional in some cases. For example, project system dynamics modelling introduced in Chapters 7 and 16 does take into account identified project risks after addressing.

Expert judgement as a standalone most ancient method is left out from the further consideration as it is of interest only from the psychological and organizational bias angle. Predetermined guidelines method—'the 10% syndrome' method (see Baccarini (2005))—is briefly discussed in Chapter 4.

As pointed out in Chapter 2, scoring method is a classical conventional method. Some simulation methods including Monte Carlo methods can be qualified as conventional if they are consistently used along with the scoring method or at least when they pay due attention to risk identification and addressing. However, some practitioners involved in Monte Carlo modelling might not pay due attention to steps to identify, address, and monitor risks. They get down to Monte Carlo risk quantification directly. Monte Carlo modelling that does not pay due attention to those steps could be re-qualified and demoted to unconventional. In fact, the author of this book started his risk management career paying attention mostly to Monte Carlo risk quantification. The other steps of the PRM process were virtually put on the back burner. This was explained by a previous good knowledge of the Monte Carlo methodology because of the background in physics and mathematics and poor knowledge of PRM at large. This could be clearly qualified as cognitive bias in a form of channel vision zealotry (see Chapters 1 and 2). A true understanding of importance of the other risk management steps came later.

It is important to point out that similar professional deformations could be observed when estimators, schedulers, safety specialists, economists, etc. begin to practise project risk management.[1] They bring methods used in their previous disciplines to risk management which often results in PRM context—PRM method mismatch. This should be qualified as a realization of a subconscious cognitive bias and in some most complicated cases as anchoring or channel vision zealotry.

Both scoring and simulation methods are reviewed in this chapter. Before technical aspects of various risk quantification methods are compared, important aspects of psychological and organizational bias related to methodology's selection should be pointed out. The reason for this is rather straightforward: there are complaints that some particular risk quantification methods are not adequate in particular situations. However, if selection of a risk quantification method is done in contradiction with project context, a wrongly selected method is not obliged to be adequate. As bias can play an active negative role in selecting risk quantification methodologies, the discussion on PRM zealotry types (see Chapter 1) is fully applicable to risk quantification methods that are part of PRM. The same is also the case in respect of the PRM

[1] For example, parametric risk quantification methods could be a good example of this as those branched out from traditional cost estimating and used mostly by good estimators turned to risk specialists.

zealotry insofar as the risk quantification zealotry may also take forms of a channel vision, anchoring, hidden agenda, a servile, or a marketing type of zealotry.

In reality there is no universal methodology that works perfectly in all circumstances and in all project phases. In addition, the following four factors should be kept in mind when selecting risk quantification methods.

First, all relatively recent risk quantification methods have been developed keeping in mind replacement of subjective empirical methods such as expert judgements and predetermined guidelines. At the same time many recent methods are still directly based on expert judgements and predetermined guidelines. This should mean only minute improvement over these early empirical methods or its illusion.

Second, some risk quantification methods may be considered ersatz-methods or even 'cargo cult' methods (see section 1.4.4) as they do not pass muster in terms of mathematics basics. This often happens when quite valuable methods in one context branch out to areas where they became inadequate. Expert judgements or references to empirical data are often used as 'band-aids' to fix low credibility of methods of this kind (see Chapter 4).

Third, some risk quantification methods simply are not robust enough as they either overlook or double-count major components of project risk exposure. They just cannot be adequate by design having apparent gaps. This is relevant to both Monte Carlo and parametric methods.

Fourth, any type of risk quantification zealotry could be reinforced by self-publishing of articles and books on the Internet or issuing them by journals or publishing houses that do not even require peer reviews. Such publications may be used to legitimize the questionable promoted methods.[2]

All considered risk quantification methods (both conventional and unconventional) should pass muster regarding the four points above before the practical applications.

3.2 Deterministic (Scoring) Risk Quantification Method

According to the PRM process of Figure 2.1, risk assessments should be undertaken twice—before and after addressing. In addition to PRM tools discussed in section 2.2.3 a scoring tool to carry out risk assessments should be introduced. Figure 3.1 demonstrates a sample risk assessment matrix (RAM) for downside risks that is typical for capital project in FEL2 or FEL3 with budgets around $1B. This contains a set of selected project objectives as well as five grades of probability and impact scores (5×5 RAM). These two parameters are to measure occurrence (section 2.1.1) and impact (section 2.1.2) uncertainties. In case of quantifiable objectives each impact is introduced as an impact range attached to a particular RAM's cell for a corresponding objective. In case of qualitative objectives short statements are attached to each RAM's cell.

[2] Recalling the definition of *bahramdipity* (see section 1.4.3), this is one of mechanisms of its realisation.

Severity colour legend	Low	Medium	High

Project Zemblanity	Downside impacts on project objectives						Probability score				
Impact score	Schedule, Mos	Cost, SM	*Quality, %*	*Reputation*	*Health & Safety*	*Environment*	< 1% Very low (1)	1 – 20% Low (2)	20 – 50% Medium (3)	50 – 90% High (4)	>90% Very high (5)
Very high (5)	>6	>50	*>10*	*International media coverage. Irreparable stakeholder impact*	*Single or multiple fatalities*	*Massive effect*	5	10	15	20	25
High (4)	3 – 6	20 – 50	*5 – 10*	*National media coverage. Substantial stakeholder impact*	*Serious personnel injury resulting in permanent disability*	*Major effect*	4	8	12	16	20
Medium (3)	1 – 3	5 – 20	*3 – 5*	*Regional media coverage. Moderate stakeholder impact*	*Serious personnel injury not resulting in permanent disability*	*Localized effect*	3	6	9	12	15
Low (2)	0.5 – 1	0.5 – 5	*1 – 3*	*Local media attention. Minor stakeholder impact*	*Medical treatment of personnel. Lost time incident*	*Minor effect*	2	4	6	8	10
Very low (1)	<0.5	<0.5	*<1*	*Slight media attention. Little stakeholder impact*	*Minor impact of personnel. First aid only*	*Slight effect*	1	2	3	4	5

Figure 3.1 A sample risk assessment matrix (RAM)

The sample RAM of Figure 3.1 shows six selected project objectives. Three of them—schedule, cost, and quality—are quantifiable hard objectives. The other three are non-quantifiable soft objectives.[3]

Three colours introduced in the RAM to reflect on severities of particular risks: high (red), medium (yellow), and low (green). In case of impacts on several objectives a top impact score multiplied by a probability score defines overall risks score and severity. A similar RAM could be developed for upside risks (aka opportunities) (Raydugin (2013)).

The discussed method is not statistical in the sense that regression-based and Monte Carlo methods are, although the expected value approach is still used to make risk assessments as products of probability and impact scores.

As one may reckon, RAM is a notoriously imprecise tool that contains scores attached to broad but rigid ranges of probabilities and impacts attached to particular RAM's cells. Sometimes there could be a need to consider probabilities or impacts that belong to two different ranges. For example, if a cost impact is expected to belong to a range from $15M to $25M, which impact score should be selected, 3 (medium) or 4 (high)? In practice more prudent higher scores are selected, although this does not eliminate infamously low RAM's accuracy. The next logical question to be asked should be why is RAM still used at all (Hubbard (2009))? This is a very fair question.

There is an ongoing discussion among risk professionals that sometimes reveal some forms of zealotry and even *bahramdipity* (see section 1.4.3). It is aimed at disavowing RAM for its critically low accuracy. As explained below, this stems from misunderstanding of the RAM's purpose, power, and applicability followed by the condemnation of this wrongly used tool.

Let's imagine a situation where a manager asks an employee to measure all physical objects in a room (binders, books, computers, chairs, desks, etc.) and prepare an inventory list that includes both their sizes and size ranking. In addition, let's imagine that the manager provides a wrongly calibrated ruler. For example, a declared standard twelve-inch ruler is, in reality, say ten-inch-long with completely wrong calibration. Its use should result in completely wrong absolute measurements by default as it has a clear systematic error. However, the task to complete size ranking is still fully attainable (Raydugin (2017a))!

The two-step risk assessment using RAM provides internal or mutual comparison of risk assessments before and after addressing. Despite both absolute assessments being quite imprecise their comparison should still be valid. It is based on assessment's differences that are actually a subject to proposed addressing actions. In other words, the two-step RAM assessment plays a supporting role in the PRM process. It supports a primary PRM goal to develop an adequate set of risk-addressing actions and controls. This comparison allows not only the elimination of doubtful absolute

[3] Only two out of three hard objectives are in the primary focus of this book: schedule and cost. However, quality objective plays an important role in non-linear knock-on risk interactions (see Chapter 8). The other soft objectives are largely left out from consideration. Despite their high importance they are completely outside the scope of this book.

risk assessments but reduces various realizations of bias: if bias of any type plays the same role in both absolute assessments it should become eliminated during the comparison. The bottom line: RAM is actually a fit not for risk quantification purpose (the second PRM goal). It is a fit for risk addressing purpose (the first PRM goal). When used properly it is a powerful and efficient tool to support risk addressing.

An importance of the scoring risk assessment method based on using RAM has another significant aspect. It allows rough assessments of risks with impacts on any selected objectives—both quantifiable and qualitative—to be carried out and development of addressing actions. These actions could address not only risks of impacts on hard quantifiable objectives but also on soft objectives including reputation, safety, environment, legal, etc. This is a huge advantage of the scoring assessment based on RAM utilization over the 'narrow-minded' Monte Carlo methodology which handles only quantifiable objectives.

To sum up this discussion, scoring RAM-based and Monte Carlo methods should not be in competition as they complement each other well. Reason for this is very practical. Scoring methodology is absolutely terrible in terms of achieving the second PRM goal—to develop schedule and cost contingencies. The Monte Carlo method should step in to fulfil this role. However, developed addressing actions should be directly used as raw inputs to Monte Carlo models to carry out risk assessments after addressing consistently.

Some practitioners modify the scoring method to estimate project schedule and cost contingencies instead of using Monte Carlo methods in an attempt to achieve the second PRM goal. It is undertaken by assigning monetary and duration values to impact scores. For example, a medium range (score 3) of Figure 3.1 is treated as \$12.5M and always has a two-month delay, which equate to mid-points of corresponding cells. In case of a very high range (score 5) which is open-ended, a mid-point cannot be defined at all. So predetermined multipliers should be established. For example, if a multiplier 1.25 is used, score 5 should always correspond to \$62.6M and 7.5 months. That is what happening when the method that is quite valuable to support development of addressing actions (the first PRM goal) is pushed behind its limits to do ersatz risk quantification (the second PRM goal). This type of risk quantification is a great candidate for qualification as a 'cargo cult method' and cannot be treated seriously.

3.3 Integrated Risk-based and Economics-based Selection of Project Alternatives

One of the main challenges in FEL2A phase of project development is a meaningful selection of a most suitable project alternative that should be further pursued. This is a bridge between a business case study in FEL1 and the rest of the project development phases. Namely, declared project objectives could be reached through several paths. For example, a capital project of a particular scope could be developed in one of several locations, may utilize one of several technologies, and might adopt one of several contracting strategies in execution phase. All viable combinations of significant parameters (locations vs. technologies vs. contracting strategies, etc) should be considered short-listed project alternatives to select from. The scoring RAM-based method discussed in section 3.2 can be applied in combination with economic evaluation in order to select a most suitable project alternative. A project RAM should contain all relevant project objectives.

To define the meaning of the word 'suitable', integrated risk-based and economic-based (IRBEB) selection is based on trade-offs as follows (Raydugin (2017b)):

- It is way more preferable to select an alternative with low risks in the first place instead of identifying and managing them successfully in later project phases,
- An alternative with the acceptable risk exposure after addressing that is the best in terms of economic returns (internal rate of return (IRR), return on investments (ROI), discounted cash flow (DCF), etc.) can be selected.

Project alternative selection is undertaken in two steps.

First, risk registers for each short-listed project alternative should be developed. A standard RAM developed for a given project in FEL1 should be used for risk assessment. Tailor-made alternative-specific addressing actions should be developed for each risk. However, if some risks are the same for all alternatives before and after addressing, they will not help out differentiate alternatives and may be ignored: only so-called risk differentiators should be kept. If an alternative has at least one critical ('red') risk after addressing related to any project objective, this alternative should be disqualified and left out from any further consideration.

Second, an economic evaluation for each alternative kept after the first step should be undertaken. Usually discounted cash flow (DCF) statements are developed for each short-listed alternative that include costs of risk addressing. An alternative with an acceptable risk exposure and the best economics can be selected. Normally, an acceptable risk exposure means that all risks after addressing associated with an alternative are low ('green'), although some alternatives with medium ('yellow') risk exposure and attractive economics could be considered too. Depending on an organizational risk tolerance, decision-makers being aware of potential threats may still consider such alternative. In other cases, an alternative with second best economics but lower risk exposure could be preferable. In both cases, it is up to decision-makers to make informed risk-based and economic-based decisions.

Besides selection of project alternatives in FEL2A, this methodology can successfully support other important project decisions. For example, it can be efficiently used for selection of engineering options in FEL3 and detail design (Raydugin (2012)) or bidder's evaluation in procurement (Raydugin (2013)).

A core advantage of this methodology is simplicity and transparency which facilitates quick decision-making. This further accentuates a power of the scoring PRM methodology when it is applied in a right way in a particular PRM context. Meaningful attempts to replace scoring risk assessments by more advanced Monte Carlo methodologies are possible only in the case of quantifiable objectives as all risks with impacts on soft objectives have to be ignored. Apparently, this would be quite one-sided decision-making that should be avoided where possible.

3.4 Probabilistic (Monte Carlo) Risk Quantification Methods

3.4.1 A Historic Overview of Monte Carlo Modelling

Monte Carlo methodology originated in the 1940s to solve some cornerstone equations of nuclear physics as part of the US Manhattan project and its Russian counterpart.[4] However, it came of age as a project schedule risk quantification method in late 1980s in combination with critical path method (CPM) and programme evaluation and review technique (PERT) (see e.g. Vanhoucke (2013)).

The initial version of PERT technique to evaluate programme schedules was developed in mid-1950 for the US Navy's nuclear submarine Polaris programme (Galway (2004)). Each task in a programme activity network diagram was assigned a duration uncertainty in a form of a triangular distribution: minimum (optimistic), maximum (pessimistic), and most likely task durations. As no occurrence probabilities were part of the consideration ($P = 1$) all considered uncertainties were general uncertainties by their nature. A version of PERT devoted to program costs was developed soon after based on consideration of project work breakdown structures (WBS).

In parallel DuPont developed a critical path method (CPM) (Galway (2004)). Despite it being based on the network representation of projects, it did not consider duration uncertainties: durations were deterministic one-point values. It was largely used to identify most critical tasks that drove project durations.

In the course of rapid computer power increase in the 1980s, PERT and CPM merged with Monte Carlo methodology. As a result, it allowed the analysing of project tasks with general duration uncertainties belonging to critical paths. Cost risk analysis was undertaken through reviewing cost distributions for WBS cost line items. Mainframe computers were used for modelling in that instance.

At that point in time, only general schedule and cost uncertainties were considered in Monte Carlo models. Uncertain events were not yet included explicitly. Moreover, the schedule and cost models were built and run separately. It was a common practice that uncertain events were factored in to general uncertainties but in an inconsistent way: each risk event contributed to higher uncertainty range around impacted normal

[4] The Manhattan project was devoted to development of nuclear weapons. Among many other challenges it was required to solve quantum Schrödinger equations for a population of neutrons to control (aka ignite) nuclear chain reaction. As required computing power was not available in 1940s, a sampling statistical technique was developed. The term 'Monte Carlo' was a code name to refer to this technique in a situation of a complete secrecy. It came up as an uncle of one of scientist was a 'practicing statistician': he was addicted to gambling and frequented Monte Carlo.

tasks. Certainly, it was well understood that in reality any normal task or cost account should be impacted by multiple risk events characterized by some specific probabilities. That said, it was a best possible technical alternative to count uncertain events in Monte Carlo models. Often it was required to add some substantial extra durations to uncertainty ranges to accommodate most severe risks, although there were no formal rules on how to accomplish this in a consistent way. As a result, this inevitably gave rise to highly skewed ranges when $(\textit{most likely} - \textit{min.}) \ll (\textit{max.} - \textit{most likely})$. As a result, it was possible to observe ranges such as (90 per cent; 100 per cent; 200 per cent) or (70 per cent; 100 per cent; 500 per cent). And, yes, this outdated type of ranging modelling is still used and unfortunately is very popular—activity ranging (AR) in the case of schedule risk analysis, and line-item ranging (LIR) in the case of cost risk analysis.[5]

The ranging risk quantification method consistently produces low and pretty much the same contingencies. This conclusion can be illustrated by comparing of three sample triangular general uncertainty distributions (80%; 100%; 150%), (90%; 100%; 140%), and (95%; 100%; 135%). It is easy to verify that all three have the same mean value of 110 per cent. If all normal tasks or cost accounts have ranges like these, a corresponding ranging Monte Carlo model will consistently produce 10 per cent contingency at the mean-value confidence level. This observation could serve as a justification for previously discussed complains by Merrow (2011) that ranging Monte Carlo methods are inadequate, which is an absolutely fair statement. This is also one reason why the predetermined guidelines and the ranging methods are treated as 'two boots that make a pair'[6] leading to the 10 per cent syndrome (Baccarini (2005)).

The next fundamental development happened in the 2000's when the first Monte Carlo tools incorporated project risk registers for uncertain risk events as separate modules.[7] This meant that risks could finally be represented in Monte Carlo models explicitly and consistently as uncertain events. Usually only a few dozen of the most severe risks make it to project risk registers with all the others still being counted as general uncertainties (see Chapter 17). Schedule and cost risk modules were built and run separately leaving out schedule-driven costs from a consistent consideration at that time.

A next break-through in modelling of schedule and cost risks occurred when a methodology to take into account schedule-driven costs in integrated schedule and cost risk analysis (SCRA) models was developed. Its principles are stated by Hulett (2011). A similar approach is put forward by Raydugin (2013). The key difference between the two cited methods is that the latter is based on a full, consistent and seamless integration of the deterministic (scoring) methodology as related to risk identification and addressing (the first PRM goal) with the schedule and cost risk analysis methodology to develop contingencies (the second PRM goal). However, in

[5] According to recent observations related to hiring, more than two thirds of risk analysts who claim that they are qualified to carry out Monte Carlo modelling actually are capable to do ranging modelling at most!

[6] A Russian proverb.

[7] It was done first in PertMaster in 2004 (Patterson (2018)).

many cases even these advanced Monte Carlo models demonstrated alarmingly low accuracy, although in some cases, Monte Carlo models can be surprisingly accurate (Raydugin (2017a), (2017c)). This encourages the further investigating of factors decreasing accuracy of the Monte Carlo modelling and finding ways to improve it.

As discussed in Chapters 8–15, non-linear risk interactions in project systems should be taken into account to drastically improve the accuracy of Monte Carlo models.

Table 3.1 summarizes the main features of various Monte Carlo simulation techniques. Four main factors of risk exposure are included in Table 3.1: general uncertainties, uncertain events, schedule-driven costs, and non-linear risk interactions. The non-linear risk interactions are included to Table 3.1 in advance anticipating the discussion in Chapters 8–15.

Despite correlations among general uncertainties and uncertain events being important factors that somewhat improve realism of Monte Carlo models (Raydugin (2013)), they are not listed in Table 3.1. As a matter of fact, any Monte Carlo technique can be used either with or without correlations. It is recommended that correlations are included to whatever version of Monte Carlo technique is used. Incidentally, correlations among risks represent a first although embryonic instance of non-linear interactions in a project system. Usually their overall contribution to contingencies does not exceed a few percentage points of baseline durations or costs.

Table 3.1 immediately leads to the following loaded question. Why is high accuracy expected from techniques that do not include some important components of risk exposure that are presumed by project context? Formal and neutral answer to this question is that corresponding techniques may be accurate enough when missed components of risk exposure are truly unimportant or irrelevant. For example, ranging

Table 3.1 Core factors included to various Monte Carlo techniques

Monte Carlo technique	Included factors of risk exposure			
	General uncertainties	Uncertain events	Schedule driven costs	Project system issues and non-liner risk interactions
Range estimating (activity and line-item ranging)	yes	no	no	no
Schedule risk analysis (SRA)	yes	yes	n/a	no
Cost risk analysis (CRA)	yes	yes	no	no
Integrated schedule and cost risk analysis (SCRA)	yes	yes	yes	no
Non-linear schedule risk analysis (N-SRA)	yes	yes	n/a	yes
Non-linear cost risk analysis (N-CRA)	yes	yes	no	yes
Non-linear integrated schedule and cost risk analysis (N-SCRA)	yes	yes	yes	yes

Monte Carlo method should be fine, if all project risks after addressing are completely avoided. Schedule-driven costs are unimportant if all planned project activities are completed on time as planned. Non-linear risk interactions can be ignored if all risks are non-interacting being truly standalone. As discussed in Chapters 8–10, this is possible in simple project systems characterized by high maturity, etc.

The next logical question to be asked is how often such conditions exist in real projects? The answer is obvious.

3.4.2 Activity and Line-item Ranging

The first practical Monte Carlo modelling method to assess project cost and schedule contingencies handled only one type of uncertainty—general uncertainty. By this reason it is still referred to as range estimating method (Curran (1989)). Schedule and cost contingencies were evaluated separately, schedule-driven costs were completely ignored. The other major risk object—uncertain events—was left out for technical reasons: earlier Monte Carlo software packages did not have a functionality to map project risks to normal tasks. Instead there were attempts to modify cost and duration ranges by using multipliers (risk factors) attributed to uncertain events. As a result, as already discussed in section 3.4.1, modified ranges became somewhat broader although probabilities of risk occurrences were completely ignored.

Opponents of Monte Carlo methodology sometimes use range estimating—activity and line-item ranging—as an easy target to criticize Monte Carlo methodology at large. The criticism is fully justifiable as range estimating does not take consistently into account three out of four major risk exposure factors (see Table 3.1). Hence, expectations that range estimating can produce accurate risk quantification results are rather preposterous.

Range estimating should be limited to introduction of general uncertainties (both types) to project schedules and base estimates. This is a risk-free situation in terms of uncertain events. Major reasons for introduction of general uncertainties should stem from a given level of project development: a stage of project development, available quotes, historic information, methodologies used for baseline development, etc. This should be understood as an uncertainty snapshot at activity (normal tasks in a schedule) and line-item (cost lines in WBS) levels in the current project phase. In a way it does not matter what are given characteristics of a project team or of a level of project complexity as the snapshot should capture resulting project general uncertainties as givens. This is a descriptive, not an inferential activity.

Range estimating could be understood as a direct match with parametric models in term of accuracy and nature of considered uncertainties: both consider general uncertainties and leave out uncertain events (see AACE (2011)). Monte Carlo range estimating could be used along with or instead of parametric modelling in early project phases when risk registers have yet to be developed and when a low accuracy of risk quantification suffices.

3.4.3 Monte Carlo Schedule Risk Analysis

Even outspoken opponents of probabilistic methods recognize a substantial value of Monte Carlo schedule risk analysis (SRA) (Merrow (2011)) when uncertain events are included to models along with general uncertainties.

There are three reasons for this:

- Higher accuracy: SRA cannot miss one of the major risk factors such as schedule-driven costs by default which is simply irrelevant (Table 3.1);
- Independency: SRA can be a standalone activity and does not require inputs from CRA;
- Project logic: It is based on project schedule logic and comprises benefits and power of PERT and CPM methods when every uncertain event is mapped to impacted normal tasks.

Although the project schedule logic is linear as opposed to non-linear logic of system dynamics discussed in Chapters 6, 7, and 16, it provides an additional realism to SRA modelling. Apparently, these benefits are not applicable to CRA.

According to Table 3.1 the only major missing factor in the case of linear SRA are non-linear risk interactions in a project system.

As soon as both general uncertainties and uncertain events are combined in a Monte Carlo model an issue of double or triple counting of risks arise: some project uncertainties could be counted more than once. The first time as general uncertainties. The second time as uncertain events. Moreover, those could be already counted as some allowances in baselines. Certainly, any instances of double or triple counting should be eradicated.[8] A simple rule is that any uncertainty factor should be counted only once as an allowance, as a general uncertainty or as an uncertain event. Most important uncertainty factors are usually counted as uncertain events. However, a practical rule is required to define when uncertainty factors can be counted as general uncertainties and when as uncertain events.

The required practical rule to carry out such delineation does exist. When an uncertainty range for a given activity or a line item is strongly skewed, there should be a reason to identify corresponding new uncertain events instead and reduce the degree of skewness (a skewness restriction rule). Practically, this should be done when

$$I_{max} - I_{ML} \geq 2 \times (I_{ML} - I_{min}) \tag{3.1}$$

Where I_{min}, I_{ML}, and I_{max} are values representing minimum, most likely, and maximum parameters of a triangular distribution for a given activity or a line item. A newly

[8] A PRM jargon term for double counting is 'double dipping'. Some risks could be already included to base schedules or estimates ('padded' baselines). 'Lean' or 'stripped' baselines should be used in Monte Carlo modelling to avoid double counting related to baselines.

identified risk event should be released from the pool of low-severity risks that are supposed to define corresponding general uncertainties.[9]

This skewness restriction rule prevents using a well-known trick when highly skewed activity or line-item ranges are imposed to augment or abate project contingencies. The direction of contingency amendment (i.e. range skewness towards higher or lower values) directly depends on a nature of a hidden agenda. The case of an abated contingency may perfectly play into the Hiding Hand discussed in the introduction.

If the skewness restriction rule is not obeyed, this would correspond to partial or full return to range estimating when general uncertainties, although counted along with uncertain events, dominate Monte Carlo modelling. The limiting case would mean very few or no risk events and highly skewed general uncertainty ranges. It is still possible to come across Monte Carlo models that have two or three risk events mapped to tasks on a critical path—often related to long-lead item's delivery—and highly skewed activity ranges of the majority of tasks.

This rule diminishes importance of general uncertainties and accentuates a paramount role of risk events in development of contingencies that take into account schedule-driven costs and non-linear risk interactions (Table 3.1). That is why general uncertainties are not the primary focus of this book. As shown in Chapter 17, general uncertainties are 'a dumping ground' for multiple low severity risks (tier-two risks) that did not get inclusion to project risk registers (tier-one risks).

3.4.4 Monte Carlo Cost Risk Analysis

Cost risk analysis (CRA) as a standalone modelling has much lower capacity to produce accurate contingencies than SRA has. Corresponding reasons are largely reciprocal to ones attributed to SRA in section 3.4.3:

- Lower accuracy: standalone CRA cannot count schedule-driven costs consistently.
- Dependency: CRA should be done after SRA to get information required for counting schedule-driven costs.
- An additional uncertainty: so-called burn rates used to calculate schedule-driven costs might have a high level of general uncertainty.
- No explicit project logic: CRA is not based on any schedule logic as CRA models are built without mapping of uncertain events to impacted WBS cost line items.

The last point is very important as cost risk impacts in linear CRA models are always local leading to one-off project budget increase. In the case of SRA modelling any delay leads to domino effect down-the-line in accordance with schedule logic.

Concerning the rule of conversion of cost general uncertainties to cost risk events discussed in section 3.4.3, it is fully applicable to CRA.

[9] The nature and origin of general uncertainties is explained in chapter 17.

Two types of cost risk impacts should be distinguished: fixed cost and variable cost impacts. It may be rightly guessed that variable costs impacts are fully attributed to schedule-driven costs. Hence, only fixed cost impacts can be taken into account in a standalone CRA modelling consistently. In terms of information collected in Table 3.1 the very major contribution—schedule-driven costs—is missed out. Instead, there could be attempts to inflate fixed-cost risk impacts implying some contribution of schedule-driven costs subjectively, although it cannot be done consistently[10] without actual running of an SRA model.

By the reasons listed above, it is not recommended to run CRA independently from SRA. Otherwise CRA results without schedule-driven costs will be quite confusing: a produced cost risk contingency will be inadequately low. From the bias-related angle, this approach could be advocated when there is a proclivity to abide project cost contingency and play into the Hiding Hand.

3.5 Accuracy of Conventional Risk Quantification Methods

Only accuracy of Monte Carlo methods is discussed in this section. The other major conventional risk quantification method—deterministic scoring method—does not need to be in the focus of this discussion. Despite it being very important to risk addressing (the first PRM goal), it has rudimental next-to-nothing quantification accuracy that does not need any additional comments.

3.5.1 When Accuracy is Low

It is easy to find complains on low accuracy of various risk quantification methods in the literature. Merrow (2011) complains that Monte Carlo simulations produce on average 9 per cent contingencies irrespectful of actual project risk exposure and comes up with fierce criticism of Monte Carlo modelling. Apparently, Merrow actually refers to range estimating Monte Carlo techniques discussed in section 3.4.2 while extending criticism knowingly or unknowingly to all types of Monte Carlo modelling. It is difficult to disagree with the criticism of ranging methods as range estimating does not count actual project risk exposure. Standalone CRA could easily produce similar results. Also, Merrow (2011) points to low accuracy of other non-Monte Carlo risk quantification methodologies including parametric modelling. The latter is discussed in Chapter 4.

Flyvbjerg (2017) points out that capital projects are traditionally characterized by severe overspendings and delays and introduces an iron law of megaprojects: 'overbudget,

[10] On several occasions it is stated in this book that a particular method cannot take into account particular components of project risk exposure consistently. In a situation when there is no required modelling functionality those components could be compensated through subjective tweaking of available modelling parameters.

over time, under benefits over and over again'. At the same time, this viewpoint on low accuracy of project risk quantification has clear bias-related or 'behaviour economics' angle. It does not matter how risks are quantified, the matter is how some decision-makers use (or mis-use) derived risk contingency information.[11]

Too often risk professionals complain about and stumble onto a fact that technically solid risk quantification results could be ignored by stakeholders and decision-makers. As mentioned in the introduction, by this reason risk quantification should be called political mathematics or political statistics as interests of many stakeholders clash around risk contingency discussions. It is extremely difficult to quantify this type of politically motivated bias, although the non-linear Monte Carlo methodology introduced in Chapter 13 does have a technical capacity to take any types of bias into account.

Previous Chapters 1 and 2 contain criticism of regression-based methodologies for using convenience or biased samples. It would be fair to point out that Monte Carlo methodology is not innocent in terms of using biased sampling too. Namely, some types of Monte Carlo models that do not take into account important components of project risk exposure (see Table 3.1) provide biased samples in each Monte Carlo iteration. If a Monte Carlo methodology does not take into account some important components of risk exposure such as risk events, schedule-driven costs or non-linear risk interaction, it plays into the Hiding Hand too: established project contingencies will be critically low.

In practice some risk analysts try to find or invent additional reasons and ways to inflate contingencies using existing lower-accuracy Monte Carlo methods. Usually it is done in the name of empirical data and experts' opinions. These are subconscious or conscious attempts to compensate methods' shortcomings and gaps. Excessive correlations of general uncertainties and uncertain events, inflated and highly skewed ranges of risk impacts and general uncertainties, inflated probabilities of risk occurrences, which play down efficiencies of risk addressing actions, etc. are typical signs of tempering of the sort.

3.5.2 When Accuracy is High

Cases when projects meet approved budgets and completion dates *do* happen, despite directly contradicting the iron law of megaprojects introduced by Flyvbjerg (2017). They are, however, treated almost as wonderful fairy tales with happy endings by the industry.

One such example is discussed by Raydugin (2017a), (2017c) and is mentioned in the introduction. A project team that worked on a carbon capture and storage project in Alberta, Canada completed it few weeks before the approved P50 date. It should be

[11] This is an allusion to understanding of democracy by the Soviet dictator Comrade Joseph Stalin: 'It does not matter how they vote; the matter is how we count.'

pointed out that an owner of this project was a top international oil and gas company with well-developed and implemented project procedures. No major scope changes were undertaken after project sanctioning. The local and federal governments were in full support of this showcase project even providing substantial grants to demonstrate efficiency of the carbon capture and storage technology. A project team assigned to this high-profile project was handpicked and very skilful. A project leadership was seasoned and considered the established completion date as a challenging but achievable stretched target. Some project risks did occur although their consequences were isolated and well managed. As discussed in Chapter 1, the system of that successful project can be qualified as almost completely inanimate. This unique combination of project attributes characterized by very low project system issues worked out successfully against the iron law of megaprojects (Flyvbjerg (2017)). Although, as they say, exceptions confirm the rules.

One important observation is that when a project team is characterized as strong, it responds well to stretched targets established by management. In other words, there is another reason why a project risk quantification would tend to be right. This reason may be labelled as a positive bias to beat the target dates and costs, although it is unrealistic to believe that weak project teams would respond well to stretched targets. The opposite might happen.

An additional discussion on the role of the positive bias is provided in Chapter 17.

3.5.3 Typical Distributions of Project Completion Dates and Costs

Any unbiased database on completed projects does contain information on both highly successful and notoriously failed projects. This means there should be projects that had adequate risk quantification methods and projects which risk quantification methods failed.

Figure 3.2 represents a sample of about 400 completed projects and information about project delays. The project schedule slippages are measured against approved project durations—the factor 1.0 means no schedule slippage. It does contain a few projects completed on or ahead of approved dates. Unfortunately, no consistent information on quality of project schedules, schedule risk quantification methods, or project system issues is available. Presumably, methods from expert judgement and predetermined guidelines, to parametric, to Monte Carlo were used in different projects. Somewhat consistent information is available on project changes. As could be expected, most notorious delays (the slippage factor 2.0 or higher) are observed in projects that had significant and multiple project changes after FEL2. Also, it could be noticed that there is a second peak on the curve in the vicinity of the slippage factors 1.7–1.8. As discussed in Chapter 17, this could be interpreted as an indicator of projects that ended in chaos.

Figure 3.3 cited and discussed by Hulett (2017) and initially produced by Hollmann (2015) shows a similar behaviour for project costs although the distribution has a

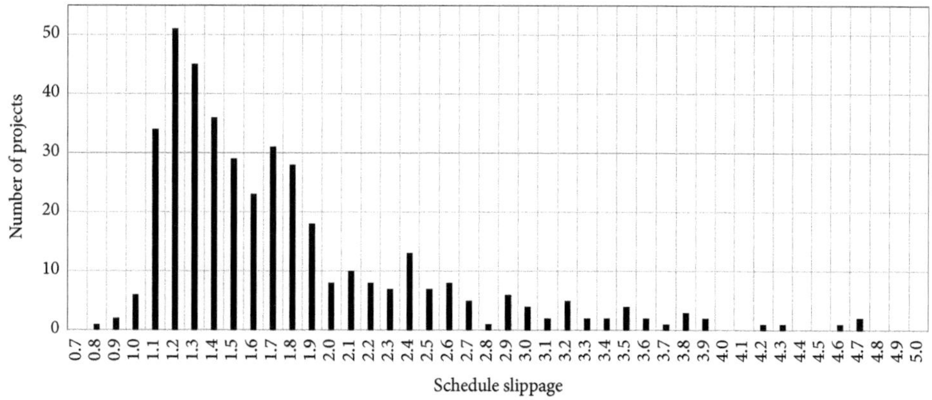

Figure 3.2 Schedule slippage for about 400 completed projects (geography: Middle East; time period: 2000–2019; project budgets: from \$1.5M to \$15B)

single peak. In a similar way as in the case of Figure 3.2, the factor 1.0 means no overspending. Actuals are compared with approved project budgets that include contingencies. Information about a number and types of projects as well as about risk quantification methods used by projects included in the sample is not available. It would be safe to presume that a variety of project system maturity levels is represented by projects included in Figure 3.3.

According to Figure 3.3, some projects were successful and their actual costs were below their base estimates. However, the majority of projects encountered overspendings of various sizes. According to Hollmann (2015) and Hulett (2017), at least 10 per cent of major projects overrun their estimated costs by 70 per cent or more. Figure 3.3 accentuates a fact that the actual cost distribution is strongly skewed in comparison to less skewed cost forecast distribution.

On the one hand, Figures 3.2 and 3.3 are not specific enough as required information about various important characteristics of included projects (risk quantification methods, maturity of project delivery systems, etc.) is not available. On the other hand, both Figures 3.2 and 3.3 point to the fact that duration and cost contingency estimating methods are often inadequate.

An inevitable blend of projects characterized by a variety of context differentiators included to a database accentuates difficulties of parametric modelling related to sampling. Using an average of just a few parameters of biased samples for inferential purposes sounds either fantastic or preposterous.[12] However, it looks as though ANN methodology introduced in Chapter 4 could eradicate this parametric method's shortcoming through much better possibilities of ANN training and testing, although quality and consistency of used data could still be a primary's method of limitation.

[12] It is difficult to expect successful diagnostics of a newly arrived patient based on comparison of his blood pressure and temperature readings with the average readings of existing hospital's patients especially if he arrived in a wrong hospital by mistake.

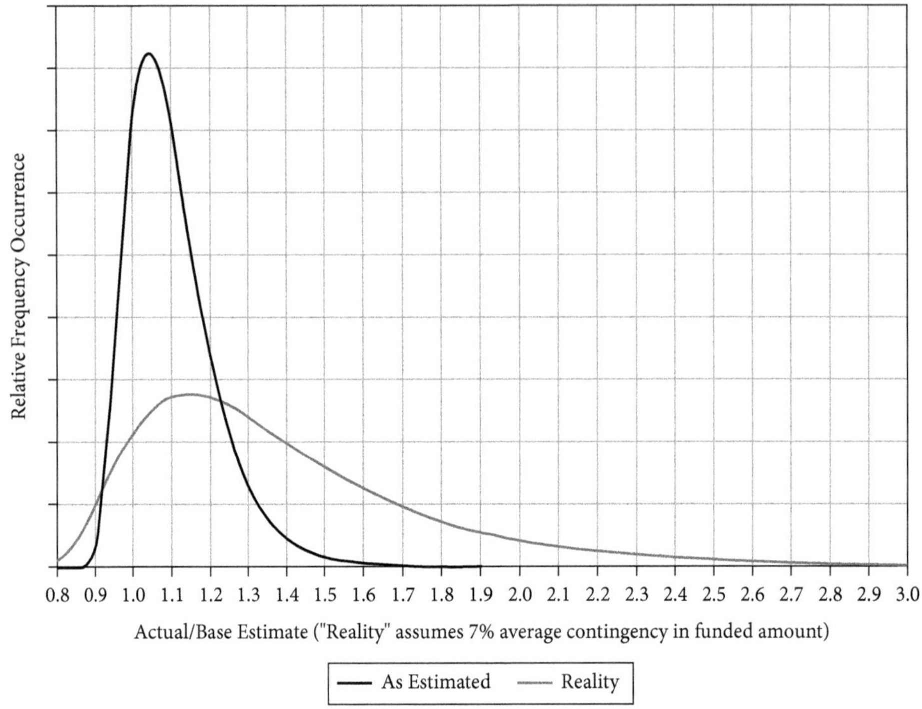

Figure 3.3 A comparison of cost risk estimate accuracy to empirical accuracy (originally published by Hollmann (2015); courtesy of IGI Global)

Currently used Monte Carlo methods miss at least one or more core risk exposure drivers. At best they can be adequate only in case of mature project delivery systems.

In addition, critics of the Monte Carlo technique argue that probabilistic methods do not use empirical data as opposed to parametric methods. First, many findings discovered by descriptive statistics can be and should be used as empirical inputs to Monte Carlo models. A new trend is to use artificial and augmented intelligence methods for providing empirical inputs to scheduling and probabilistic schedule risk methods (Patterson (2018)). This should be considered a promising attempt to better incorporate empirical data to scheduling and schedule risk management and break an imaginary self-proclaimed monopoly of parametric modellers on empirical data. Second, if a method is based on some empirical data it is not a guaranty of its adequacy (see Chapter 4): empirical data could be biased and an empirical method could be inadequate. Regarding Monte Carlo methods, even after using correct empirical data Monte Carlo models should still comprise all major components of risk exposure.

The bottom line is straightforward: there is no risk quantification methodology overviewed in this chapter capable of explaining the distributions like those of Figures 3.2 and 3.3 in their entirety for both successful and failed projects. In other words, all overviewed methodologies have their shortcomings and limitations. This suggests the need for a new Monte Carlo risk quantification methodology that is capable of explaining all features of the distributions represented by Figures 3.2 and 3.3. Even the most notorious cases such as the Sydney Opera House or the Boston Big Dig

Artery/Tunnel projects should be explainable and predictable. Such methodology should take into account all important risk exposure factors overviewed in Table 3.1. And the last missing factors are non-linear risk interactions.

3.5.4 Statistical Bias and Systematic and Random Errors

Various realizations of psychological and organizational bias are discussed in section 2.3. Another type of bias—statistical bias—is discussed in this section that has a direct link with the other discussed types of bias.

Statistical bias in the context of this discussion describes mathematically reasons for low accuracy of risk quantification methods. It comprises multiple reasons that include various aspects of PRM context—PRM method mismatch as well as relevant realizations of psychological and organizational bias. In turn, the PRM context—PRM method mismatch means that some critical components of project risk exposure including schedule-driven costs or non-linear risk interactions are missed out. In addition, banal reasons such as missed (unidentified) risks that stay unknown could contribute to statistical bias too.

A general discussion on statistical bias is provided by Spiegel and Stephens (1999). In a language of inferential analysis, an estimator or predictor is a statistic derived from evaluation of selected samples in order to infer a value of a population parameter of interest. An estimate is a one-point value produced by each sample.

In cases of Monte Carlo simulation each iteration can be treated as a sample that produces a one-point duration or cost estimate. Together all samples (aka iterations) yield a duration or cost distribution that can be understood as an estimator. This estimator can be used to infer a future project duration or cost outcome. Two main outcomes of the inference are as follows.

- Confidence interval: estimator's parameters (statistics) can be used to evaluate an interval of duration or cost values that should contain the expected project outcome with some probability.
- Confidence level: the statistics can be used to elaborate a specific duration or cost value that is expected to be higher or equal to the outcome of the upcoming project with some probability.

For example, an 80 per cent confidence interval should contain the actual project outcome which is always a one-point value with probability 80 per cent, although there is 20 per cent probability that the outcome would be outside of that interval. An example of the confidence level application that is used much more often could be a P70 duration or cost value which means that with probability of 70 per cent the project outcome should be lower or equal to this value, although there is 30 per cent probability that the outcome would be higher. Confidence level could be selected at any level from 0 per cent (P0) to 100 per cent (P100) which usually reflects on a corporate contingency development criterion. As a case in point, a mean value of the estimator μ_s could be adopted as the criteria although in practice it is seldom used.

A serious practical issue with application of the confidence concept is that sampling methodology used for the inference could be inadequate itself. In simple terms, a risk quantification model might have intrinsic deficiencies: either some major components of risk exposure are overlooked (the Monte Carlo modelling case) or a sample used for inference is biased (the parametric modelling case). This phenomenon is called a statistical bias.

It could be imagined for a moment that there is an ideal statistical distribution that perfectly includes all components of the project's risk exposure in accordance with project context.[13] A practically used estimator to approximate that ideal distribution is based on available samples. Two high-level parameters can be introduced to evaluate quality of the approximation—a systematic error and a random error.

A systematic error Δ (aka statistical bias) of the estimator used for the inference can be defined as

$$\Delta = \mu_p - \mu_s, \tag{3.2}$$

where μ_p is a mean value of that ideal imaginary distribution, and μ_s is a mean value of the predictor—a distribution produced by Monte Carlo sampling (see examples of Monte Carlo distributions in Chapters 5 and 16). In case of parametric models, μ_s is a mean value resulting from parametric modelling that is used for inference (see Chapter 4).

An estimator is called unbiased in a statistical sense if

$$\mu_p \approx \mu_s. \tag{3.3}$$

This corresponds to an ideal situation of highly accurate estimator when a systematic error is negligible $(\Delta \rightarrow 0)$.

In addition, a standard deviation of the estimator σ_s or its variance σ_s^2 could be introduced and treated as a measure of a random error around the mean value μ_s as samples (aka iterations in case of Monte Carlo modelling) which produce results (estimates) are scattered around the mean value μ_s. If estimates are scattered significantly around μ_s, the corresponding estimator is called inefficient. However, as discussed below, very high or very low random error might not be a bad thing. Simply, the imaginary distribution might have a particular (not necessarily low) random error which is a characteristic of the project risk exposure.

Hence, four situations could be told apart based on four types of estimators: biased vs. unbiased depending on the presence of a systematic error and efficient vs. inefficient depending on scattering of estimates.

[13] This thinking experiment with the ideal distribution required as an upcoming project is not completed yet. The imaginary population (all possible project outcomes) to approximate is defined as that ideal distribution. This is the difference with typical inference tasks when the population and samples that are drawn from the population to approximate its statistics are all existing.

If estimates produced by samples do not differ from each other much when the condition (3.3) is obeyed, a corresponding estimator can be called both unbiased and efficient having both low systematic and random errors. This is an ideal if not utopian situation which corresponds to a narrow distribution that does cover an actual project duration or cost outcome. This case is characterized by both high accuracy and high precision.

If estimates are scattered significantly around μ_s when the condition (3.3) is obeyed, a corresponding estimator is called inefficient although still unbiased as this allows a broad range of outcomes that covers the actual outcome. This case is represented by a wide distribution and could be understood as having high accuracy although low precision.

If

$$\mu_p >> \mu_s, \tag{3.4}$$

or

$$\mu_p \ll \mu_s, \tag{3.5}$$

the estimator produces perfectly wrong results systematically and it is called biased due to the high systematic error $(|\Delta| \gg 0)$.

Despite the expressions (3.4) and (3.5), some samples (aka iterations) might produce estimates that could be close enough to the actual project outcome if the precision is low (high σ_s). However, the vast majority of estimates will stay far away. This could be tagged as 'an inference paradox' when low precision of an inefficient estimator saves an inference model. The low accuracy 'supported' by the low precision still might cover an actual outcome! In essence the inference paradox states that 'the worse the precision, the better (if accuracy is low)'. This has a direct military analogy with the 'carpet-bombing' during aerial bombardment using powerful although low-precision ammunition.

The inference paradox explains a known propensity of some risk analysts to compensate the known low intrinsic accuracy of some particular Monte Carlo methods through indiscriminate inflating of random errors. In practice this is done through inflating risk probabilities, risk's and general uncertainty's impacts, and imposing excessive correlations among Monte Carlo model's parameters (see Chapter 2). Sometimes it is done in the name of empiricism. These actions do increase random errors without much justification making distributions less precise.

In the case of parametric models that produce one-point mean values only without any spread or random error, a spread around the produced mean values is imposed forcefully just because there should be some such spread! The controversy of this method is demonstrated in Chapter 4.

However, if systematic error is too large, inflated random errors might not save the flawed imprecise methodology anyway. The examples of project outliers—Sydney Opera House project (1,400 per cent cost overrun) and Boston Big Dig Artery/Tunnel project (220 per cent overspending)—should be drawn again to illustrate this. So far, neither traditional Monte Carlo nor existing parametric models are capable of predicting such outcomes.

Figure 3.4 illustrates the role of statistical bias (aka systematic errors) and random errors in defining method's accuracy and precision (Raydugin (2018a), (2018b)). It recaps the discussion provided in this section.

Unbiased estimators ($\Delta \rightarrow 0$, quadrants A and B of Figure 3.4) are characterized by high accuracy of inference on average as actual project outcome certainly belongs to the interval covered by a Monte Carlo distribution. However, situation represented by quadrant B corresponds to an inefficient estimator that has high random error (low precision). This makes the situation represented by quadrant A ideal (high accuracy and high precision) although in most cases utopian: the real risk exposure might point to higher random error.

Estimators depicted by quadrants C and D are clearly biased as they systematically produce wrong results (low accuracy): an actual project outcome is outside of the interval covered by the distribution. It does not matter that an estimator of quadrant C produces consistent estimates (high precision) as all these estimates are precisely wrong. Monte Carlo ranging methods are a good example of this. They consistently produce contingencies in the vicinity of 10 per cent although rarely are accurate (Merrow (2011)).

It is worthwhile to point out that accuracy of an estimator takes precedence over precision. If an estimator has low accuracy (i.e. high systematic error Δ), high precision (low σ_s) will not help to predict a project outcome at all (quadrant C). As discussed above, a situation when both accuracy and precision are low has more chances of predicting a right outcome due to the inference paradox: a broad distribution (low precision) might compensate low accuracy. A weather forecast example could be drawn to illustrate this. If a weather man on BBC World Service stated that the daytime temperature would be in the interval from −50 C up to +50 C he should be quite accurate despite the ridiculously low precision of the forecast.[14] The practical value of this forecast is a different topic despite the 99 per cent confidence interval. The practical values of corresponding inferential methodologies could be comparable.

3.5.5 Root Causes of Statistical Bias and Random Errors

In general, a systematic error of a distribution that reflect on statistical bias of a methodology reflect on a degree of bias of a used sampling. In case of the Monte Carlo methodology this mostly stems from elements of overall project risk exposure that are missed out. In case of the parametric modelling this is a direct product of convenience and judgement samplings.

[14] As the author writes this chapter, the outside temperature in Dhahran, Saudi Arabia, is +52 C. This is another reminder about the role of outliers in forecasting as well as about the flaw of averages (Savage (2009)).

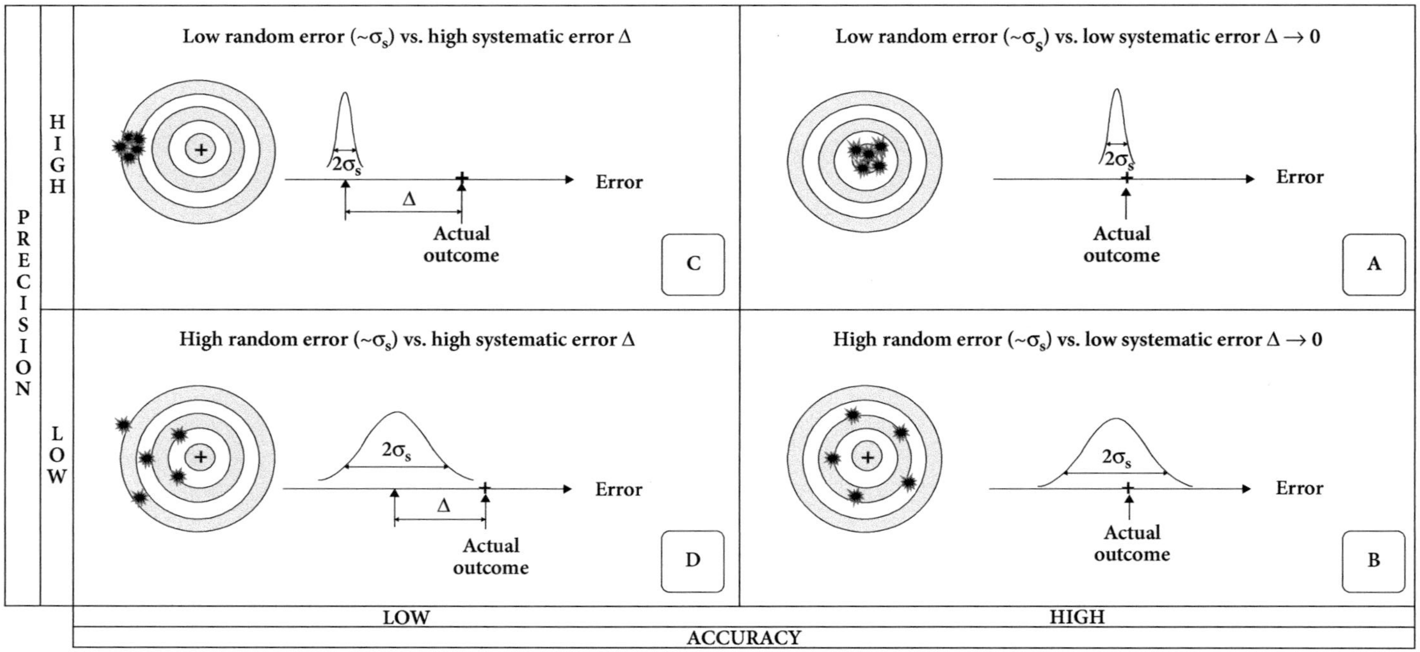

Figure 3.4 Systematic and random errors as characteristics of accuracy and precision

A distribution's random error is a reflection on the overall project schedule or cost risk exposure.[15] This means that high random error is not necessarily an indicator of a flawed Monte Carlo methodology if all relevant risks and their interactions are consistently identified, assessed, and addressed. If a project is characterized by multiple interacting risks of high severity that are possibly highly correlated, a width of the resulting schedule or cost distribution and the associated random error should inevitably be high. A low random error could correspond to the low level of risk exposure but not only. This could simply indicate that some risks are not identified, not assessed or not addressed correctly which inevitably gives rise to systematic error too. Another reason for low random error is that some significant correlations and other non-linear interactions among risk are missed.

As demonstrated in Chapter 15, the newly introduced non-linear Monte Carlo SCRA methodology does produce highly skewed risk distributions with 'fat tails' to cover actual outcomes. This is a feature of distributions pertaining to complex projects with a rich risk exposure that includes non-linear risk interactions. This methodology allows us to predict outcomes even of projects with abnormally high levels of delays and overspending comparable with those for Sydney Opera House or Boston Big Dig Artery/Tunnel projects.

This discussion is continued in Chapter 4 where the concepts of the estimator's accuracy and precision are tried on parametric methods.

3.6 Chapter's Markup and Takeaway

- Scoring and Monte Carlo methodologies are two main conventional risk quantification methods.
- Selection of right risk quantification methods can be impeded by various realizations of bias and even by PRM zealotry.
- RAM is used in the scoring method for high-level assessments of risk consequences through evaluation of probability and impact scores.
- The scoring method has a notoriously low accuracy and actually should not be used for development of risk contingencies (the second PRM goal). It should be used to support development of risk addressing actions (the first PRM goal).
- The scoring method can be adopted for selection of project alternatives in FEL2A if used along with economic evaluation.
- The Monte Carlo methodology may achieve critically low accuracy if major components of risk exposure presumed by project context are not taken into account.
- Monte Carlo ranging methods—activity ranging (AR) and line-item ranging (LIR)—miss out risk events and schedule-driven costs (CRA case) handling only

[15] A word 'error' could sound confusing. In case of Monte Carlo modelling a random error being a characteristic of the overall risk exposure does not bring any negative connotation. Projects with substantial risk exposure usually have a higher random error. In case of parametric modelling, it directly depends on a used sample and not actual risk exposure of the upcoming project (see Chapter 4).

general uncertainties. They consistently produce critically low contingencies (the 10 per cent syndrome).

- If ranges are abnormally skewed, new risks should be identified and released from the ranges. The project risks should be taken into account in a separate risk register module of Monte Carlo models.
- The traditional CRA method that handles both project cost risks and cost general uncertainties misses out schedule-driven costs.
- The traditional SCRA method misses out risk interactions.
- The traditional SCRA models can be accurate if risk interactions are negligible. In all other cases a more comprehensive Monte Carlo SCRA methodology is required to take risk interactions in to account consistently.
- Duration and cost distributions are usually significantly skewed.
- A concept of statistical bias explains systematic errors and, hence, low precision of risk quantification methods if important components of risk exposure are missed out.
- The inferential paradox demonstrates how a methodology with critically low accuracy can cover an actual outcome. It is possible if methodology's precision is critically low too.

4
Overview of Unconventional Risk Quantification Methods

This chapter provides a high-level overview of unconventional risk quantification methods despite them being actually outside the scope of this book: these are not risk quantification methods per se being project outcome forecasting methods. In terms of the traditional PRM process these methods are reduced to the project outcome assessment step in place of the risk assessment step. The other steps—identify, address, monitor, etc.—are not used largely as irrelevant. The power and limitations of the unconventional methods in particular situations are discussed in this chapter.

4.1 Regression-based Risk Quantification Methods

4.1.1 Historic Overview of Regression-based Methods

The empirical methods could be labelled as retrospective methods (Galway (2004)). In turn, retrospective methods could be broken down to descriptive and prescriptive (inferential). The first group is traditionally based on simple bivariate regression analysis when one independent and one dependent parameter are considered. For example, the RAND corporation was actively involved in such studies in 1950s–1970s although it effectively terminated them in the early 1990s (Galway (2004)) due to the emergence of the Monte Carlo methods. The second is referred to as cost estimating relationship (CER) or parametric method based on multiple regression analysis. It is used for inferential purposes. This apparently has birthmarks of estimating and utilizes regression-type equations to predict project cost outcomes.

4.1.2 Descriptive and Prescriptive Regression-based Methods

A paramount role of regression-based methodology to describe statistically significant interdependencies between predictors and expected schedule and cost outcomes is praised in Chapter 1. However, when the successful descriptive methodology encompasses the inference area, it stumbles across several difficulties.

Parametric modelling is usually designed in a way that takes into account findings of bivariate regression analyses. These findings are facts and givens pertaining to

Modern Risk Quantification in Complex Projects: Non-linear Monte Carlo and System Dynamics Methodologies. Yuri G. Raydugin, Oxford University Press (2020). © Yuri G. Raydugin.
DOI: 10.1093/oso/9780198844334.001.0001

previously completed projects which were discovered as part of the descriptive regressing exercises. There is always an underlying assumption that upcoming projects should have pretty much the same features as completed projects. This assumption plays down uniqueness of upcoming projects that can be tagged as project context differentiators. Implicitly this assumption states that the upcoming project should have the same traits as previous projects. Although this is in line with the 'iron law of megaprojects' introduced by Prof Flyvbjerg (2017), reasons for project failures could differ significantly. Unfortunately, the depersonalization of projects and risks deprives project teams from a possibility of addressing relevant project risks. On the other hand, risk addressing step is not part of unconventional risk methods anyway.

Moreover, to be on the level, the parametric risk quantification modelling is not a risk quantification method per se.[1] By design, it is an attempt to quantify overall project outcomes of upcoming projects based on known overall outcomes of a number of completed projects included to a sample set without identification or addressing of risks. Depending on project context, actuals for completed projects and for upcoming projects besides baselines may include:

- occurred risk events,
- allowances,
- general uncertainties,
- cost of risk addressing,
- cost escalation and exchange rate risks and cost of hedging,
- costs of change orders,
- costs of claims,
- penalties,
- loan interest rates,
- etc.

The mix of these items is specific to each completed project and will be unique to the upcoming project constituting project context differentiators. No particular risk events are identified, assessed, or addressed when running a typical parametric model as there is no way to take into account the context of the upcoming project consistently anyway. All above-listed items are represented in parametric analyses undistinguishable on both sides: completed projects and an upcoming project.

4.1.3 Direct Parametric modelling

Parametric modelling has two distinctive stages. The first stage is to derive an analytical formula for mean values of a dependent variable using information from an

[1] The same is true in case of the other two discussed unconventional risk quantification methods—system dynamics and ANN. It is difficult to quantify risks that are not identified and addressed.

available database of previously completed projects. Usually those dependent variables are a cost growth or a schedule slippage. Several selected independent parameters (predictors) are used in the multivariate regression analysis. For the purposes of this discussion, this stage could be labelled as a direct regression analysis and it pertains to the traditional descriptive statistics. Several models for cost growth and schedule slippage were developed in the past (see e.g. Hackney (1965), Merrow (1981), or AACE (2011)).

The outcome of a direct multivariate regression analysis would be a formula for a mean value of a dependent parameter represented as a sum of selected predictors with weighting coefficients as follows.

$$MV = c + \varepsilon_1 \times F_1 + \varepsilon_2 \times F_2 + \cdots + \varepsilon_n \times F_n, \tag{4.1}$$

where MV is a mean value of a dependent parameter of interest, n is a number of selected predictors, F_i is a selected predictor ($i = 1, 2, \cdots n$), ε_i is an i-th predictor's weighting coefficient, and c is a fitting constant (Hollmann (2017a)). Factually, expression (4.1) should include all relevant factors listed in section 4.1.2 including all occurred risks factored in to actual outcomes of completed projects. Using similar formula for duration-related parameters is technically possible although less credible as each project from a database has its own unique schedule logic along with the other uniqueness characteristics which would be perfectly ignored.

It should be clearly stated outright that the produced mean value characterizes only the set of projects included to the used database. Any context differentiators of upcoming projects including their project risk exposure cannot be adequately represented. As discussed in section 4.1.4, all attempts to do this are reduced to tweaking weighting coefficients ε_i.

4.1.4 Reverse Parametric Modelling

The second stage of parametric modelling is to make the regression analysis relevant to an upcoming project which cannot be part of the used database by default. This may be labelled as reverse regression analysis. This step belongs to the inference statistics.

For this purpose, weighing coefficients ε_i of selected predictors F_i in formula (4.1) are amended to reflect on specifics of an upcoming project. Expert judgement and some empirical data are used to assign discrete weighing scores, say, from one to five to condition the predictor's weighting coefficients (Hollmann (2017a)). Each score gets rules on how weighting coefficients ε_i ('the tweaking coefficients') should be modified depending on an assigned score. This resembles the deterministic scoring methodology discussed in section 3.2. For example, a formula may have a project complexity as an independent predictor F_{compl}. Scores from 1 to 5 should be introduced to define rules on how the weighting coefficient ε_{compl} should be amended. Most complex projects could get the top score 5 and significant multiplier to increase ε_{compl}, and a very simple project should get score 1 and small or no increase in ε_{compl}, etc. Developing rules for tweaking of weighting coefficients and assigning them to selected predictors could be quite subjective although it is usually done in the name of empiricism.[2]

As discussed in Chapter 1, tweaking the weighting coefficients in formula (4.1) reveals a special type of biased sampling—judgement sampling—that is imposed on top of convenience sampling. It can be justified as an apparent attempt to compensate the degree of bias brought up by the convenience sampling by the bias of judgement sampling to improve relevance of the used sample to the upcoming project. As discussed in section 4.4.2, these coefficients are objectively re-evaluated and re-assigned by the adopted ANN tool as a result of a mathematical error minimization process in the case of the ANN modelling.

As a rule, any available databases contain a mix of projects with a spectrum of characteristics, that is, their unique context differentiators. Normally the number of predictors (see formula (4.1)) rarely reaches ten whereas the number of important project characteristics reflecting on project context (the context differentiators) is much higher. For example, any legacy database contains projects with various levels of complexity, skills attributed to project teams, execution strategies, locations, levels of stakeholder's alignment, level of engineering development, robustness of business case, regulations, technologies, features of local markets, etc, although there is no possibility of attaching multiple project characteristics to a few predictors. This means that many project characteristics have to be ignored. Apparently, if another database is selected as a source of data for parametric modelling, the results could differ for the same upcoming project. The situation may get much worse if the actual number of samples in the database used in the parametric modelling is low which, according to the small sampling theory (see Spiegel & Stephens (1999)), requires additional conditioning (see Figure 1.2).

4.1.5 Introduction of Uncertainty to Deterministic Results

The amended formula (4.1) with project-specific weighting coefficients yields a one-point mean value for a schedule slippage or a cost growth. This is a deterministic one-point value result obtained through averaging of characteristics of completed projects and tweaking of the weighting coefficients.

Obtaining one-point mean values strongly resemble deterministic project cost estimating or schedule development: the results do not bring forward any uncertainty of outcomes or confidence intervals. The absolute level of certainty does not add credibility to the obtained results. An introduction of uncertainty is usually attempted as the next step to make the results look less certain and more credible. This could be done through assigning uncertainty ranges around the derived one-point mean values. The approach is based on the following two assumptions (Hollmann (2016)).

[2] Does this not resemble a blind wine testing party? Various expert opinions about a possible year of grape production along with speculating if it was grown on a southern or western slope of a hill could be expressed based on years of heavy empiricism. This example does not contradict the notion of the judgement sampling.

First, it is assumed that schedule slippage or cost growth data are normally distributed around a mean value derived from the regression formula. Using standard deviation data for the sample set and its mean value from the available regression analysis model it is advised to derive all confidence level values from P0 to P100 for the symmetrical bell-shape normal distribution.

This is another reason why the parametric modelling is highly database specific: parameters of the normal distribution depend only on previously completed projects included to a database, not on features of an upcoming project. This distribution cannot be skewed for any given sample set being symmetrical (bell-shaped) by design.

More intrigue is added by Hollmann (2016) in the same publication: it is argued that schedule slippage and cost growth distributions must be substantially skewed. This is a right concept to expect that these distributions should be highly skewed (see for example figures 3.2 and 3.3). The apparent discrepancy related to the type of distribution (bell-shaped vs. skewed) cannot be explained mathematically.

Second, it is argued that one standard deviation of a normal distribution can be proclaimed a required project contingency empirically based on expert experience and judgement (Hollmann (2016)). It is well known that expert judgements even in the name of empiricism are too often subjective and biased. This makes parametric risk quantification strongly dependent on a used database as well as on expert judgements, not on characteristics of the upcoming project. If the sample is changed, a different contingency would be derived.

Moreover, if the assumptions on a shape of a distribution (bell-shaped vs. skewed) and a contingency definition (one standard deviation) are combined, another ambiguity is revealed. A standard deviation of a normal distribution and a standard deviation of a skewed distribution should yield significantly different contingency results.

This discussion explains why parametric modelling is highly sample and expert judgement specific. All in all, these discussed shortcomings are an equivalent to the statement that 'the upcoming project does not have a right to differ much on average from previously completed projects included to the database'.

4.1.6 Systematic and Random Errors of Parametric Modelling

If in the case of the Monte Carlo methodology a major source of systematic errors are missed risk exposure factors, in the case of parametric modelling it is a product of convenience and judgement biased samplings.

Figure 3.4 advises on method's accuracy and precision. It is developed initially to review the Monte Carlo methodologies. Both accuracy and precision of parametric methods depend directly on characteristics of selected project databases used for inferential analysis. Moreover, a systematic error directly depends not only on a selected database but also on assumptions and judgements used to derive weighting coefficients (see expression (4.1)).

If a database used in parametric modelling is actually a convenience sample set that is tweaked by judgement sampling, a substantial systematic error due to statistical bias is guaranteed. It could also be expected that the random error should be high enough if a high dispersion of outcomes is observed in the sample set. This is normally the case when irrelevant projects are part of the convenience sample, which is one possible definition of the convenience sampling. Both factors are characteristics of the available database and a way it was tweaked, not of the upcoming project. As discussed in Chapter 1, the situation would be further exacerbated if the actual number of relevant used samples is low (small sampling theory). In terms of Figure 3.4, this is an additional source of both systematic and random errors (see e.g. Spiegel and Stephens (1999)).

If a database contains a number of projects that are very similar to an upcoming project in all aspects and characteristics, a systematic error would be minimal. Actually, a random error should be minimal too as there should not be much dispersion of actual outcomes. This case is slightly exotic as it describes a situation of simple alike projects without any distinctive features or context differentiators. By this reason it does not require any parametric modelling: a direct expert judgement could be used to predict the same result as before without any analysis.

In practice specific parametric models have both high systematic and random errors. Their value may be justified by the inference paradox discussed in Chapter 3. The lower the precision, the better (see quadrant D of Figure 3.4). A derived low-accuracy—low-precision parametric modelling should cover almost any possible project outcome. However, outcomes comparable with outcomes of project outliers such as Sydney Opera House or Boston Big Dig Artery/Tunnel projects are still out of reach.

The situation could be more difficult if a low-accuracy model is precise enough. This happens when the dispersion of results for completed projects is low as projects are quite similar but irrelevant to the upcoming project as the sample set is highly biased. This is represented by quadrant C of Figure 3.4.

4.1.7 A Role and Place of Parametric Modelling

The discussion above sheds a slightly pessimistic light on parametric analysis methodology's role in project risk quantification. Actually, as discussed in Chapter 1, the opinion on limited applicability of the parametric modelling is well understood in the industry. For instance, an AACE recommended practice 42R-08 (AACE (2011)) does prescribe using parametric risk models only in earlier project phases when Monte Carlo modelling is not viable yet (no scope defined prior to the end of FEL2A) and when an expected accuracy of estimates is very low. In this situation parametric modelling is used as a rough estimate for a class of projects an upcoming project belongs to, not as an estimate of the individual upcoming project.

It is often claimed by proponents of regression-based methods that the parametric methodology consumes empirical data and does not succumb to subjective or biased judgements. This is not aligned with points discussed in this chapter as related to inferential statistics. However, this would be fair to underscore a natural and important descriptive role of regression-based methods in descriptive statistics.

The shortcomings of parametric inference methods overviewed above demonstrates that if a method uses empirical data as inputs it is not a guarantee that the method is robust. Such situation could be labelled as an empirical irrelevancy when empirical data are treated and interpreted in a questionable and subjective way. The subjective tweaking of weighting coefficients in expression (4.1) in the name of empiricism demonstrates this.

On the other hand, the same empirical data could be successfully used to better define inputs to Monte Carlo models (Galway (2004)). This includes recent trends in artificial intelligence applications to more precisely define inputs to scheduling and schedule risk analysis models (Patterson (2018)).

It should also be kept in mind that parametric models were developed way before simulation analysis modelling came of age (see e.g. Hackney (1965)). In essence they were developed to diminish power of previously dominating subjective methodologies—expert judgement and predetermined guidelines. However, parametric method does require and consume expert judgements itself.

Intentionally or unintentionally, regression analysis methods stand behind widely used predetermined guidelines methodology. As discussed in Chapter 1, these methods were used for development of AACE cost estimating classes that in practice support and feed predetermined guidelines methodology: many companies still apply 10 per cent cost contingency across their project portfolios because AACE class 3 stipulates minimum contingency of 10 per cent. This is what Baccarini (2005) calls the '10% syndrome'.

Simulation (aka Monte Carlo) methodologies started taking over from regression-based models in late the 1980s and early 1990s, although this process was not smooth. In particular the Monte Carlo technique, especially its earlier versions had its own limitations: range estimating Monte Carlo models were introduced first that were actually no better than parametric modelling in terms of accuracy (see Chapter 3). That is why parametric and ranging methods could be used interchangeably in early phases of project development such as FEL1 and FEL2A.

An esoteric ANN methodology may become a reinforcement and an upgrade to parametric methods to improve accuracy drastically (see section 4.8).

4.2 A Hybrid Model

A recognized role of regression-based methodology in descriptive statistics is discussed and praised on several occasions in this book. However, several shortcomings of parametric modelling that belong to inferential statistics are discovered in section 4.1 relating to quality of samples used for regression and features of direct and reverse regression steps.

In addition to regular parametric modelling, attempts and efforts to combine parametric and Monte Carlo methods were committed (Hollmann (2016)). According to the developed method, all predictors that are used in parametric modelling are called 'systemic risks'. In addition to 'systemic risks', so-called 'project-specific' risks are introduced.

Schedule or cost contingencies derived from parametric models are put as single additions to Monte Carlo SRA or CRA models to take into account 'systemic risks'. In case of schedule risks is it added as a single dummy normal task at the very end of a project schedule used in SRA.

This eclectic approach has two limitations as follows.

- Double-counting. The parametric part is based on actual outcomes of completed projects. All occurred risk events labelled 'project-specific' should be part of the actuals already. However, they are counted again in Monte Carlo modules of hybrid models.
- Low intrinsic accuracy. Overall accuracy of any model is defined by its modules of lowest accuracy. As discussed in section 4.1, parametric modelling has its own accuracy limitations which define low accuracy of entire hybrid models.

These shortcomings should be even more prominent in case of schedule risk modelling when schedule logic is ignored.

Whereas parametric modelling should be used in early project phases when a project scope is not finalized and risk registers are not yet developed, the hybrid model does not seem to have much room for application at all. When risk registers are not available—pure parametric models without Monte Carlo modules should be used in FEL1 and FEL2A. As soon as risk registers are available, Monte Carlo modelling can take over from parametric models starting with FEL2B.

Needless to say, neither parametric nor hybrid models can accommodate schedule-driven costs not to mention non-linear risk interactions in project systems.

4.3 System Dynamics Risk Quantification Methods

This section provides a high-level overview of system dynamics at large and its possible applications to project management in particular including project risk quantification.

A value of system dynamics to model real-life processes was well known for decades (Forrester (1961, 1968), Ferrari (1978), Coyle (1996)), although even now it is considered largely an esoteric activity. Very few project practitioners use it and mostly in the area of dispute resolutions and claims. At the same time, system dynamics offers valuable insights on projects as systems and can provide strong support to project risk quantification. As a case in point, change orders that are in focus of dispute resolutions and claims activities may be treated as disruptions or perturbations to a project system. As such they have the same nature as the other project risks with one exception. Change orders are treated in dispute resolutions and claims as actuals *post factum*. Treating project risks including possible change orders as future events has to include probabilistic angle on risk occurrence.

The purpose of this section is to:

- introduce system dynamics to project practitioners as a non-esoteric project control and risk quantification methodology,
- demonstrate its power to support risk quantification, and
- discover a significantly non-linear nature of interactions in project systems including non-linear interactions of risks.

4.3.1 Applications of System Dynamics to Project Management

A first practically significant system dynamics application to project management was developed by Cooper (1980) to resolve disputes and claims in a US Navy shipbuilding project. A challenge was to demonstrate that multiple changes introduced to the project by a project owner led to so-called ripple effects: disruptions (reworks, out-of-sequence works, lower productivity, etc.) to those parts of the project that were not directly impacted by changes and were not paid for. This pioneering work introduced a project rework cycle representing a project work as a flow fulfilled with a variable productivity impacted by multiple conditions including changes. The rework cycle included categories such as 'work to be done', 'work really done', 'undiscovered rework', and 'known rework'. These categories were 'depleted' or 'replenished' by 'work progress' and 'rework discovery' rates that depended on work quality and productivity. The purpose of the modelling was to demonstrate impact of changing parts of a project impacted directly by owner's change orders on seemingly unchanging but actually impacted parts. Along the way, it was demonstrated that the rework models had clearly non-linear relations between inputs and outputs.

A complete model developed by Cooper (1980) is extremely complex. It has several modules and multiple interrelated factors influencing project system dynamics. Actually, it took several years to develop it. The model has a sufficiently high level of realism and detail to be presented and defended in court.[3] This project system dynamics approach is reiterated by Cooper and Lee (2009) providing additional valuable insights. It seems that the formidable level of complexity, detail, and efforts pushed most project practitioners to turn project system dynamics down.

Besides four major work categories—'work to be done', 'work really done', 'undiscovered rework', and 'known rework'—that are called 'stocks' or 'levels' in

[3] Eventually that case was settled out of court based on convincing system dynamics modelling results.

system dynamics, the following factors called 'variables' or 'parameters' are taken into account to establish mathematical relations among stocks (Cooper (1980)):

- work quality
- engineering revisions
- productivity
- out-of-sequence work
- schedule pressure
- perceived progress
- staff on project
- overtime
- staffing request
- staff hiring
- staff turnover
- worker experience
- worker morale
- etc.

Full replication and debugging of the Cooper's model would take months if not years. Simplified but practical versions of project system dynamics models based on the rework cycle approach are introduced in Chapters 7 and 16 that serve risk quantification purposes. System dynamics basics are discussed in Chapter 6. It is shown in Chapter 6 how established relations among stocks and variables can be converted to differential equations that stand behind system dynamic modelling. Analytical solution of these equations are also provided.

Another informative example of successful application of system dynamics to a project litigation case is provided by Williams et al. (1995). The effect of design changes on project costs leading to significantly non-linear dynamic interactions in a project system is investigated using a system dynamics model comprising the following categories:

- design changes,
- cross-impacts among design components,
- enforced work on unfrozen items,
- increased rework,
- approval delays,
- tight timescale,
- more work to do,
- etc.

It is also pointed out that the traditional linear network planning of projects might not be able to describe and forecast the discussed non-linear dynamic interactions. It looks like two models—a traditional schedule-based risk management model and a system dynamics model—should be run in parallel for any project. This approach is demonstrated in this book.

One recent example when a system dynamics model is accepted by a court in a litigation case is provided by Goodchild (2018). One of the reasons why courts are still reluctant to accept system dynamics models seems to be a missing proactive project control approach. Namely, the collection of evidence data related to contested claims' cases usually starts after a claim is filed. Namely, a suspicion on a biased collection of evidence data *post mortem* might result in lower credibility of otherwise robust system dynamic models. If related data are collected systematically and used in system dynamics models built in the course of the project development and execution along with project cost estimates and schedules, this should dramatically increase credibility of discussed models. This practical and proactive project controls approach based on system dynamics methodology is discussed by MacGuinnes (2019).

A cornerstone book by Sterman (2000) lays out a theoretical foundation for system dynamics modelling in general and in project system dynamics in particular. It has a section on the project system dynamics work by Cooper (1980).

A work by Lyneis and Ford (2007) provides a comprehensive and relatively recent overview of application of system dynamics to project management. The following main factors should be considered when applying the system dynamics approach:

- rework cycle,
- out-of-sequence work,
- ripple and knock-on effects,
- perception gaps between perceived and real progress,
- delays in recognizing the gaps,
- etc.

Cooper (2004) puts forward a unifying theory for project risks and changes based on system dynamics. Compounding effect of impacts from multiple risks and changes is investigated. It is shown that timing and magnitude of the individual impacts are important to define the compounding effect: the higher the impacts and the later in a project they occur, the higher the compounding effect.

Gray and Shahidi (2011) also accentuate similarities of project changes and risks. A project system might react in seemingly disproportionate ways to risk events including project changes due to the interconnected and interdependent nature of the risks themselves. A quest for moving from considering risks and changes standalone events (an isolated approach), to considering their non-linear interactions (a domino effect), is discussed.

Following the approach by Gray and Shahidi (2011), multiple intertwined drivers and causes of risks are considered in this book. These include project system issues as additional (systematic) causes of risk events. Interactions among project risks are kept in mind: knock-on and compounding effects in a family of project risks are introduced. A special type of risk interactions—intra-risk interaction or internal risk amplification—is also introduced. This is responsible for amplification of regular standalone project risks due to the presence of static project system issues called chronic in this book. Its origin is discussed in Chapter 17.

4.3.2 Suitability of System Dynamics to Project Risk Quantification

Although the idea that risks handled by project risk management and changes discussed in dispute resolutions and claims have many similarities there is one technical difference between the two. Dispute resolutions and claims handle occurred risks (mostly project changes initiated by project owners) *post mortem* upon project completion. This is a backward-looking activity. On the contrary, risk management handles future risk events including project changes that might (or might not) occur (Cooper (2004)). By this reason, actual delays and extra costs are subject to claims, whereas any risk quantification modelling should be based on evaluation of expected values of possible schedule and cost risk consequences. This is a forward-looking activity. A product $P \times I$ of a risk occurrence probability P and risk impact I is subject to risk quantification, not just actual impact I. This probabilistic factor is taken into account in this book.

Three primary conclusions can be drawn and used in risk quantification modelling including Monte Carlo modelling based on system dynamics experience as follows.

- Due to similarities of project changes and risks, system dynamics modelling is capable of enriching project risk quantification.
- The non-linear nature of interactions in project systems should be taken into account in project risk quantification explicitly.
- In place of considering risks as standalone isolated events, non-linear interactions among them should be taken into account in risk quantification explicitly.

All future risks including possible change orders initiated by a project owner should be treated as interacting perturbations to a project system. These may result in several interacting factors as follows.

- Rework cycle including:
  - adding a new work (extra delay and cost),
  - cancelling a part of unfinished work (sunk costs),
  - scrapping a part of finished work (sunk costs),
  - adjusting a part of finished work (extra delay and cost).
- Out-of-sequence work including:

- delays in fulfilling a work due to lack of required resources or materials when required (so-called work constraints),
- lower work quality,
- lower labour productivity,
- schedule pressure to accelerate works,
- etc.

Full-scale system dynamics models comprise dozens of relevant factors. For simplicity purposes only a few of the most important factors are included into high-level models discussed in Chapters 7 and 16. As discussed by Lyneis and Ford (2007), Cooper (1980), Williams et al. (1995), and Gray and Shahidi (2011), the factors listed above have complex cause-effect relations organized in non-linear feedback loops which are a foundation of system dynamics approach. Some of these loops are listed below.

First, a rework cycle often involves parts of a project that seemingly are not impacted directly (a ripple or knock-on effects).

Second, if two or more risks occur simultaneously or near simultaneously they may collectively exacerbate their standalone impacts (a compounding effect).

Third, some risks might impact down-the-line risks (a knock-on effect) that do not seem to be impacted directly.

Fourth, ripple, compounding, and knock-on phenomena could occur in projects simultaneously leading to additional highly non-linear interactions in a project system.

As a result, traditional linear cause-effect relations utilized in a bowtie diagram approach for standalone risks (see Chapter 2) should be substantially modified to reflect on non-linear loops and interactions in affinities of project risks which is done in Chapter 9. These affinities are introduced and visualized through dynamic risk patterns in Chapter 14.

Project system dynamics models developed in this book (Chapters 7 and 16) clearly demonstrate the role of non-linear knock-on and compounding risk interactions.

4.4 ANN Risk Quantification Methods

This section overviews a concept that can be successfully used in project risk quantification—ANN. ANN methodology has been used for decades to resolve problems related to image recognition, stock market prediction, sales forecasting, etc. with a variable level of success (Vargas (2015)) based on analysis of big strays of data. Despite there being dozens of publications on ANN applications in project management (see e.g. review papers by Waziri et al. (2017) or Sarcia et al. (2007)), it is still considered an esoteric activity by project management practitioners. At the same time, it could become an indispensable replacement of various low accuracy regression-based models. Multiple examples of predictive superiority of ANN methodology over parametric models in project management (cost estimating, construction scheduling, decision-making, risk assessment, and dispute resolution) are pointed out by Waziri et al. (2017).

This section provides a very simple version of ANN modelling just to overview the ANN concept and establish its link with the parametric modelling.

4.4.1 The ANN Concept

Artificial neural networks mimic brain activities to recognize complex relations and patterns in large arrays of data. This capability can be used for predictions related to new sets of data. In simple terms, ANN methodology might be considered an upgrade to the parametric methodologies that also use multiple independent parameters for inference. Two main categories of problems could be addressed using ANN: classification and numeric (Palisade (2015)). The former provides a definition into which category an unknown item should fall. For example, an upcoming project could be categorized as a success or failure against some specific criteria or thresholds (e.g. 25 per cent schedule slippage or 25 per cent overspending), although a numerical value of a delay or an overspending for a particular upcoming project is not inferred. The latter handles specific numeric forecasts, for example, an expected actual duration or cost of an upcoming project.

Key elements of artificial neural networks are 'brain cells' or 'neurons' that are connected to networks through 'dendrites'. Neurons can receive impulses from other neurons and, depending on the strength of received impulses, may pass signals to other neurons. Eventually the entire network produces a single output that is a function of inputs and their interactions and transformations.

Neurons are organized in layers. A first layer directly receives inputs as values of independent parameters. Independent parameters in case of project management applications may include project characteristics as follows:

- budget,
- duration,
- complexity,
- type of contract,
- location,
- level of scope development,
- project team skills,
- leadership skills,
- stakeholders,
- etc.

Inputs could be defined either as categories (e.g. low, medium, or high complexity) or as numeric values (e.g. a project budget). Dependent parameters also could be defined as either categories or numerical values and may include as follows:

- project success vs. failure (a classification problem),
- actual duration,

- actual cost,
- duration variance,
- cost variance,
- etc.

Information from a first layer neuron is passed to neurons belonging to internal or 'hidden' layers. The number of ANN hidden layers may be higher than one. All outputs form neurons belonging to the last hidden layer are passed to the neurons that generate the final outputs of the entire ANN. It is well known that single-output ANNs are more accurate than multiple-output ANNs (Palisade (2015)). Apparently, independent parameters get into ANN through the first layer whereas dependent parameters are associated with the last layer.

Figure 4.1 represents a simplistic ANN with two independent parameters and one hidden layer of three neurons. It is known that one-layer ANN models are sufficient enough in the majority of project management applications (Palisade (2015)). Multi-layer ANN models could, however, be selected to improve accuracy of modelling.[4]

According to Figure 4.1, two selected independent parameters are represented as inputs I_1 and I_2. Each link from input to hidden-layer neurons has a weighting coefficient W_{ij}, where $i = 1,2$ and $j = 1,2,3$ in this example. A summary signal S_j received by a hidden-layer neuron can be defined by the following general formula.

$$S_j = c_j + \sum I_i \times W_{ij}, \tag{4.2}$$

where c_j are fitting constants called bias.

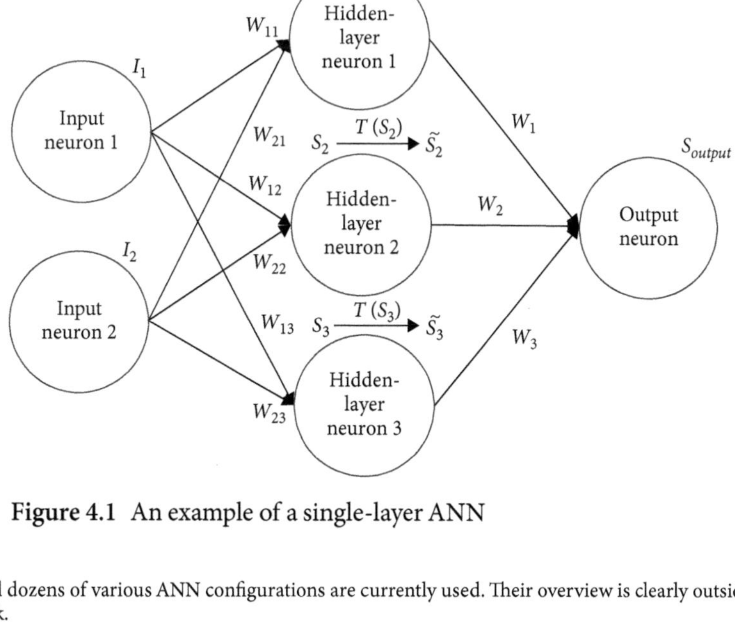

Figure 4.1 An example of a single-layer ANN

[4] Several dozens of various ANN configurations are currently used. Their overview is clearly outside the scope of this book.

For example, in the case of the hidden-layer neuron 2, the received signal can be defined as:

$$S_2 = c_2 + I_1 \times W_{12} + I_2 \times W_{22}. \tag{4.3}$$

A signal received by a hidden-layer neuron is transferred to down-the-line neurons with amendments. Namely, a so-called transfer (or activation) function $T(S_j)$ is used that transforms an incoming signal S_j to a signal $\tilde{S}_j$ received down-the-line from a neuron *j*.

There are several options for transfer function's selection. For instance, in early ANN's models a simple step-function was used. It is equal to zero below a certain threshold and to one above the threshold. If a threshold is established at 1/2 level, this could be directly used as a binary-type outcome to be used in classification problems. In terms of the ANN outputs this could lead to two outcomes: zero (no signal) or one (full-strength signal). The step transfer function plays a role of a trigger or an activation function. In a simple language this may correspond to a human-brain-neuron heuristic situation when one says about an idea or finding that 'it struck me'.

Some other smoother transfer functions could be used in hidden layers, for example, a sigmoid function[5]

$$T(S) = e^S / (1 + e^S), \tag{4.4}$$

Where e^S is the exponential function. Apparently, a sigmoid function that has *S*-shape, approaches 1 when $S \to +\infty$; it is equal to 1/2 when $S \to 0$; it approaches zero when $S \to -\infty$. In contrast with the binary-type step function, sigmoid function is active and works for all values of a signal S_j except a limiting case of $S_j \to -\infty$. Apparently, the smooth sigmoid function does not play a role of a trigger: a signal *S* is always transferred down the line although it has varying strength unless a specific threshold is established. In human-brain-neuron heuristic terms this could mean slow 'flirting with an idea that was put in somebody's head'.

In case of one-layer ANN, the output is defined as a sum of inputs from a hidden layer as follows:

$$S_{output} = \Sigma \tilde{S}_i \times w_i = \tilde{S}_1 \times w_1 + \tilde{S}_2 \times w_2 + \tilde{S}_3 \times w_3 \tag{4.5}$$

It would be worthwhile to establish a connection between ANN and parametric modelling. First, parametric models have a trivial transfer function $T = 1$ as they do without any hidden layers and neurons: 'the transfer valve' is always fully open and does not modify inputs. From this angle, the linear parametric modelling is a limiting case of ANN modelling when all hidden layers fold up and there is no need for:

- a transfer function $(T = 1)$,
- cross-coefficients W_{ij} between neurons $(W_{ij} = 0)$, and
- weighting coefficients w_i for signals $(w_i = 1)$.

[5] Another S-shaped transfer function that is often used instead of the Sigmoid functions in some applications is a hyperbolic tangent function. Other popular types of transfer function are linear, piecewise linear, Gaussian, etc.

In this case, as expected, expressions (4.2)–(4.5) are degenerated to formula (4.1) for the linear parametric modelling that describes the mean value of a project outcome.

In comparison with the parametric modelling, ANN modelling allows much higher flexibility in handling inputs and converting them to highly accurate results. It takes into account more sophisticated relations among independent parameters in comparison with the linear parametric modelling. Certainly, more that just a few predictors can be used by ANN models. However, this imposes very high requirements on databases used in ANN modelling.

Also, ANN input–output relations are substantially non-linear when non-linear transfer functions are used.[6]

4.4.2 The ANN Modelling Process

A foundation of ANN methodology is a learning process (Vargas (2015)): an ANN model can recognize complex relations and patterns in a given array of data. It utilizes the identified patterns to forecast future results based on similar results obtained in similar circumstances.

As mentioned in the previous section, several ANN configurations are currently used depending on a nature of solving problems (classification vs. numerical) and required accuracy. For example, in case of classification problems probabilistic neural networks (PNN) and multi-layer feedforward networks (MLFN) are applicable. Numerical problems can be solved using MLFN and generalized regression neural networks (GRNN) (Palisade (2015)). Modern ANN software packages offer a best network search functionality to avoid any guesswork when several network configurations are tried simultaneously. In turn, these main types of ANN configurations have several subtypes. Their overview is outside the scope of this book. The ANN example of Figure 4.1 represents a MLFN type with just a single layer.

The initial ANN step starts with selection of data used for modelling. It is similar to the selection of sample sets (databases) for the parametric methodologies and has the same limitations discussed in section 4.3, although ANN requires much higher quality and granularity of data collected in the used database as it uses many more parameters. In some rare cases, it could be possible that some high-quality databases that have been used for the parametric modelling can be used for ANN modelling too.

Similar to parametric modelling a set of independent parameters should be selected. Descriptive regression analysis results can be used as a starting point for this. As opposed to the parametric modelling, the number of independent parameters could

[6] Without getting into details, it should be pointed out that only non-linear transfer functions with a simple derivative and non-vanishing gradients can be used in ANN modelling. This is a technical requirement for application of a gradient descent algorithm used for minimization of errors during a training step—a backward propagation—of ANN modelling (see section (4.4.2)).

be quite high. One or two dependent parameters should be identified. Usually project durations or costs are selected as dependent parameters.

A type of model structure should be selected when independent and dependent parameters have been selected. Review of ANN models that have applications in project management are discussed by Waziri et al. (2017). The main issue with ANN modelling is that recognized standards for ANN modelling have yet to be developed. This is a sign of exploding development of this research area which is also an obstacle to its practical implementation at this point in time.

When selection of a sample set, independent and dependent parameters, and an ANN structure is completed, there are three other steps to build and run ANN models as follows (Palisade (2015)).

- Training:[7] adjustment of weighting and bias parameters of equation (4.2) to fit known outcomes from a selected sample set.
- Testing: using the model after training to verify how well it predicts known past outcomes from the sample set.
- Inference: finally, the trained and tested model can be vetted and used to predict unknown outcomes.

Detailed database records that contain complete information not only about inputs but also about outputs are used for ANN training.

The same as in case of the parametric modelling, ANN has direct modelling steps and reverse modelling steps: feedforward and backward propagation. The purpose of training is to evaluate and optimize a set of the coefficients W_{ij} and w_i and the bias parameters c_i (see expressions (4.2), (4.3) and (4.5)) to match known outcomes from a database and, hence, minimize a difference (an error) between the known actual outcomes and their prediction. After each instance of the feedforward propagation the error is propagated backward. It is apportioned to each neuron in ANN through corresponding weighting coefficients in accordance with the amount of the total error for which a particular neuron is responsible. In place of mean squared error evaluation method used in the parametric modelling, more sophisticated methodologies are used in ANN, for example, gradient descent algorithm. Although, essentially these are based on the same principles of error minimization.

A residual error threshold can be established as a criterion for acceptable ANN training results.

A cardinal improvement pertaining to the ANN modelling in comparison with the parametric modelling is that it does not treat the weighting coefficients as subjective tweaking coefficients defined by experts (see expression (4.1) for the parametric modelling). These coefficients are objectively re-evaluated and re-assigned by an adopted

[7] Two types of ANN training are used: supervised and unsupervised. Only simple supervised training is discussed here. More advanced unsupervised training presumes that ANN selects information and method for optimization self-consistently on its own.

ANN tool as a result of the error minimization process (see expressions (4.2), (4.3), and (4.5)).

As soon as training provides acceptable results, ANN testing is undertaken to check accuracy of outcomes. This step also utilizes information about database records with known outcomes. Error tolerance for good vs. bad evaluations is set up to judge when ANN yields accurate enough results: these should belong to an established tolerance range around actual results.

It is not unusual for about one-quarter of available database records to be used for training purposes and a comparable number of records used for testing.

When both training and testing ensure acceptable results, actual inference can be undertaken to predict unknown outcomes.

4.4.3 ANN Modelling Difficulties

Despite ANN modelling looking like a promising substitute for the low-accuracy parametric modelling, it has several intrinsic shortcomings discussed above.

First, the ANN modelling is sample specific. The same as in the case of parametric modelling, quality of databases used for the ANN modelling directly impacts quality of ANN modelling outcomes. In biased databases, the results cannot be accurate.

Second, there is no single ANN method or even a group of ANN methods that is recognized and used by researchers and practitioners as a standard. New ANN methodologies or amendments of existing ANN methodologies appear quite regularly. This makes ANN results method specific.

Third, the ANN methodology is considered by many as 'a black box'. In reality this is an update of the parametric modelling based on multiple 'non-linear relays' built in 'hidden layers'. They just provide more flexibility and more possibilities or degrees of freedom to better fit information collected in large arrays of data and carry out more precise inference.

When ANN comes of age in project management it might become a powerful supporting tool to Monte Carlo risk quantification methods the same way as system dynamics may do.

4.5 Accuracy of Unconventional Risk Quantification Methods

Multiple complaints on low accuracy of unconventional risk quantification methods could easily be found in literature (see e.g. Merrow (2011), Hulett and Whitehead (2016), Hollmann (2015, 2017a), etc.). Each unconventional methodology has its reasons for lower accuracy discussed in this chapter besides the fact that they do not pay due attention to risk quantification, addressing, and monitoring in the first place.

At the same time, some unconventional methods are quite valuable in many situations. For instance, low accuracy of parametric models is a perfect match to low accuracy requirements in early project phases when Monte Carlo modelling does not yet work.

System dynamics modelling may provide amazingly accurate results for some types of risks related to change orders, low quality, and reworks. Moreover, it is possible to upgrade system dynamics risk models to conventional status if risk addressing is taken into account in modelling inputs. At the same time, some other project risks might require an excessive level of sophistication to be adequately modelled. Integration of system dynamics with Monte Carlo methods especially SRA methods might be an answer to this difficulty. This should require a development of new generation of risk quantification tools that integrate Monte Carlo and system dynamics methodologies. As these tools are not yet developed, this is done in this book using those tools which are available.

ANN methodologies have yet to become a project risk quantification mainstream. On the other hand, learning about practical ANN modelling does not demand the possession of advanced mathematical skills. It could be comfortably done when using commercially available ANN software packages.[8] As discussed above, a mandatory condition to undertake meaningful ANN modelling is the availability of well-developed databases of previously completed projects. This is one of the reasons why a temptation to include an ANN modelling case in this book has been suppressed.

As soon as ANN modelling becomes a part of project risk quantification mainstream, the traditional low accuracy, parametric methods may peacefully retire: ANN modelling has a clear capacity to predict project outcomes with much higher accuracy.

Any types of unconventional risk quantification methods may play a good supporting role through providing ammunition (inputs) to Monte Carlo methods in order to make them fully empirical. Modelling of Monte Carlo inputs using the system dynamics or ANN methods is capable to replace some instances of expert judgements. However, the traditional Monte Carlo methodology requires a significant overhaul itself to increase its accuracy. The non-linear Monte Carlo methodology discussed in this book is the primary goal and the direct answer to this challenge.

4.6 Chapter's Markup and Takeaway

- The main unconventional risk quantification methods focused on risk assessment steps only are the parametric modelling, system dynamics, and ANN.
- Strictly speaking, the unconventional risk quantification methods are not risk quantification methods *per se*: it is difficult to quantify risks without their identification and addressing. Actually, these methods estimate overall project durations and cost outcomes.

[8] For example, the user-friendly NeuralTools by Palisade Corporation (2015) or its more sophisticated analogues.

- The first step of the parametric modelling is to obtain a multivariate regression formula that links dependent variable—duration and cost outcomes—with independent parameters obtained from descriptive regression analysis.
- The second step of the parametric modelling is to define one-point mean value for project's outcomes which is somewhat achieved through tweaking of weighting coefficients of the regression formula to make it more relevant to the upcoming project.
- As the one-point mean value bring up an absolute certainty, the next step is to bring uncertainty back through imposing dispersion characteristics of the used database. A normal (bell-shape) distribution is used for this.
- Using symmetrical bell-shape distribution to establish uncertainty is contradictory to empirical data that the distribution should be highly skewed.
- Based on an expert judgement, one standard deviation of the bell-shape distribution should be used as a project contingency, which is a highly subjective recommendation.
- According to the inferential paradox, a low-accuracy of parametric results may still cover an actual project outcome if the precision of the parametric method is also low. However, this fully depends on characteristics of the used database.
- If a database contains similar projects with similar outcomes (low dispersion) that are irrelevant to the upcoming project (a biased sample set), a narrow distribution will be far apart from the actual outcome due to the systematic error.
- A hybrid model that combines parametric module with Monte Carlo module succumbs to the double counting of project risks that are taken into account twice in parametric and Monte Carlo modules of a hybrid model.
- If the parametric modelling is a good fit for purpose in early project phases FEL1 and FEL2A, the hybrid model does not seem to have much room for applicability.
- System dynamics modelling allows the description of project rework cycle through taking into account non-linear knock-on and compounding risk interactions in a project system.
- The ANN method is a natural upgrade to the low-accuracy parametric modelling as it opens more possibilities for non-linear modelling of interdependencies among the model's independent parameters.
- Both system dynamics and ANN methods can become effective modelling tools for providing some empirically vetted inputs to Monte Carlo models.

5

Project Zemblanity Business Case (I): Linear Monte Carlo Schedule and Cost Risk Analysis (L-SCRA)

This chapter introduces an example of a traditional integrated schedule and cost risk analysis (SCRA) model. Namely, this demonstrates probabilistic risk quantification methodology that has been used by advanced project risk analysts for more than a decade up to now. This methodology is referred to as linear Monte Carlo methodology in this book and it is quite adequate in case of mature project systems. Two pertinent books on linear Monte Carlo methodology by Hulett (2011) and Raydugin (2013) should be cited. A modified non-linear methodology that takes into account risk interactions is developed in this book in consecutive chapters as an upgrade to the traditional linear SCRA methodology. This newly developed non-linear methodology is based on its linear version as a starting point and a limiting case. By this reason the traditional methodology should be reviewed in detail prior to branching out to its non-linear version. In addition, this chapter may be considered a crash course on the traditional L-SCRA. A workable model discussed in this chapter is available on the book companion website at **www.oup.co.uk/companion/Raydugin.**

This chapter introduces an imaginary business case of an oil and gas project Zemblanity undertaken in the Arctic by an international consortium, Serendipity United PLC.

5.1 A Business Case

An international consortium Serendipity United PLC of European, Middle-Eastern, Asian, North American, and Russian major oil and gas companies agreed to develop a huge newly discovered offshore oil field near Novaya Zemlya in the Russian Arctic. This project became viable due to a favourable political situation encouraging close international cooperation and accelerated melting of the Arctic ice cap. Among various facilities to develop there is a gas-oil separation plant (GOSP). A project to develop the GOSP received a name—project Zemblanity. A business case study (FEL1) has been successfully completed although several business issues among consortium members are still unresolved. The project has entered a scope developing phase (FEL2). A phase FEL2A devoted to selection of a business alternative in

Modern Risk Quantification in Complex Projects: Non-linear Monte Carlo and System Dynamics Methodologies. Yuri G. Raydugin, Oxford University Press (2020). © Yuri G. Raydugin.
DOI: 10.1093/oso/9780198844334.001.0001

terms of location, technology, transportation methods, served markets, etc. has been completed recently although some important components of scope have yet to be settled among the consortium members. In addition, it is expected that quite significant changes of scope could result from coming environmental and safety studies as well as from permitting process.

Risk Services & Solutions Inc. based in Calgary, Alberta, Canada, has been selected as a project management consultancy (PMC) to support FEL2B and FEL3 activities including risk management activities. Selection of engineering, procurement, and construction (EPC) contractor(s) is planned at the end of the project definition phase FEL3 when a final investment decision (FID) should be made.

It has been decided that the GOSP should be able to process 450,000 barrels of crude oil a day. It also has been decided that the GOSP should be located in the northern part of Novaya Zemlya in an on-land point that is closest to the developed off-shore oil field located to the north. A base estimate developed upon selection of the preferable project alternative is $1B (Table 5.1). The overall duration of the project since the start of the FEL2 phase is estimated at about 44 months (Table 5.2, Figure 5.1).

The integrated project team (IPT) has held a risk identification and assessment session upon selection of the preferable project alternative based on qualitative (scoring) methodology. A bowtie diagram (see Figure 2.2) and risk assessment matrix (RAM) (see Figure 3.1) have been used for initial risk identification and scoring assessments. However, only two quantifiable objectives—schedule and cost—have been in the focus of risk identification. Two assessments have been done: before and after risk

Table 5.1 A base estimate of project Zemblanity

ID	Summary Activity	Total Cost, $'000
SD00	Scope Definition (FEL2)	6,200
PD00	Project Definition (FEL3)	21,400
DE00	Detail Engineering (incl. Procurement)	466,000
CN00	Construction	448,500
CS00	Commissioning and Startup	57,900
Total	Project Zemblanity	1,000,000

Table 5.2 Project activity durations of project Zemblanity

ID	Description	Duration, d	Start	Finish
MS01	Project Start	0	1-Apr-2020	
SD00	Scope Definition (FEL2)	140	1-Apr-2020	18-Aug-2020
PD00	Project Definition (FEL3)	226	19-Aug-2020	1-Apr-2021
MS02	FID	0	2-Apr-2021	1-Apr-2021
DE00	Detail Engineering	270	2-Apr-2021	27-Dec-2021
CN00	Construction	550	28-Dec-2021	30-Jun-2023
CS00	Commissioning and Startup	173	12-May-2023	31-Oct-2023
MS03	Project Completion	0		31-Oct-2023

Project Zemblanity
Serendipity United PLC

RISK SERVICES & SOLUTION Inc. ™

Activity ID	Description	Early Start	Early Finish	Duration
<NONE>		01-Apr-2020	31-Oct-2023	1309 d
MS01	Project start	01-Apr-2020	01-Apr-2020	0
SD00	Scope definition (FEL2)	01-Apr-2020	18-Aug-2020	140 d
PD00	Project definition (FEL3)	19-Aug-2020	01-Apr-2021	226 d
MS02	FID	02-Apr-2021	02-Apr-2021	0
DE00	Detail engineering (incl. procurement)	02-Apr-2021	27-Dec-2021	270 d
CN00	Construction	28-Dec-2021	30-Jun-2023	550 d
CS00	Commissioning & startup	12-May-2023	31-Oct-2023	173 d
MS03	Project completion	31-Oct-2023	31-Oct-2023	0

Figure 5.1 A schedule of project Zemblanity

Table 5.3 Project Zemblanity risks

Risk ID	Risk Title (Tag)	Cause(s)	Risk event	Impact(s)
R01	Design Basis Change (Proponent-Driven)	A. On-going reservoir simulations B. Changing stakeholders' requirements C. Overlooked components of scope	Design basis may be changing by request of the proponent (after FEL2)	A. Rework and extra cost (engineering) B. Rework and schedule delay (engineering) C. Quality (engineering)
R02	Overlooked External Stakeholder's Requirements	A. Lack of communication with external stakeholders B. Recommendations of environmental impact assessment (EIA) and HAZOP studies	Overlooking of major requirements of external stakeholders (government, media, etc.) and opposition from their side could take place	A. Rework and extra cost (engineering) B. Rework and schedule delay (engineering) C. Quality (engineering)
R03	Errors and Omissions (Engineering)	A. Interface management failure (among engineering disciplines) B. Inconsistent implementation of QA/QC procedures in engineering	Errors and omissions in engineering could take place	A. Rework and extra cost (engineering) B. Rework and schedule delay (engineering) C. Quality (engineering)
R04	Poor Work Quality (Construction)	A. Interface management failure B. Contractor's coordination difficulties C. Inconsistent implementation of QA/QC procedures in construction	Poor quality of construction works could take place	A. Rework and extra cost (construction) B. Rework and schedule delay (construction) C. Quality (construction)
R05	Delays in Material Delivery	A. Several delivery items on project's critical path B. Limited capacity of qualified vendors and competing projects C. Logistics difficulties	Some critical items might not be delivered to the construction site on time	A. Schedule delay (construction)
R06	Commissioning and Startup (CSU) Failure	A. Poor planning and unavailability of required CSU skilled labour B. Poor planning and unavailability of required spare parts C. Failed CSU equipment tests	Timely CSU might not be possible	A. Rework and extra cost (CSU) B. Rework and schedule delay (CSU) C. Quality (operations)

addressing based on proposed addressing actions. The outcome of the session has been collected in a project risk register containing six project risks. A list of identified risks is shown in Table 5.3. A risk register that includes after-addressing risk assessments as well as generic addressing actions is shown in Table A.1 of Appendix A.

5.2 L-SCRA Method Statement

As a next step, the IPT has been asked to develop schedule and cost contingencies to ensure a required confidence level for a project duration and a cost estimate. A post-mitigated confidence level P70 has been established by the consortium as a contingency selection criterion for both.

The IPT has reviewed various risk quantification methodologies to select a most appropriate one.

First, it has been decided to reject any methodology based on pure expert judgements as they could be quite subjective and biased. It has also been decided to reject a methodology based on predetermined guidelines as that methodology does not take into account actual project context and uniqueness (context differentiators) whatsoever. Other methodologies based on analyses of empirical data including parametric modelling have been rejected too. They are based on empirical analysis of completed projects of the past that had nothing to do with complexity and uniqueness of project Zemblanity. In addition, they cannot take into account an actual project risk exposure and miss out an important risk-addressing angle completely, although the parametric methodology was used in FEL1 phase to have high-level rule-of-thumb assessments of project cost when the project scope was not yet defined. Several projects to develop GOSPs completed in 1960–2010 in various geographies—from the Gulf of Mexico to Alberta, to the North Sea—have been used as a sample set to estimate a possible cost outcome of project Zemblanity. Actually this was an assessment for the project class, not for project Zemblanity. Initially it has been argued that the accuracy of the base estimate assessment (see Table 5.1) in FEL1 should be –30 per cent/+50 per cent following the industry practice and the AACE guidelines. However, when various unique project uncertainties related to the project location have been thoroughly considered, it has been argued that the accuracy of the assessment should be –35 per cent/+90 per cent. Due to the wide uncertainty range, it has been recognized that the parametric assessment of this kind could not be practically useful for decision-making.

Second, after selection of a preferred alternative for project Zemblanity has been made in FEL2A, it has been decided to use probabilistic Monte Carlo methodology to estimate both schedule and cost contingencies. It has been proposed by one of the consortium members to use activity ranging (AR) and line item ranging (LIR) Monte Carlo methodologies that have been used by the industry for several decades. One additional reasoning has been that decision-makers should easily understand such a straightforward and traditional methodology and digest the results. Being based on evaluation of general schedule and cost uncertainties only, this proposal has been rejected as this could not take into account real project risk exposure including risk addressing. Initially it has been decided to carry out two standalone schedule and cost risk analyses: SRA and CRA. This has been opposed by a construction manager: according to his previous experience so-called schedule-driven costs usually make significant contributions to cost risks and should not be missed out.

As a result, it has been decided to carry out integrated SCRA using the existing base estimate (Table 5.1), the schedule (Table 5.2 and Figure 5.1), and the risk register (Tables 5.3 and A.1) as initial raw inputs. This preliminary developed information collected prior to the SCRA workshop have been reviewed, validated, and endorsed by all internal stakeholders. It has also been decided to conduct the SCRA modelling for the after-risk-addressing case only to accentuate the value of proactive risk addressing as an important project control instrument.

5.3 A SCRA Workshop

A four-day SCRA workshop to collect input information for SCRA modelling has been held in a project Zemblanity field office in Rogachevo, Novaya Zemlya, Russia. Representatives of the integrated project team including representatives of all consortium partners (project owners) and employees of the PMC contractor have attended the workshop to ensure a quorum. A risk management facilitator from the PMC contractor's organization has been appointed to run the SCRA workshop.

General schedule uncertainties have been reviewed and defined as a first step of the SCRA workshop (Table A.2 of Appendix A). As a second step schedule-driven costs have been discussed. To accommodate the schedule-driven costs in risk quantification models, project schedulers and estimators have been asked to break down each cost account of Table 5.1 to two components using activity durations of Table 5.2—fixed (time-independent) costs and variable (time-dependent) costs. The summary of this exercise is collected in Table A.3. Table A.3 contains so-called burn rates measured in dollars per day so that any possible schedule delay associated with schedule risks will inevitably result in extra schedule-driven costs.

As a concluding part of the SCRA workshop, a project risk register has been developed to specifically define schedule and cost uncertain events. The initial project risk register (Table A.1) has been used including addressing actions of identified risks. Participants of the workshop converted the initially collected risk data of Table A.1 to SCRA inputs to obey probabilistic (Monte Carlo) input data requirements. Workshop participant's judgements based on their experience and available historic data have been used to complete the risk assessment. As a result, the very raw scoring risk assessments of Table A.1 have been cardinally revised, mapping of risks has been added, and main correlations among risks have been established (Tables A.4 and A.5).

The collected risk register data have been reviewed, validated, and endorsed by all interested internal stakeholders upon completion of the SCRA workshop. They have been used to build and run a linear L-SCRA model.

Results of the L-SCRA modelling are discussed in section 5.4.

5.4 L-SCRA Modelling Results

Key results of the integrated linear L-SCRA after addressing are presented by Figures 5.2–5.10. As decision-making should be done based on after-addressing modelling results, the before-addressing results are not discussed at all.

Figures 5.2 and 5.3 demonstrate two versions of SRA outcomes: a distribution of completion dates and its equivalent for project durations, both after risk addressing. Expected project completion date at P70 confidence level is 25 April 2024 which

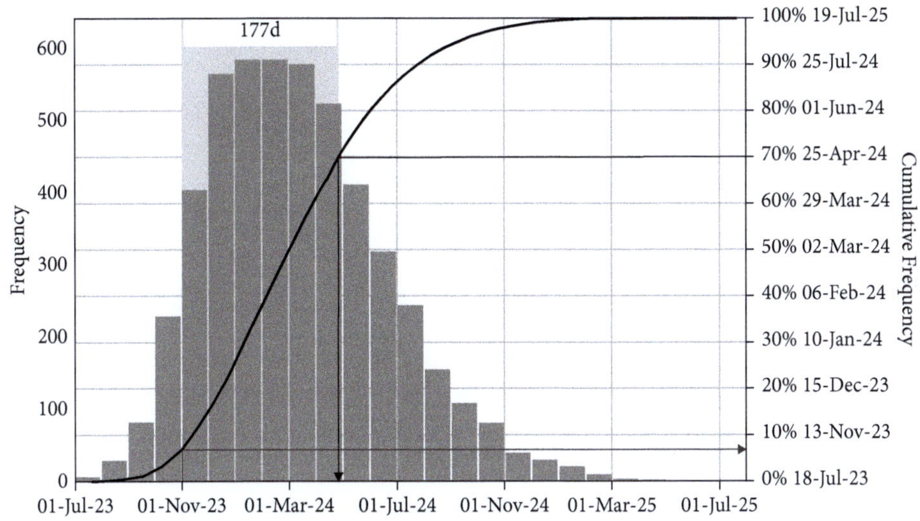

Figure 5.2 A distribution of project Zemblanity project completion dates (a linear L-SCRA case)

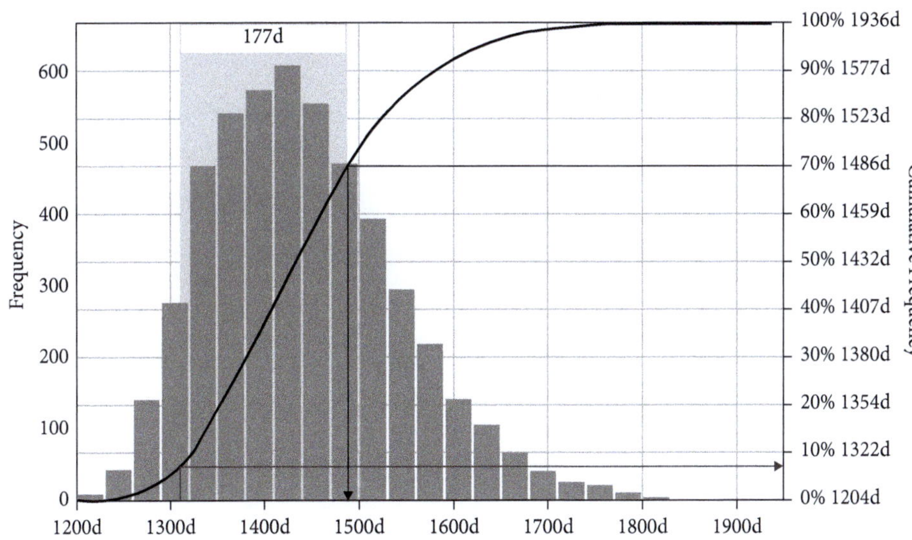

Figure 5.3 A distribution of project Zemblanity project durations (a linear L-SCRA case)

corresponds to a project duration of 1486 days (Figure 5.3). A required schedule contingency is estimated as 14 per cent of the deterministic schedule duration of 1309 days.

Figure 5.4 shows L-SCRA cost outcome as a total project cost after risk addressing. Total cost contingency at the P70 confidence level would be $148M which corresponds to 15 per cent of the $1B base estimate.

Total schedule-driven costs are estimated as $37M at P70 level of confidence which corresponds to about 3.7 per cent of the base estimate (see Figure 5.5).

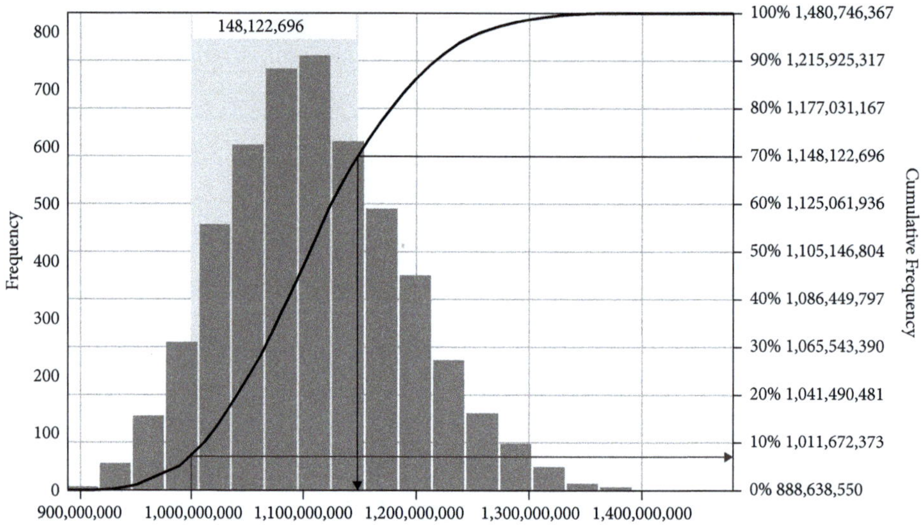

Figure 5.4 Project Zemblanity cost distribution including schedule-driven costs and cost general uncertainties (a linear L-SCRA case)

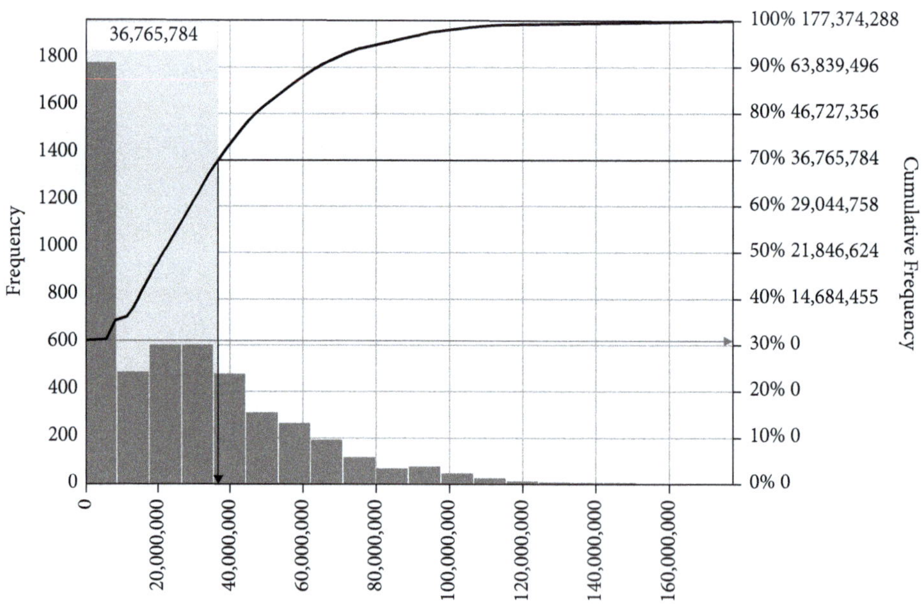

Figure 5.5 Project Zemblanity cost distribution for schedule-driven costs (a linear L-SCRA case)

Figure 5.6 provides a project duration distribution when only general duration uncertainties are counted. At P70 level it is 1395 days which corresponds to 7 per cent duration contingency. Cost contingency of $115M at P70 level required due to the cost of general uncertainties corresponds to about 12 per cent (Figure 5.7). This cost contingency includes schedule-driven costs stemming from schedule general uncertainties.

The general uncertainty results of Figures 5.6 and 5.7 are in a good compliance with previous observations on schedule and cost contingencies produced by activity and line-item ranging methods (Merrow (2011)) discussed in Chapter 3. However,

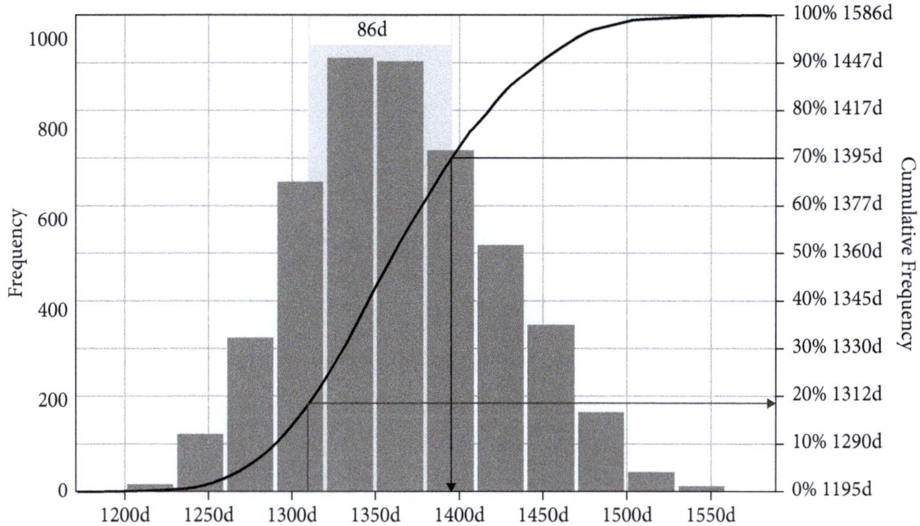

Figure 5.6 Project Zemblanity duration distribution if only general uncertainties are included (a linear L-SCRA case)

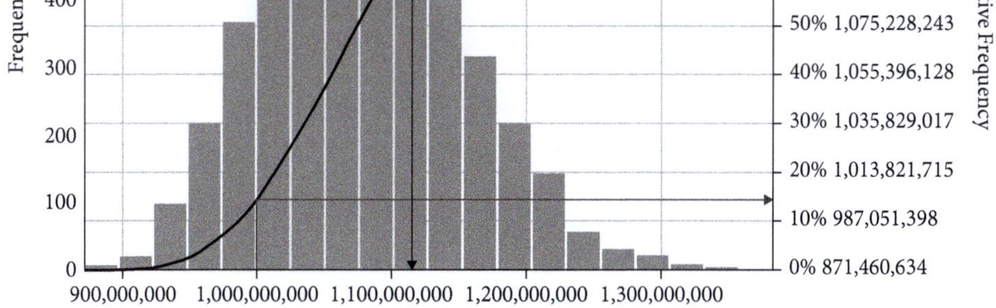

Figure 5.7 Project Zemblanity cost distribution if only cost general uncertainties as well as schedule general uncertainties are included (a linear L-SCRA case)

it should be kept in mind that Figure 5.7 contains schedule-driven costs stemming from general schedule uncertainties.

Figures 5.8 and 5.9 demonstrate schedule and cost distributions when only risk events are counted with general uncertainties being excluded. A required schedule contingency at P70 level is about 9 per cent (Figure 5.8),[1] whereas cost contingency at

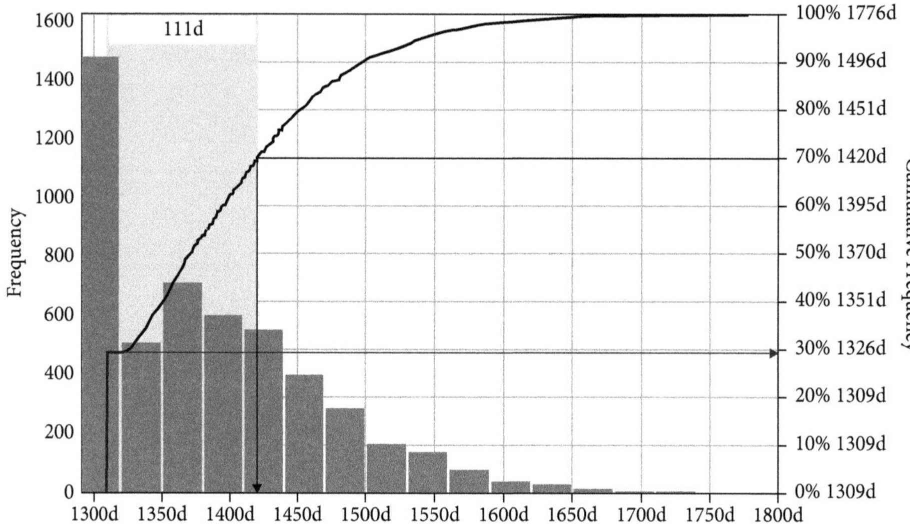

Figure 5.8 A project duration distribution when only risk events are included (a linear L-SCRA case)

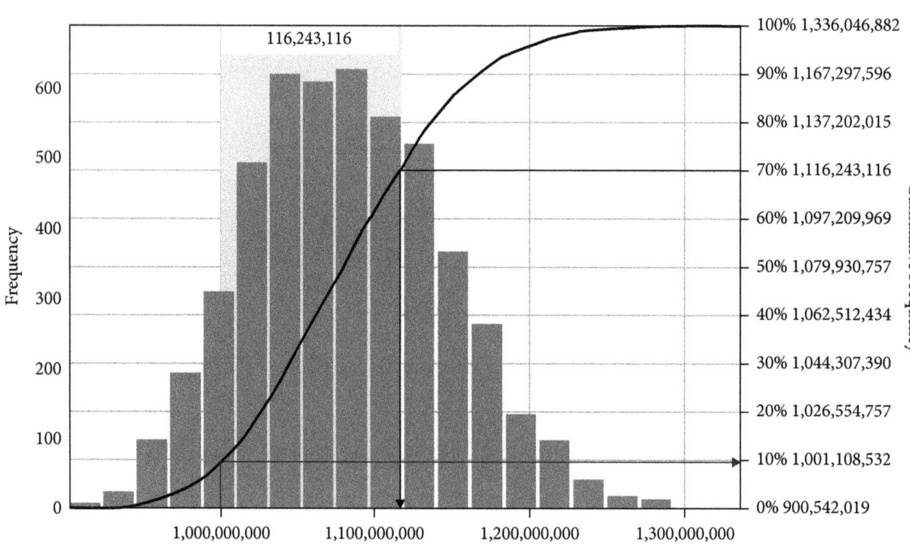

Figure 5.9 A project cost distribution after addressing when only risk events are included (with schedule-driven costs) (a linear L-SCRA case)

[1] It should be noted that the duration distribution of Figure 5.8 has a mean value of 1388 days. This information is used in Chapter 7 to compare L-SCRA results with results of system dynamics modelling.

P70 level due to cost risks is 12 per cent (Figure 5.9) which includes schedule-driven costs associated with uncertain events only.

5.5 Discussion of L-SCRA Results

The results of the undertaken SCRA analysis have been reviewed by representatives of the consortium. Several project specialists including engineering, construction, commissioning and startup, quality, environment protection, etc. managers have attended the review session. It has been pointed out that considering schedule-driven costs explicitly has improved understanding of the overall project risk exposure. At the same time, reviewers of the SCRA results have expressed a concern that both schedule and cost risk contingencies would be too low for a project of a given level of structural and delivery complexity. Two major observations have been put forward.

First, an opinion has been expressed that discussed SCRA results are a fit to projects:

- with scopes fully agreed by all stakeholders,
- with well developed and implemented project delivery procedures,
- undertaken in well-developed areas, and
- delivered by highly-experienced project teams and contractors.

An opinion has been expressed that in reality identified risks would have much higher severities than factored into modelling.

Second, speaking from their practical experiences, several project specialists attending the review meeting have attracted the participant's attention to situations when risks interact either through their compounding or due to 'domino', 'ripple', 'snowballing', or 'knock-on' effects.

Based on these observations some more specific points of concern have been raised.

- Keeping in mind recent heated unsettled debates among the project stakeholders on the project scope and an execution strategy, it has been pointed out by the stakeholder management specialist that late scope changes would be imminent in later project phases. This pessimistic conclusion has been drawn along with a statement that scope development and change management procedures have yet to be finalized. Several consortium members have been trying to impose their own conflicting procedures related to scope development and change management. As has been pointed out by a quality manager, any possible late changes requested by consortium members after completion of FEL2 phase would have unproportionally high or 'non-linear' impacts on the project duration and cost due to several factors. These factors could include re-work, discarded previously completed work, work constraints, out-of-sequence work, sunk costs, etc.
- According to the environmental protection manager, expectations of external stakeholders have yet to be fully understood. It seems to be very likely that some

of the key stakeholders among local and federal branches of government have yet to be identified. Stakeholder management procedures have been developed and implemented although have not been followed on several occasions. If late proponent-driven changes become compounded with changes initiated by external stakeholders, the overall impact on the project duration and the budget would be unproportionally high or, again, 'non-linear'.

- Due to high complexity of the project including the project location and multiple interfaces, it has been pointed out by a procurement manager that selecting and mobilizing of EPC contractors that could deliver the project flawlessly would be challenging. It may be anticipated that engineering errors and omissions and lower quality in construction works would be much higher than factored in to the currently developed risk quantification models. Current risk assessments related to quality of EPC works seem to imply that the best EPC contractors who successfully completed several oil and gas projects in the Arctic have been selected. Real contractors might have much lower experience and skills especially in the light of cost pressure and a propensity of some consortium members to select low bidders.
- The engineering manager has predicted that any late scope changes after FEL2 by any reason can induce a knock-on effect on engineering and procurement activities. Moreover, these could instigate possible engineering errors and omissions or at least amplify them. Such amplification, compounding, and knock-on effects should be reflected in risk quantification modelling in a way.
- The quality manager has accentuated a possibility that engineering errors and omissions are not always detected before engineering is completed. A situation when those are discovered in construction usually leads to significant rework and unproportionally high delays and extra costs for the entire project. Some errors and omissions like this can be fixed without returning flawed drawings to engineering. Corresponding amendments could be done by the construction team and reflected in as-built drawings. In more notorious cases, engineering rework is inevitable. As the construction manager has argued, both cases could result in rework, work constraints, and out-of-sequence work leading to lower productivity, delays, and extra costs, including schedule-driven costs.
- It has been pointed out by the procurement manager that delivery of materials to Novaya Zemlya could be challenging for two reasons. First, selected vendors might not be capable of producing technological equipment for working in a harsh Novaya Zemlya environment on time, on specification, and with acceptable quality. Second, logistics issues could be higher than factored in to the risk quantification models. Also, a delay in accessing material can easily lead to construction work constraints and out-of-sequence work leading to additional delays and extra costs. It has been suggested that compounding of material delivery delay with risks to construction activities should be also reflected in risk quantification modelling.
- The construction manager has pointed out that out-of-sequence work might be needed due to so-called work constraints when required materials or required

labour are not available simultaneously. He has referred to work constraints using an old Russian construction industry joke: 'today bricks – yea, though mortar – nay; tomorrow bricks – nay, though mortar yea'. These work-constraint factors should somehow be reflected in a realistic risk modelling.

- It has been argued that overall commissioning and startup delays and extra costs might be much higher than anticipated in the risk quantification model. On the one hand, if the identified commissioning and start up risks are treated and evaluated as standalone risks, the agreed inputs to the risk models can be considered somewhat adequate. However, it has been pointed out by the experienced commissioning and start up manager that any quality issues occurred in earlier project phases and not identified could result in abnormally high and sort of 'non-linear' impacts on commissioning and startup risks. The possibility of the delayed knock-on impacts has not been taken into account in the risk quantification models but should be to increase model's realism.

A strong opinion has been expressed that the proposed risk quantification methodology seems to miss out some important factors related to features of project development including its level of complexity, leadership, team members' skills, adopted procedures, location, etc. Apparently, it has been argued by participants that the proposed Monte Carlo model does not seem to produce accurate and credible results. There are several hints that the proposed SCRA Monte Carlo model has substantial systematic error (see Figure 3.4) due to several important risk factors that have not made it to the model.

Based on the provided feedback and comments it has been recommended to revise the risk quantification methodology to better reflect on project realities. Hence, the challenge is now in the court of the project risk manager.

5.6 Further Improvements of the L-SCRA Modelling

The initial intent has been to improve the existing Monte Carlo model using known factors and methods discussed below in this section. Artificial inflation of impact ranges of general uncertainties as well as of probabilities and impact ranges of uncertain events have been ruled out as this cannot address expressed issues and concerns in a structured and consistent way.

As discussed in this section, none of the contemplated improvement measures can address concerns expressed during the SCRA result's review.

5.6.1 Before-addressing and After-addressing Modelling

The considered L-SCRA model has been run for the after-addressing scenario only. All decisions related to project contingencies are done using outcomes of after-addressing risk quantification. Although, in case of complete failure or ignoring of proposed risk-addressing actions the two assessments—before and after addressing—are supposed to be identical. For this reason, before-addressing assessments still have a value and may be undertaken: they delineate a range of possible risk quantification outcomes if addressing completely fails.

Some modern Monte Carlo modelling software packages allow the possibilities of risk-addressing failure to be factored in. As assessments after addressing directly depend on this, it should be reasonable to utilize this functionality to produce more realistic assessments and contingencies after addressing. In a limiting case of failure of all addressing actions after-addressing risk assessments should simply be replaced by risk assessments before addressing. For simplicity purposes the possibility of risk addressing failure is not considered in this book. In any case it would not be able to address issues and concerns put forward during the modelling result's review.

5.6.2 Burn Rates

The burn rates used to define schedule-driven costs in Table A.3 are one-point numbers. To avoid the absolute burn-rate certainty, their distributions could be introduced instead. For example, a triangular distribution could be used to introduce minimum, most likely, and maximum burn rate's values. This is not done for simplicity purposes, although if skewed burn-rate distributions are used this could lead to somewhat higher cost contingency.

5.6.3 Jointed Confidence Level (JCL)

The JCL concept can also be used for increase of project contingencies. Mathematical foundation of JCL is explained in detail in the sections below. Practical applications and examples of the JCL concept are explored by Hulett (2019).

5.6.3.1 JCL: a general concept

So far individual confidence levels P70 are being considered separately for schedule and cost risk distributions. An advanced decision-maker might ask a loaded question why these confidence levels are not considered jointly. It is easy to imagine that a project spends a lot of extra money to accelerate a schedule (or crash it on the risk jargon) in an attempt to meet an established completion date. Cases like this are not unheard of. What if the decision-maker asks about a case when both schedule and cost contingencies have the P70 confidence level simultaneously or the joint confidence level JCL70. Apparently individually developed P70 values and JCL70 values should differ as JCL70 is based on an additional strong condition

$$JCL70 = (Schedule = P70 \cup Cost = P70) = (P_S 70 \mid P_C 70) = (P_C 70 \mid P_S 70), \qquad (5.1)$$

where sign $\cup$ indicates unification of two conditions—P_S70 duration values for the schedule distribution, and P_C70 cost values for the cost distribution—that must be obeyed simultaneously. In a general case any confidence levels or their combinations could be established in decision-making, if required. For instance, a $(P_C50 \mid P_S90)$ condition could be established that answers a question 'what would be a cost P_C50 value if the schedule P_S90 value is fixed?' It is said that the cost P_C50 value is conditioned by the schedule P_S90 value. It could be called a joint confidence level JCL50(C)/90(S).

As a next step, these one-point value conditions could be expanded to a situation when an entire schedule or cost distribution is conditioned by a fixed cost or schedule one-point value. For instance, a conditioned distribution could be established $(Cost \mid P_S70)$ for the entire cost distribution with the fixed P_S70 value. The cost distribution will be different if it is conditioned differently: the conditioned cost distribution will be unique for each fixed schedule-confidence-level conditioning value. The same is applicable to conditioned schedule distributions.

If following this path, a most general condition could be established between two distributions: a cost distribution conditioned by a schedule distribution (*Cost* | *Schedule*) or a schedule distribution conditioned by a cost distribution (*Schedule* | *Cost*). Such combined distribution is called bivariate, whereas two individual distributions are called marginal.

Apparently, both marginal distributions can be represented as two-dimensional charts (see e.g. Figures 5.2–5.4): values that are realized most frequently form the peaks of the marginal distributions. However, their bivariate distribution should be three-dimensional. It should look like a hill which peaks when formed by most frequent pairs of schedule-cost values occurring in Monte Carlo iterations. Actually, Figure 5.10 does represent the schedule-cost bivariate distribution which is supposed to be three-dimensional. For simplicity purposes, it was reduced to the two-dimensional chart resulting in the 2D 'American football' chart instead of the 3D hill chart. Although the hints on the third dimension—the hill's height reflecting on the schedule-cost pair's frequencies—is kept in a way, which is visible through the density of the schedule-cost pair dots on Figure 5.10.

The conditioned cost distribution $(Cost \mid P_S70)$ mentioned above can be imagined as a cross-section of the hill for the fixed P_S70 value. A set of cross-sections representing conditioned cost distributions can be obtained for any other conditioning schedule values through slicing the hill up. The same is applicable to conditioned schedule distributions.

The theoretical and analytical foundation for schedule-cost JCL approach is developed by Garvey et al. (2016) based on a study of bivariate normal and lognormal probability distributions as developing analytical formulas for a general case is not possible. However, schedule and cost distributions can be roughly treated as near-normal (see Figures (5.2)–(5.4)) if the resulting schedule and cost distributions have not been altered too severely.[2]

[2] As shown in Chapter 15, schedule and cost distributions are substantially skewed if project system issues and risk interactions are taken into account. By this reason the American football chart takes the form of a plume.

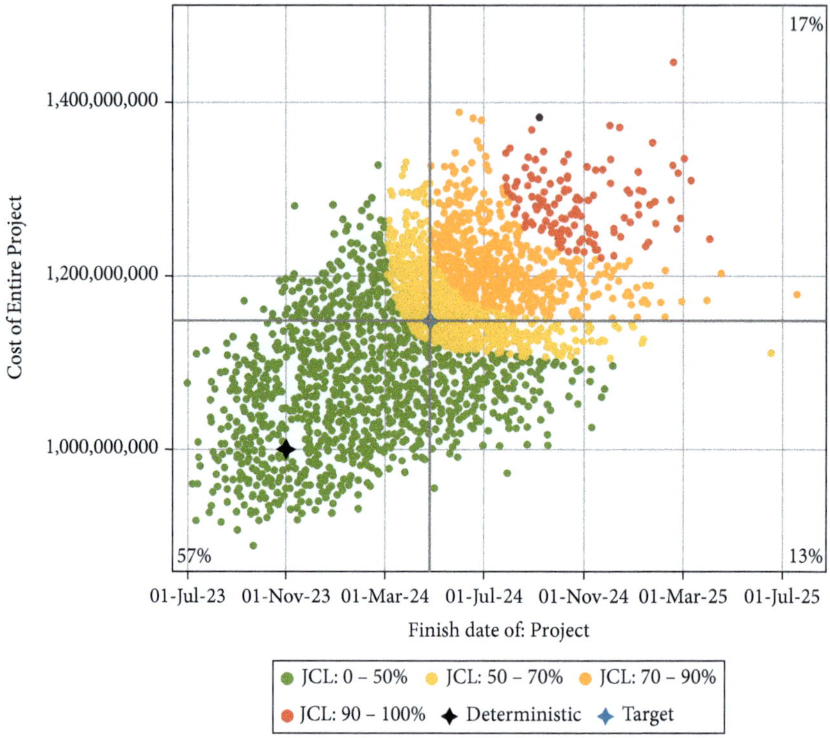

Figure 5.10 Joint confidence level (JCL) chart for project Zemblanity (a linear L-SCRA case): the crosshair shows a P70 completion date (25 April 2024) and a P70 cost ($1.15B).

According to Garvey et al. (2016), in case of two marginal normal distributions, the conditional expected values (mean values) for the bivariate schedule-cost distribution are

$$E(Schedule \mid P_C 70) = \mu_S + (\sigma_S / \sigma_C) \times K_{SC} \times (P_C 70 - \mu_C), \tag{5.2}$$

$$E(Cost \mid P_S 70) = \mu_C + (\sigma_C / \sigma_S) \times K_{CS} \times (P_S 70 - \mu_S), \tag{5.3}$$

Where μ_S and μ_C are mean values of the marginal schedule and cost distributions, $-1 \le K_{SC} \le 1$ is a schedule-cost correlation coefficient ($K_{SC} = K_{CS}$), σ_S and σ_C are standard deviations of marginal schedule and cost distributions.

For the purpose of this discussion, expressions (5.2) and (5.3) can be interpreted as a conversion of random errors in one marginal distribution to amendment of systematic errors in the other marginal distribution through taking into account schedule-cost data dispersion and correlation.

Besides conditional mean values the bivariate distribution is described by conditional standard deviations as follows.

$$\sigma(Schedule \mid P_C 70) = \sigma_S \times \sqrt{1 - K_{SC}^2} \tag{5.4}$$

$$\sigma(Cost \mid P_S 70) = \sigma_C \times \sqrt{1 - K_{CS}^2} \tag{5.5}$$

Certainly, expressions (5.1)–(5.5) can be easily generalized to any conditioning schedule and costs confidence levels.

5.6.3.2 JCL: applicability

Apparently, when correlation coefficient $K_{SC} = 0$, any individual confidence levels of marginal distributions are equal to JCL by default as two distributions are simply independent.

When $K_{SC} \neq 0$, and if in place of P_S70 and P_C70 conditioning values used are mean values μ_S and μ_C, expressions (5.2) and (5.3) yield the following results.

$$E(Schedule|\mu_C) = \mu_S \tag{5.6}$$

$$E(Cost|\mu_S) = \mu_C \tag{5.7}$$

Despite expressions (5.6) and (5.7) looking quite trivial, they actually imply two remarkable conclusions.

First, the JCL concept is important only if both conditioning confidence level criteria significantly differ from the mean values. For instance, if the corporate schedule and cost criteria are based, say, on P50 or P_{mean} confidence level, which are usually quite close to each other, the JCL concept is to produce JCL50 or JCL_{mean} that is close to individual P50 or P_{mean} values. This means that the JCL is not terribly useful in this case as it does not help to increase contingencies. It makes sense to use the JCL criteria if the individual and JCL values are based on much higher confidence levels far away from mean values P_{mean}. Only in this case the difference between individual and JCL results would be significant and worth playing with (see expressions (5.2), (5.3), (5.6), and (5.7)).

For instance, NASA (2015) adopted the joint confidence schedule-cost level 80 per cent or JCL80 for all its projects. This actually corresponds to abnormally high individual confidence levels of schedule and cost distributions that are higher than P90. Among other things, this could be explained by the novelty of NASA projects as well as features of NASA funding process in the US Congress. Apparently, the adopted JCL80 criterium reflects on a comfortable political NASA–US Congress joint confidence level. In a hypothetical case if NASA used JCL criteria around JCL50 or JCL_{mean}, it would not be a useful tool to require more money for projects.

Second, being based on correlations between marginal schedule and cost distributions, JCL correction factors impact both random and systematic errors of the Monte Carlo SCRA methodology (see Figure 3.4). However, if individual confidence levels are selected in the vicinity of mean values, only random error is affected which depends on the correlation coefficient only (see expressions (5.4) and (5.5)). In this case systematic error is not addressed or changed at all.

Actually, formulas (5.1)–(5.5) and other expressions provided by Garvey et al. (2016) can be used for analytical calculations of JCL values if marginal schedule and cost distributions are not skewed. The resulting oval-shape bivariate distributions are similar to the one of Figure 5.10. The only practical difficulty for this approach is deriving right correlation coefficients K_{SC}. Good news is that most of modern Monte Carlo SCRA software packages allow the JCL calculations to be easily performed behind the scene which is already demonstrated by Figure 5.10. The JCL charts are standard by-products of Monte Carlo modelling when modern Monte Carlo software packages are used.

Although the importance of the JCL concept is not denied, it should be considered a side-effect of a regular Monte Carlo modelling. In contrast with other risk factors that could be addressed, JCL correction factors—the differences between JCL and individual confidence level values—are givens and cannot be addressed directly. At the same time, the JCL correction factors do depend on the other risk factors and their addressing indirectly. The JCL concept should be used when decision-makers require a higher level of confidence in project outcomes based on a trade-off between project durations and costs.

By this reason JCL correction factors can be used to condition Monte Carlo modelling results after the main Monte Carlo modelling is completed. This should be the case when a schedule-cost trade-off is of high importance.

5.6.3.3 JCL correction factors for the L-SCRA model

In the L-SCRA case considered in this chapter, along with the two P70 values—a project due for completion on 25 April 2024 (Figure 5.2) with a cost $1.148 B (Figure 5.3)—Figure 5.10 does provide a two-dimensional bivariate distribution of the American-football shape. Figure 5.10 clearly shows the border that tells apart JCL50–JCL70 and JCL70–JCL90 areas which exactly correspond to JCL70. The independently developed P70 values (represented as a crosshair by Figure 5.10) actually correspond to JCL57.

Apparently, the JCL70 band of Figure 5.10 represents a continuum of date–cost combinations. In practice considered for decision-making are date–cost pairs that are located in the middle of the football chart or that are closest to the cross-hair, although there is a slight ambiguity which pair to select. In case of Figure 5.10 a selected date–cost pair could be 1 June 2024 (duration of 1523 days, see Figures 5.2 and 5.3) and $1.182B (see Figure 5.4). Getting back to Figures 5.2–5.4, it could be mentioned that this pair corresponds to about P80 individual confidence levels. In turn, the difference between P70 and JCL57 confidence levels—the JCL correction factor—is measured as about 2.8 per cent $=\left(\frac{1523d-1486d}{1309d}\right)\times 100\%$ for the schedule contingency, and 3.4 per cent $=\left(\frac{\$1.182B-\$1.148B}{\$1B}\right)\times 100\%$ for the cost contingency.

As JCL correction factors in the case of the discussed model are insignificant they are ignored in the decision-making.

5.6.4 Sensitivity Charts

Various sensitivity charts can be used to identify most important risk factors in a project schedule or base estimate (Raydugin (2013)). The primary idea behind using sensitivity analysis is that a small change in an uncertain event or a general uncertainty may lead to an unproportionally large change to overall project duration or cost. A sensitivity of a particular risk depends on its probability and impact, its belonging to project critical or near-critical paths, correlations with other risk factors, etc.

In the traditional Monte Carlo case, all risks factors are considered standalone. The only exceptions are correlations. Results of sensitivity analyses (so called 'tornado charts') are used for further risk addressing. Addressing of top sensitive risks is most efficient: according to the sensitivity concept, even small change of their sensitivity should lead to drastic improvements of project outcomes. Addressing of low sensitivity risks would not make much difference and sense.

Despite this being an important and practical concept it is not directly relevant to the core topic of this book. Initially developed sensitivity charts for the models L-SCRA (this chapter) and for the models N-SCRA-A and N-SCRA-B (Chapter 15) are left out to save space. However, the sensitivity charts are back in favour in Chapter 17 where they are directly used as a primary tool to investigate project failure modes.

5.6.5 Fighting Systematic Error

The initially contemplated improvements of L-SCRA modelling pointed to in this section are useful and relatively important although not cardinal. They are certainly recommended for use in practical Monte Carlo models. They may somewhat amend (increase) derived contingencies as they increase random errors of the L-SCRA methodology making distributions wider (see Figure 3.4). However, these cannot reduce systematic errors. One exception is the application of the JCL correction factors that might also reduce systematic errors slightly, as discussed in section 5.6.3.

The influence of the discussed factors on project schedule and cost contingencies is the same as with correlations among risks: contingencies could be increased by just a few percentage points due to higher random errors (aka wider distributions). Apparently, this would not help in developing adequate contingencies for notoriously delayed and overspent projects. Needless to say, application of these factors has nothing to do with counting of risk interactions or project system issues.

Concerning project Zemblanity, more cardinal modifications are required to increase accuracy of the Monte Carlo methodology through suppressing systematic errors. The need for this cardinal modification is a major outcome of the first project Zemblanity SCRA modelling exercise. The efforts to develop the modified Monte Carlo modelling referred to as non-linear SCRA or N-SCRA are made in consecutive chapters step by step by taking into account risk interactions.

5.7 Chapter's Markup and Takeaway

- Following the discussion in Chapters 1–4, it has been decided to select integrated Monte Carlo SCRA modelling approach for development of schedule and cost contingencies for project Zemblanity as the most advanced method. All other risk quantification methods do not pass muster.
- The proposed approach resulted in 14 per cent duration and 15 per cent cost contingency after addressing.
- Several concerns have been expressed that the derived project contingencies are critically low for a complex project like this in the Arctic area.
- Among the multiple concerns there have been points on partners' misalignments related to the project scope, some poorly developed and implemented project procedures, possible low competencies of selected contractors due to the low-bidder-award approach, potential snow-balling, knock-on, and ripple effects of interacting risks, etc.
- An opinion has been expressed that the proposed model has substantial systematic error as several factors of risk exposure have been missed out.
- Several measures have been contemplated to improve (increase) project contingencies: from taking into account a possibility of risk-addressing failure, to increasing burn-rate uncertainties, to adopting JCL approach.
- It has been recognized that the contemplated measures cannot address fundamental risk exposure factors such as risk interactions or specific internal inefficiencies.
- A need to cardinally modify the traditional Monte Carlo SCRA methodology has been fully recognized.

6
A Brief Introduction to System Dynamics (SD)

This chapter introduces general principles and terminology of system dynamics (SD) using a basic but highly informative model called a bathtub (BT) model. This model is considered a forerunner of many modern system dynamics models that are used in various areas. Although this chapter is not related to project risk management directly, it establishes a required background for project system dynamics cases discussed in Chapters 7 and 16.

Besides the primary goal to introduce system dynamics principles and terminology, this chapter is a reminder about a direct link between modern object-based system dynamics modelling applications and a bit of archaic last-century calculus and analytical solutions of differential equations. The latter stand behind modern system dynamics applications. It would be great not to lose this valuable link in the age of apps.

This chapter should be especially beneficial to students of advanced risk management and system dynamics programmes and any project specialists who are interested in mathematical implications of system dynamics for project risk management. 'Mathematical gourmets' could find this chapter enjoyable. Otherwise it might be ignored.

6.1 A System-dynamics Crash Course: An Overview of a BT Model

Although a model introduced in this section is not used in project SD, it overviews all key relevant principles and language of SD modelling.

A BT model (Hines (1996)) may describe a water level in an actual bathtub depending on water inflows (through 'a tap') and outflows (through 'a sink'). It might be proclaimed an ultimate parent to many project SD models. This model and its versions have multiple applications in economics, demographics, marketing, medicine, zoology, etc. (Sterman (2000)). Five versions of the BT model are introduced in this chapter through considering various interactions and boundary conditions. The corresponding versions referred to as models BT1–BT5 in this chapter are supported by Appendix B. Workable models BT1–BT5 can be downloaded from the book's companion website at www.oup.co.uk/companion/Raydugin.

In a generic version of the model boundary conditions are not applied that leads to unrealistic solutions when unlimited growth of water level or its unlimited decay including negative level values are allowed (a model BT1). In another case, boundary conditions are applied as limiting factors to cut off unrealistic solutions (a model

Modern Risk Quantification in Complex Projects: Non-linear Monte Carlo and System Dynamics Methodologies. Yuri G. Raydugin, Oxford University Press (2020).
DOI: 10.1093/oso/9780198844334.001.0001

BT2). In practical terms, one of these boundary conditions is a must to avoid home water flooding incidents. In the other three cases the boundary conditions are obeyed automatically due to the nature of dynamic interactions (models BT3–BT5) resulting in smooth growth or delay. Strictly speaking, some of these models deviate from the traditional definition of the BT model. Although observing models BT1–BT5 all together and tracking the evolution of their solutions provide valuable insights for project SD modelling.

A general approach adopted in SD modelling is based on resolving of systems of first-order differential equations describing time dynamics of key parameters. The beauty and the elegance of the bathtub model is that it could be resolved analytically for all five discussed versions through analytical integration of its differential equations (see Appendix B). All aspects of analytical integration are hidden in Appendix B as the differential equation's calculus could make some readers uncomfortable or disinterested.

Good news is that modern SD software packages[1] hide explicit differential equations in a user-friendly way behind verbal expressions ('objects'). These expressions represent an equation's variables and parameters through adopting object-based programming.

Any SD model has two major types of parameters. The first type is called 'a level' (or 'a stock'). The second type is called 'a flow' (or 'a rate'[2]) referring to a speed of inflows and outflows. All discussed versions BT1–BT5 of the BT model are shown by Figures 6.1–6.3. Flows, as expected, are indicated as the valve signs. The small cloud signs delineate borders of a model: an inflow source and an outflow sink are ignored in these BT models. Additional parameters ('variables') are used to establish required mathematical relations and calculate levels and flows. Levels and flows may be linked through so-called feed-back loops.

Apparently, a bathtub accumulates a difference between an inflow and an outflow. This process is described by a simple first-order differential equation. Relations between a level in a given moment of time with inflows and outflows could be from trivial when the flows are time-independent (Figure 6.1), to more complicated when the flows directly depend on the levels themselves due to introduced time-dependent feed-back loops (see Figures 6.2 and 6.3). The nature of the dependency could vary and may be either linear or clearly non-linear.

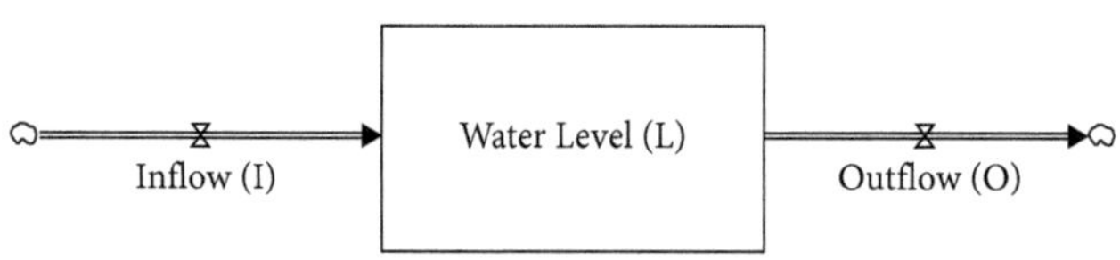

Figure 6.1 A bathtub model without feedback loops (model BT1)

[1] A software package Vensim PLE Plus v. 7.3.5 by Ventana Systems Inc. is used in this book. A free version of Vensim PLE as well as required user manuals could be uploaded from https://vensim.com/free-download/ that has all required functionalities. A more sophisticated and modern alternative to Vensim is its Russian analogue AnyLogic that also supports agent-based and discrete-event modeling and has an extremely user-friendly interface.

[2] Terms 'a level' and 'a flow' are used in this section paying a tribute to the bathtub model. Terms 'a stock' and 'a rate' are also used in consecutive sections interchangeably with 'a level' and 'a flow'.

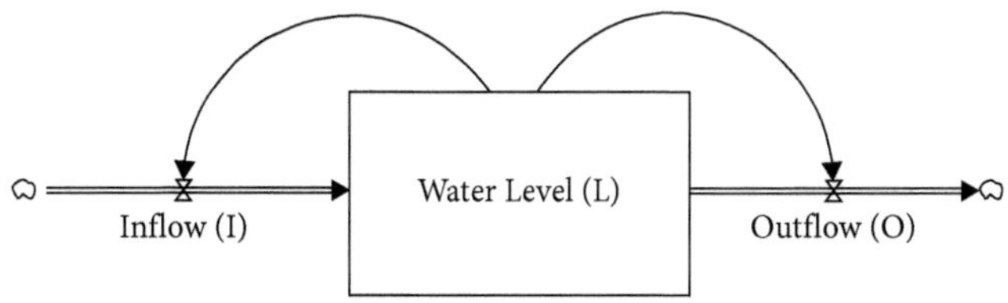

Figure 6.2 A bathtub model with boundary condition's feedback loops (model BT2)

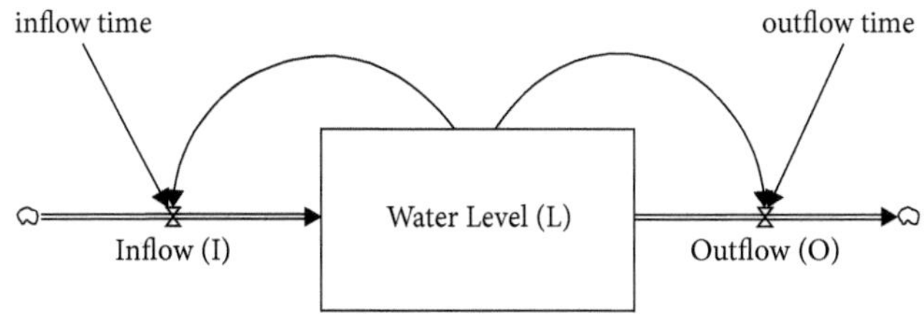

Figure 6.3 A bathtub model with feedback loops (models BT3–BT5)

Instead of mathematical expressions used in differential equations written in analytical form, word phrases ('*Water Level (L)*', *water inflow (I)*', "*water outflow (O)*, '*time*', etc.) can be used for object-based programming and numerical solving of differential equations.[3]

Intuitively, the following expression should be right based on a common sense in the case of the BT model.

$$\text{'}Water\ Level\ (L)\text{'} = (\text{'}water\ inflow\ (I)\text{'} - \text{'}water\ outflow\ (O)\text{'}) \times \text{'}time\text{'} \tag{6.1}$$

Although a more general and specific equation could be written instead:

$$\Delta(\text{'}Water\ Level\ (L)\text{'}) / \Delta(\text{'}time\text{'}) = \text{'}waterinflow\ (I)\text{'} - \text{'}wateroutflow\ (O)\text{'}, \tag{6.2}$$

or simply

$$\Delta L(t) / \Delta t = I(t) - O(t), \tag{6.3}$$

Apparently, equation (6.3) can be rewritten as a typical differential equation (see appendix B)

$$dL(t) / dt = I(t) - O(t). \tag{6.4}$$

Equations (6.3) and (6.4) use the following variables: $\Delta L(t)$ symbolizes a change of the level $L(t)$ during a period of time Δt in (6.3) that is replaced by a first derivative of a level over time $dL(t)/dt$ in (6.4).

[3] Level/stock word phrases start with capital letters by a tradition to tell them apart from expressions for flows/rates and other variables.

An initial condition for equation (6.4) is $L(t=0)=L(0)$. Expression (6.4) being a first-order differential equation is easily analytically solved in Appendix B for five discussed versions of the BT model as follows.

- A generic and rather unrealistic BT model with time-independent inflows and outflows that allows any negative levels $L(t)<0$ and any positive levels even above allowed maximum level $L_{\max}$ ($L(t)>L_{\max}$) in a bathtub: no boundary conditions are imposed whatsoever (a model BT1).
- A more realistic BT model that sharply cuts off any negative levels and levels that exceed an allowed maximum level $L_{\max}$ ($0<L(t)<L_{\max}$): two boundary conditions are imposed (a model BT2).
- A BT model that allows an exponential decay to zero level ($L(t)\to 0, L(t)\geq 0$) but still allows exponential growth of the level *L(t)* above the maximum level $L_{\max}$ ($L(t)>L_{\max}$): one boundary condition is obeyed and the other is not imposed yet (a model BT3).
- An improvement of the model BT3 that smooths the exponential growth up to the maximum level $L_{\max}$ ($L(t)\to L_{\max,}$ $L(t)\leq L_{\max}$): the second boundary condition is also obeyed (a model BT4).
- A modification of model BT4 through introduction of a non-linear (square) dependency of a level change over time $\Delta L(t)/\Delta t \sim L^2(t)$ resulting in a smooth growth: the second boundary condition is also obeyed (a model BT5).

The term 'imposed' is used for situations when solutions of equation (6.4) are cut off 'manually' for values that are outside the boundary conditions. The term 'obeyed' means that corresponding solutions do not cross boundary conditions naturally.

In parallel to analytical solutions for models BT1–BT5 their SD models are run and their solutions are visualized in this chapter. In all cases the models have an initial level $L(0)=0.5$ metres (an initial condition for *t*=0). One exception is the non-linear model BT5 as it has two different classes of solutions depending on initial conditions. To tell apart these two different classes two different initial levels are considered: 0.5 metres and 0.05 metres.

In a limiting case when inflow and outflow rates are equal, the initial water level stays unchanged over time for all models $(L(t)=L(0))$.

In those models where boundary conditions are set up, a minimum water level is presumed $L_{min}=0$ metre, and a maximum water level $L_{max}=1$ metre.

Besides being a useful tool to demonstrate SD principles, the BT model has a direct relevance to project management. Levels could be interpreted as a project work either to be done or already completed. Flows may be understood as workflow rates related to productivity, etc., although many more sophisticated feedback loops are required to describe project SD processes. Chapters 7 and 16 are devoted to SD applications in project management.

6.2 Results for Models BT1 and BT2

According to equation (6.3) the behaviour of the models BT1 and BT2 is fully defined by the difference between an inflow and an outflow (see sections B.2 and B.3 of Appendix B). Both flows of the model BT1 are constants being independent of levels in any given moment of time. If an inflow is higher than an outflow ($I > O$), a linear unrestricted growth of the level $L(t)$ should be observed. If L < O, a linear unconstrained decline of $L(t)$ is expected. In case $I = O$, the initial level L(0) will be kept over time. Figure 6.4 represents sample results for the model BT1 for the three discussed combinations of inflows and outflows.

The unconstrained level's behaviour in the model BT1—either growth or decline—is explained by the absence of any feedback loops (see Figure 6.1). In mathematical terms, the corresponding differential equation does not have any imposed boundary conditions. Imposing two natural boundary conditions $L_{min} < L(T) < L_{max}$ cuts off all solutions outside the established interval (Figure 6.5). This is a model BT2 (see Figure 6.2).

6.3 Results for Models BT3 and BT4

It would be worthwhile mentioning that the feedback loops of Figures 6.2 and 6.3 are trivial and called first-order loops as they involve only one level. A more complex second-order or third-order loop involves two or three levels correspondingly (see Chapter 16).

In contrast with the trivial behaviour of the models BT1 and BT2, models BT3 and BT4 demonstrate more sophisticated level–flow relations: flows become directly

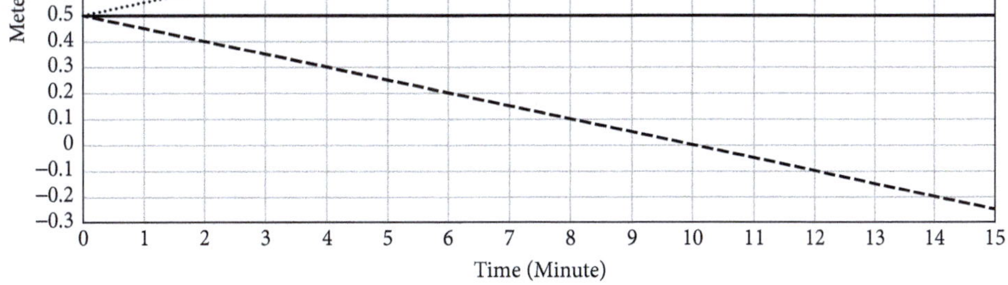

Figure 6.4 Sample results for the model BT1 (unconstrained linear growth or decay)

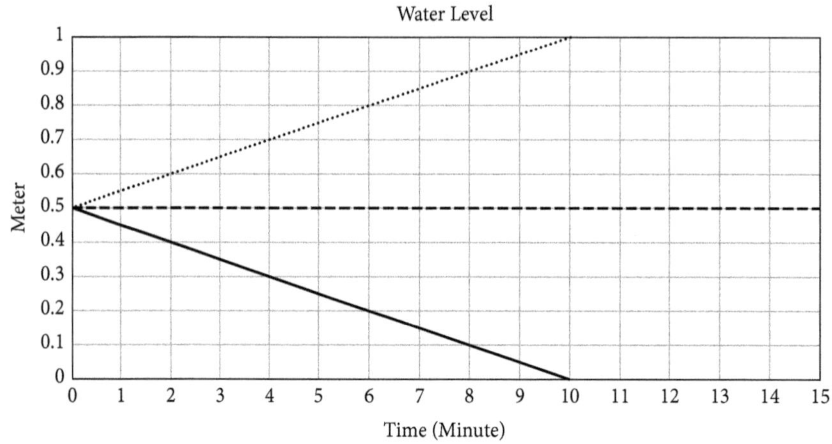

Figure 6.5 Sample results for the model BT2 (constrained linear growth or decay)

depended on a level in each given moment of time (see sections B.4 and B.5 of Appendix B).

Apparently, flows being proportional to $\Delta L(t)/\Delta t$ should be measured in metre-per-minute units which corresponds to flow units. As shown by Ferrari (1978), inflows *I(t)* and outflows *O(t)* could be represented simply through levels *L(t)* and corresponding characteristic times t_{in} and t_{out} as follows.

$$I(t) = L(t)/t_{in} \tag{6.5}$$

$$O(T) = L(t)/t_{out} \tag{6.6}$$

These characteristic times define delays required to reach certain levels. For instance, in the case of the exponential growth (model BT3) with zero outflow, $L(t) = L(0) \times exp(t/t_{in})$. At the moment $t = t_{in}$ the level of interest is $L(t_{in}) \sim 2.72 \times L(0)$. In case of exponential decay (model BT3) with zero inflow, $L(t) = L(0) \times \exp(-t/t_{out})$. At the moment $t = t_{out}$ the discussed level is $L(t_{out}) \sim 0.37 \times L(0)$. In simple terms, expressions (6.5) and (6.6) for flows are based on an assumption of exponential solutions (exponential growth or decay) of equation (6.4).

The characteristic times t_{in} and t_{out} are shown by Figure 6.3 for models BT3 and BT4 as variables 'inflow time' and 'outflow time'. Sample results for the models BT3 and BT4 are represented by Figures 6.6 and 6.7. As Figure 6.6 demonstrates, the model BT3 allows unconstrained growth which is replaced by smooth growth in the model BT4 shown by Figure 6.7 (see also sections B.4 and B.5 of Appendix B).

Strictly speaking, differential equations (B1.3), (B1.8), and (B1.10) discussed in Appendix B for above-discussed models BT1–BT4 are linear as they contain none (in case of BT1) or linear expressions (in case of BT2–BT4) for levels in the right part of equation (6.4). However, even in their linear forms, models BT3 and BT4 have clearly non-linear solutions (see Figures 6.6 and 6.7). This points to a need for a shift from a linear to a non-linear paradigm when studying SD including project SD. One

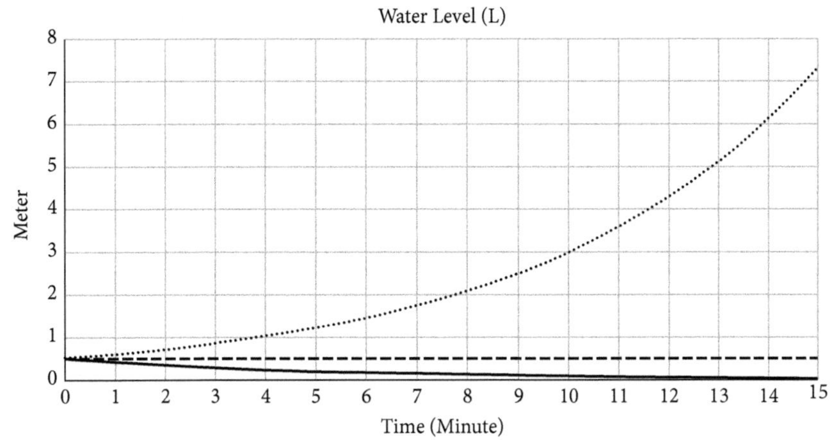

Figure 6.6 Sample results for the model BT3 (decay and unconstrained exponential growth)

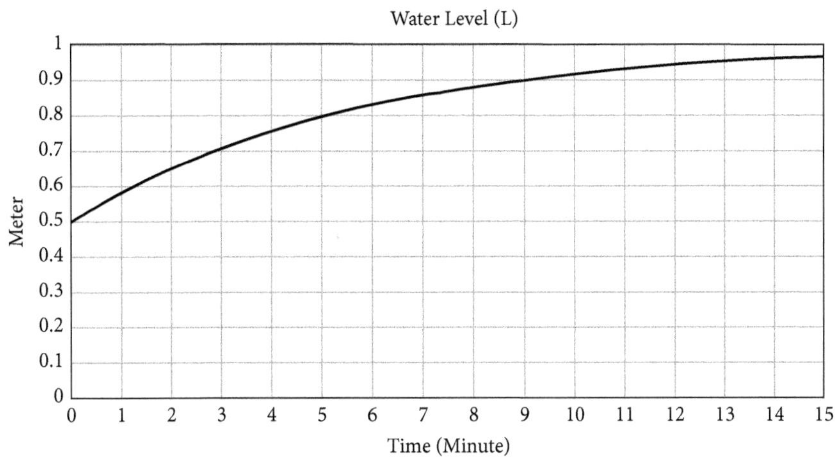

Figure 6.7 Sample results for the model BT4 (smooth growth)

additional truly non-linear model (a model BT5) is discussed in section 6.4 that could be considered a truly non-linear modification of the models BT4.

6.4 Results for Model BT5

Some BT model applications have quite sophisticated dynamics when inflow and outflow rates depend explicitly on levels in a non-linear manner. One example is a population growth model developed by Verhulst (1845). It is developed to challenge the Malthus theory of geometric-progression population growth. Apparently, the model by Malthus simply had no boundary conditions and allowed unconstrained growth similar to the model BT3. Its outcome truly scared some scholars and politicians of the nineteenth century.

According to equation (B1.10) of Appendix B, the modified inflow and outflow rates in the non-linear case of the model BT5 are represented as

$$I(t) = (L(t)/t_{in}) \times (L_{max} - L(t)) \tag{6.7}$$

$$O(T) = (L(t)/t_{out}) \times (L_{max} - L(t)) \tag{6.8}$$

When expressions for inflow and outflow rates are combined in equation (6.4) (see equation (B1.10)), its solution could be easily obtained by direct integration yielding so-called logistics distribution or sigmoid function (Strogatz (2015)) shown by Figure 6.8 for two characteristic situations.[4] As a matter of fact, the behaviour of the model BT5 strongly depends on the difference between $L(0)$ and L_{max}. If the difference is not that big ($L(0) \preceq L_{max}$) the solution approximates and resembles a solution for model BT4 (a smooth growth) shown by Figure 6.7. If the difference is substantial enough ($L(0) \ll L_{max}$), the solution sags and takes a form of a typical S-shaped sigmoid function.[5] A switch between these two classes of solutions occurs around $L(0) = 0.5$ metres. So, the solution with $L(0) = 0.5$ metres shown by Figure 6.8 belongs exactly to the border separating the two classes.

It would be reasonable to speculate a bit on the nature of the feedback loops in the model BT5. Formally speaking, they are still first-order loops as shown by Figure 6.3 representing one level $L(t)$. However, the same level $L(t)$ plays a role in the loops twice due to quadratic nature of the equation (6.4) in the case of the model BT5. This is a

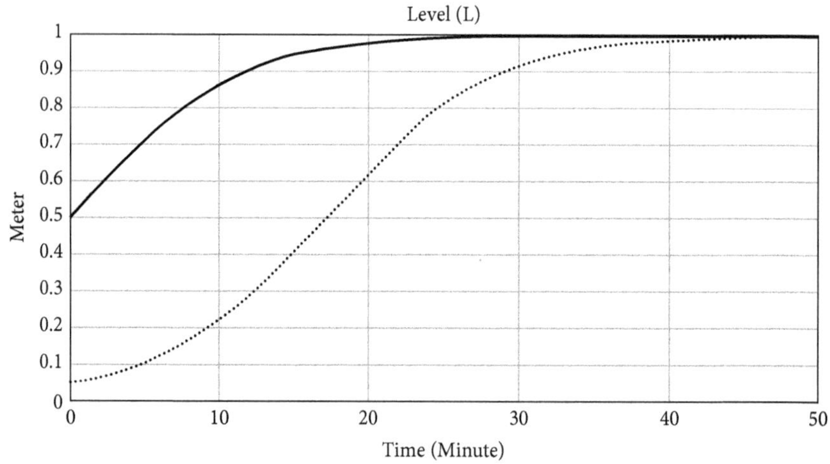

Figure 6.8 Sample results for the model BT5

[4] As an alternative to integrating, the validity of the derived solution (B.13) for the model BT5 can be checked by direct substituting it back into equation (6.4) when expressions (6.7) and (6.8) are used.

[5] By the way, the sigmoid function is discussed in Chapter 4 in a different context relating to ANN modelling (see expression (4.4)).

direct result of inflow's and outflow's dependence on the level (see expressions (6.7) and (6.8)). This effectively redefines (elevates) the feedback loop's orders.

It would be worthwhile to point out that if $L(0)=0$, $L(t)=0$ at any time t, which is a trivial solution. Although it does not contradict the purpose of Verhulst (1845) to describe the population growth—no initial population, no population growth. Although this does contradict the nature of the BT model.

Strictly speaking, model BT5 drifts away from the traditional bathtub model by showing its boundaries. Actually, it is included under the BT model umbrella in this book for demonstration purposes only simply because the model BT5 still allows analytical solutions.

6.5 Chapter's Markup and Takeaway

- A BT model bridges a cognitive gap between classical differential equation calculus that may produce analytical solutions and SD modelling that is based on numerical calculations of the equations. All five discussed versions of the BT model do have analytical solutions.
- In more sophisticated cases analytical solitons are impossible. This is when SD applications should step up.
- The BT model directly associates relevant differential equations with feedback loops used in SD modelling. Only first-order loops that include only single levels are parts of the BT model.
- When several levels are covered by feedback loops a system of several first-order differential equations should be solved. In the vast majority of cases it cannot be done analytically.
- The BT model shows that even in case of linear differential equations their solutions may be clearly non-linear reflecting on non-linear relations of parameters and variables included in feedback loops. Due to feedback loops, SD models cannot treat system's parts independently, and they cannot ignore interactions among system's parts either.
- The BT model allows the introduction of main ideas, concepts, and terminology of SD to link it with non-linear nature of interactions in project systems. In turn, this should allow reveal major gaps in current risk identification and quantification methods.
- The BT model could be valued as a foot in the door to project SD.

7

Project System Dynamics (SD): A Linear Case

Following the general SD introduction in Chapter 6, this chapter displays a role of SD modelling in project risk quantification. As the SD modelling undertaken in this chapter is devoted to mirroring the linear L-SCRA model developed in Chapter 5, the corresponding SD model can also be conditionally referred to as linear.

Project system dynamics was successfully used in modelling of project disputes and claims for years. At the same time, it is considered a highly esoteric activity by many project practitioners. The primary purpose of this and Chapter 6 is to push project system dynamics to risk quantification mainstream. The secondary purpose is to use it for demonstration and better understanding of non-linear nature of interactions in project systems including non-linear risk interactions.

However, it should be kept in mind that there are no one-to-one relations between sets of parameters used in Monte Carlo and SD models. By this reason it is difficult to expect that results of two models would be fully aligned automatically. However, it is relatively easy to align results of two types of modelling through imposing some reasonable assumptions on the model's parameters to reduce the number of loose modelling ends. This also opens possibilities for proper mutual calibration of two types of models using historical data. As this book puts forward a conceptual approach that unites Monte Carlo and SD modelling the calibration aspects are mostly ignored.

One key misalignment between Monte Carlo and SD models is that SD models discussed here do not take into account general uncertainties. For this reason abridged results of Monte Carlo models that factor in only risk events (see e.g. Figure 5.8) should be used for direct comparison with SD modelling results. Counting of general uncertainties in SD models is possible although it drastically increases the SD model's complexity.

Only schedule aspects of SD are considered in this chapter and Chapter 16. Technically it is not that difficult to incorporate cost aspect into SD modelling too, although corresponding models are too complicated to discuss here.

7.1 Project System Dynamics Modelling Approach

There is neither need nor desire to re-build a full-scale project model in a comprehensive form proposed by Cooper (1980), which might take many months if not years. That highly complex model has been used for forensic purposes to demonstrate main non-linear interactions in a project system. A high-level project SD model could be

Modern Risk Quantification in Complex Projects: Non-linear Monte Carlo and System Dynamics Methodologies. Yuri G. Raydugin, Oxford University Press (2020).
DOI: 10.1093/oso/9780198844334.001.0001

built relatively quickly instead that is simple enough and at the same time informative enough for risk quantification. This inevitably reduces the number of used parameters and links among them to a relatively low manageable level.

As discussed in Chapter 4, the project SD model introduced by Cooper (1980) is based on a concept of a rework cycle which in turn is defined by quality and productivity of project works. Dozens of parameters have been identified that may influence rework cycle features: from available stuff, overtime, hiring, worker's experience and morale, etc. to schedule pressure, out-of-sequence work, vendor design information availability, etc. (Cooper and Lee (2009)).

To build a lean version of a project SD model, the following assumptions are used to consider risks as perturbations to project works. Occurred risks can:

- impact quality of work,
- add work to initially planned work, and
- impact productivity (aka workflows).

According to the first point, two quality-related risks of project Zemblanity discussed in Chapter 5—R03 (errors and omissions (engineering)) and R04 (poor work quality (construction))—should be treated differently from the other four discussed risks (R01, R02, R05, and R06). Standalone risks R03 and R04 can be counted in a rework cycle through poor engineering and construction quality parameters. These are introduced as shares of drawings or construction units that require rework. In turn, it is assumed that the other four risks impact add additional work and may reduce productivity (work flows).

Only schedule-related part of SD modelling is discussed in this chapter. For example, sunk costs stemming from cancelling a part of unfinished work or scrapping a part of finished work that are taken into account in a cost module of a Monte Carlo model are left out for simplicity purposes.

All SD models discussed in this chapter and Chapter 16 are built based on two combined engineering and construction re-work cycles following the modelling approach by Williams et al. (1995). This is a certain simplification of the mammoth model developed by Cooper (1980).

Three models SD1–SD3 are built and discussed in this chapter starting with a trivial risk-free model SD1 (Figure 7.1). This simplistic model is developed for calibration purposes to directly replicate the schedule of project Zemblanity introduced in Chapter 5 for the Monte Carlo L-SCRA model (Figure 5.1): it simply reproduces the engineering duration (21.2 months) and the project duration (43.6 months) for a risk-free case.

A model SD2 (Figure 7.2) comprises two rework cycles although they are still fully independent. The construction rework cycle is triggered as soon as engineering work is completed, although a poor quality in engineering cannot impact quality in construction. This model SD2 includes explicitly two quality risks R03 and R04 of the L-CSRA model whereas the other four risks are left out: these two quality risks are fundamental for setting the two rework cycles. Figures 7.3–7.6 demonstrate results for the model SD2 as well as for the model SD3.

A model SD3 (Figure 7.7) does include the other four risks R01, R02, R05, and R06 too. This directly corresponds to the scope of the model L-SCRA developed in Chapter 5.

Two additional SD models SD4 and SD5 are built in Chapter 16. These take into account non-linear internal risk amplification and cross-risk (knock-on and compounding) risk interactions. The models SD4 and SD5 are built to mirror non-linear Monte Carlo N-SCRA model developed in Chapter 15.

All workable SD models SD1–SD5 discussed in this book could be downloaded as supplementary materials from the book's companion website at www.oup.co.uk/companion/Raydugin. It should be kept in mind that graphical representations of SD models provided in the main text of this book (see Figures 7.1, 7.2, 7.7, and 16.1) are highly simplified just to demonstrate main stocks, rates, variables, and feedback loops. Quite a few auxiliary components of these models are not even shown.

7.2 Project System Dynamics Model SD1: A Risk-free Case

Figure 7.1 represents a trivial project SD model referred to as the model SD1 that consists of two semi-independent modules: engineering and construction. Procurement is considered part of engineering whereas commissioning and startup is part of construction. Capital letters 'E' and 'C' are used to point to engineering and construction works, correspondingly. These two modules are connected only through a usual schedule-logic-type link indicated by a dotted arrow: construction starts as soon as engineering is completed. The role of the first-order feedback loops is to impose boundary conditions as the workflows should be stopped as soon as all planned engineering drawings and construction units are completed.

The proposed SD1 model can be considered a direct equivalent to a deterministic risk-free project schedule. All normal engineering and construction tasks of the L-SCRA model (see Figure 5.1) are fully aligned with the model SD1 by default. Namely, engineering normal tasks of Figure 5.1—SD00 (scope definition (FEL2)), PD00 (project definition (FEL3)), DE00 (detail engineering (incl. procurement))—are included to the engineering module of the model SD1. Construction normal tasks—CN00 (construction) and CS00 (commissioning and startup)—are covered by the construction module.

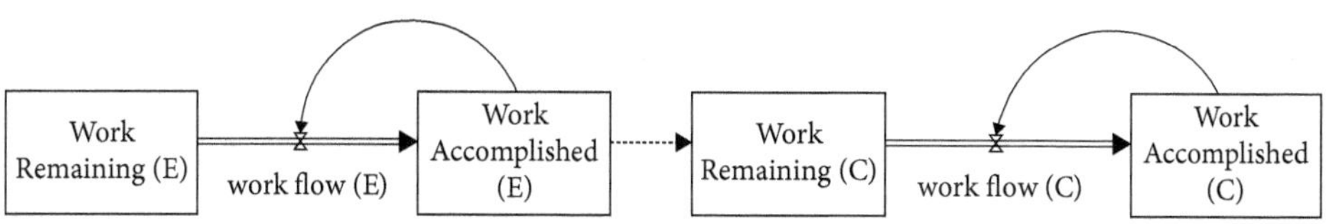

Figure 7.1 A project system dynamics model SD1 for a risk-free case

No surprise that the model SD1 yields the same deterministic duration as the L-SCRA model: 21.2 months of engineering duration and 43.6 months of overall project duration. This trivial case is required to ensure a full alignment of the L-SCRA and SD1 models in the risk-free case.

Initial engineering and construction quantities are established in all models SD1–SD5 in the following way. The number of engineering drawings is 2000, and number of construction units is 200. This allows the establishment of the base undisrupted engineering workflow rate as 94.5 drawings per month. The number of completed construction units per month is established as 8.9 in all models.

7.3 Project SD Model SD2: Two Standalone Rework Cycles

The trivial model SD1 discussed in section 7.2 does not contain any risks whatsoever. As a next step two quality risks—R03 and R04—are added.

This generic two-rework-cycle scenario is referred to as the model SD2. It is based on the L-SCRA model of Chapter 5 but with two quality-related risks R03 and R04 only. Risk R03 that impacts three normal tasks in the L-SCRA model—SD00, PD00, and DE00—is introduced through 'poor E-work quality' variable (see Figure 7.2). Similarly, risk R04 mapped to a normal task CN00 is associated with 'poor C-work quality'. These two quality risks—R03 and R04—result in forming and activating of the two rework cycles.

It usually takes some time to discover engineering drawings with errors and omissions and construction units with deficiencies. Following Cooper (1980), there should be additional stocks 'Undiscovered Rework (E)' and 'Undiscovered Rework (C)', correspondingly. These stocks are being depleted in the course of deficiencies' discovery, which depends on 'rework discovery rate (E)' and 'rework discovery rate (C)', correspondingly. When discovered, drawings with errors and omissions and construction units with deficiencies go back to square one: they enter 'Work Remaining (E)' and 'Work Remaining (C)' stocks, respectively. When 'Work Accomplished (E)' stock

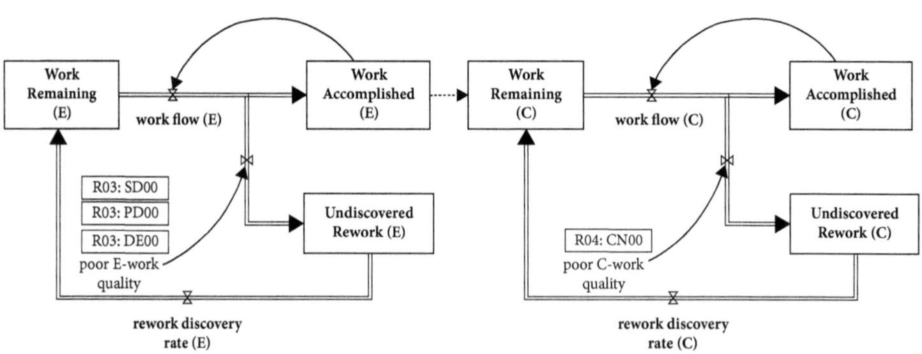

Figure 7.2 A project system dynamics model SD2 for two separated rework cycles

becomes equal to planned engineering work (i.e. 'Work Remaining (E)' is equal to zero), construction starts. Eventually, when 'Work Accomplished (C)' stock becomes equal to planned construction work (i.e. 'Work Remaining (C)' is equal to zero), the project is considered completed. In both cases corresponding rates become equal to zero thanks to the feedback loops established in the model (see Figure 7.2).

Any occurred risks should lead to extra work—additional drawings and construction units. This should correspond to delays that could be recalculated in terms of these extra drawing or construction units re-entering these rework cycles. This has two important aspects.

First, probabilities of risk occurrences P should not be forgotten. They should be factored in the recalculation as risks might happen or might not. Because of this, expected values $P \times I$ of risks should be used in the recalculation, not just impacts I. This is a fundamental difference with SD models that came from dispute resolutions: incurred changes that inflicted harm to project durations or costs are considered retroactively as actuals, which have probabilities $P = 1$ by default.

Second, multiple revisions of drawings that are routinely made in the course of engineering development should not be counted in the poor-quality category. Only overlooked actual errors and omissions that are discovered with a delay should be counted as unplanned rework.

A key assumption used in converting risk impacts of the model L-SCRA to risk impacts of the models SD2 and SD3 is that all drawing associated with risks require 100 per cent rework. Similar assumption is done for construction risks. This is a very rough simplifying assumption as in reality a share of drawings and units that require really full rework is much lower. Some of them could be slightly adjusted, a few discarded, etc. This means that several drawings or units that require partial rework can be counted as one fully reworked drawing or unit. This is a full rework equivalent or a new work equivalent which is presumed in this chapter and in Chapter 16.

As shown below, the SD models discussed in this chapter are not that sensitive as all considered risks are standalone and do not take part in interactions. This is the same situation as in the case of the model L-SCRA. Therefore no big difference is observed if 100 per cent rework or, say, 10 per cent rework is assumed for engineering drawings and construction units in the models SD2 and SD3. However, very intensive risk interactions are intrinsic to the models SD4 and SD5 (see Chapter 16) that mirror the non-linear Monte Carlo models SD4 and SD5 (see Chapter 15). Because of this, a more realistic approach is used in Chapter 16 where assumptions on a share of drawings or units requiring full rework are done. Simply speaking, if the 100 per cent rework assumptions are kept in the highly non-liner models SD4 and SD5, they will produce unrealistic results: the project would never be completed as more rework generates even more and more rework.

Appendices C and E lay out steps to convert risks of Monte Carlo models to inputs of SD models. Appendix C is devoted to recalculation between the model L-CSRA and the models SD2 and SD3. Sections C.1–C.4 describe these steps in general whereas section C.5 provides recalculation results for the models SD2 and SD3. For simplicity purposes all recalculations are done for after-risk-addressing situations of schedule risks only.

The detailed recalculation process outlined in Appendices C and D might be found to be a bit onerous. The key reason for this is that required software applications that can do this in a user-friendly way have yet to be developed. Hence, the recalculations have to be done manually for existing software tools.

As discussed in section C.5, numbers derived for risks R03 and R04 can be used in the models SD2 and SD3 as 'poor E-work quality' and 'poor C-work quality' variables. Risks R03 and R04 are shown by Figure 7.2 next to these variables symbolically: they

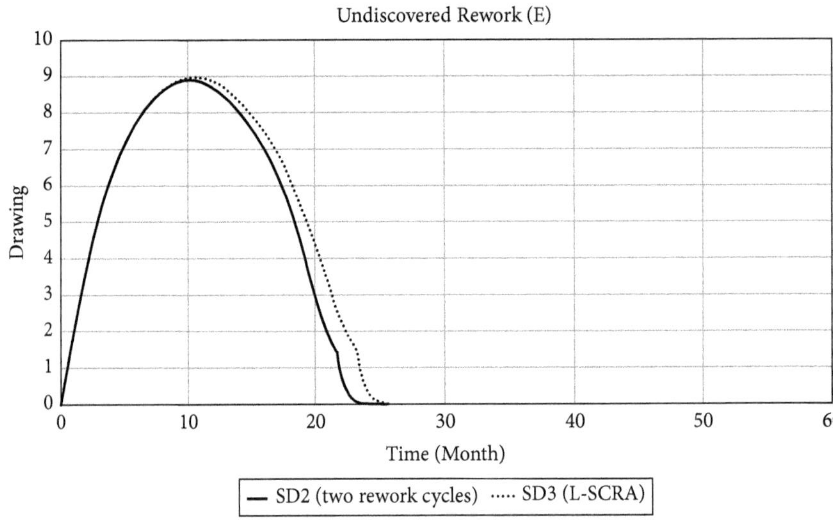

Figure 7.3 Undiscovered engineering rework (E) for the models SD2 and SD3

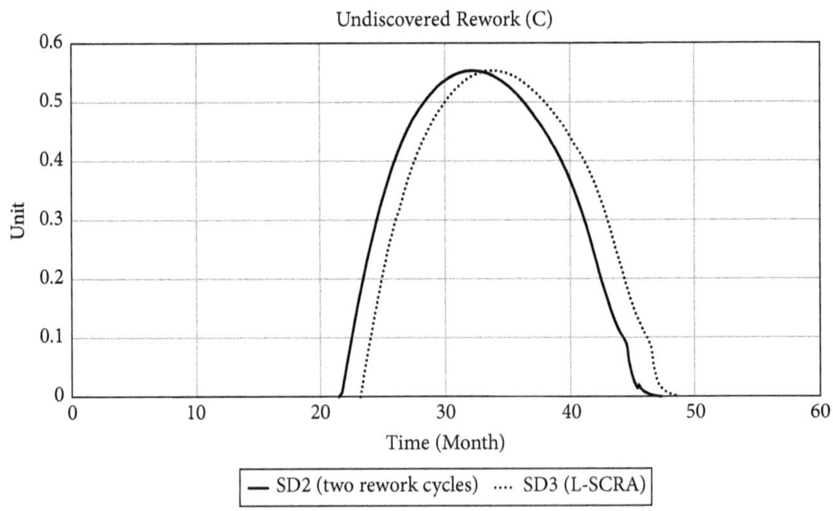

Figure 7.4 Undiscovered construction rework (C) for the models SD2 and SD3

are actually taking part in the model SD2 through the poor-quality variables only. Table C.3 yields required poor quality parameters for the models SD2 and SD3. The 'poor E-work quality' parameter is equal to 2.3 per cent, the 'poor C-work quality' is equal to 1.5 per cent.

Results for model SD2 are shown in Figures 7.3–7.6. These figures also contain results for the model SD3 discussed in section 7.4 below. Figures 7.3 and 7.4 demonstrate clearly non-linear behaviour of undiscovered engineering and construction rework stocks. Figures 7.5 and 7.6 show engineering and construction workflows. They are still linear in the models SD2 and SD3 as occurred risks do not impact them.

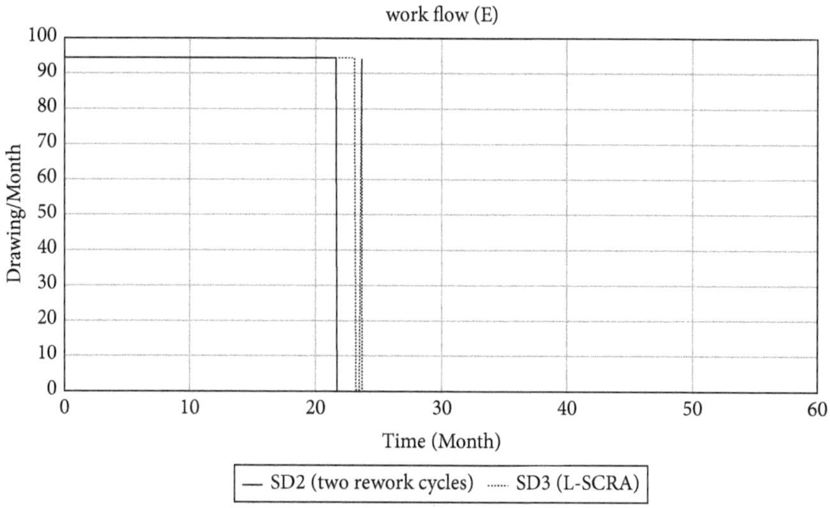

Figure 7.5 Engineering work flows for the models SD2 and SD3

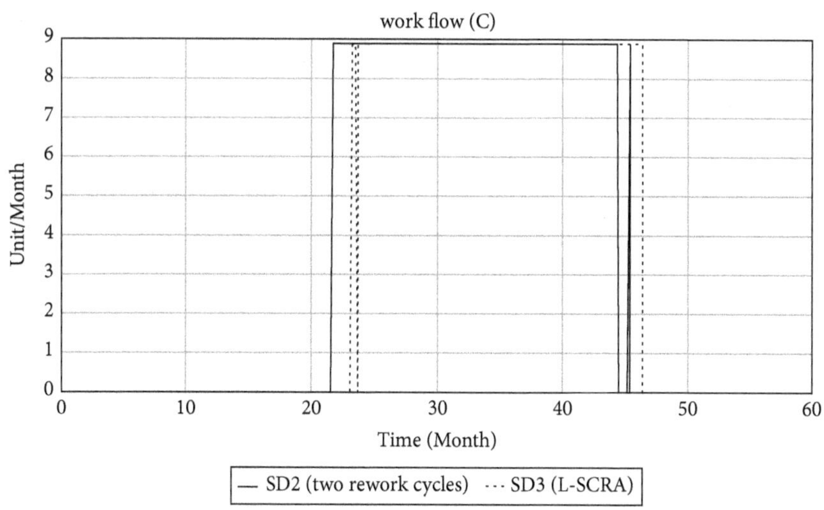

Figure 7.6 Construction work flows for the models SD2 and SD3

The case when occurred risks do impact both workflows in a non-linear way is demonstrated in Chapter 16.

Figures 7.3–7.6 advise that engineering work requires 21.6 months and the entire project Zemblanity should need 45.5 months in case of the model SD2. These durations are defined by moments when workflows stop. This corresponds to longer engineering work duration for about 2.0 per cent and overall project duration for 4.4 per cent in comparison with the risk-free model SD1. Apparently, the project duration increase associated with the model SD2 is minute.

7.4 Project SD Model SD3: Mirroring the L-SCRA Model

The linear Monte Carlo L-SCRA model discussed in Chapter 5 is converted to the corresponding SD model in this section as a comparison. Actually, the poor-quality risks of the L-SCRA model have already been taken into account by the model SD2 in section 7.3. Now it is time to add the remaining L-SCRA risks R01, R02, R05, and R06. The role of the additional risks is to add extra work through redefining quantities of drawing and construction units. Risk impacts on productivity of works are still ignored in this linear case.

Figure 7.7 introduces a SD model with all six project risks included in the model L-SCRA. This model is referred to as the model SD3. It also presumes that possible poor quality of engineering and construction works (risks R03 and R04) should become fully resolved inside engineering and construction, correspondingly, the same way as in the model SD2. The same as in case of the model SD2, risks R03 and R04 trigger the two rework cycles for 2.3 per cent of drawings and 1.5 per cent of construction units, correspondingly.

The two rework cycles are still completely separated. The other risks—R01, R02, R05, and R06—are attached to engineering and construction 'Work Remaining (E)'

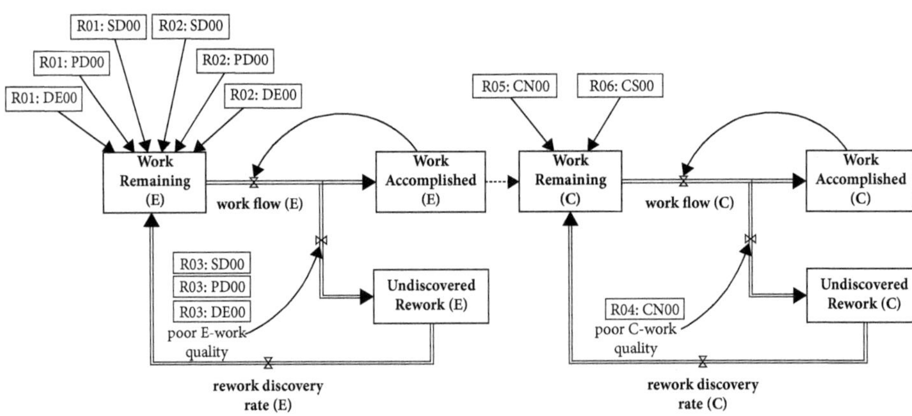

Figure 7.7 A project two-rework-cycle model SD3 that mirrors the L-SCRA model

and 'Work Remaining (C)' stocks. They are considered extras to initial project quantities for engineering (2000 drawings) and construction (200 units) (see Table C.3). For example, risk R01 should add 34.3 drawing to each of the engineering activities SD00, PD00, and DE00, or 102.3 drawing in total, although these three additions are done in different moments in accordance with Table C.2. The exact timing of risk occurrences (see Table C.2) is selected in accordance with their mapping in the model L-SCRA (Table A.4 of Appendix A). Risk R02 would add 14.8 drawings to each of these three impacted engineering activities. Similarly, construction risks R05 and R06 should add 2.5 construction units together to the initially planned 200 construction units (see Table C.3).

Although risks R01, R02, R05, and R06 bring forward additional work they do not influence productivity represented by 'work flow (E)' and 'work flow (C)' rates, as no non-linear risk interactions are considered in the model SD3 yet. The model SD3 is still linear in term of absence of non-linear risk interactions.

The model SD3 produces the following results.

Undiscovered engineering and construction rework graphs for the model SD3 are almost identical to the ones of the model SD2 (see Figures 7.3 and 7.4). Small additions of work quantities induced by risks R01, R02, R05, and R06 are easily processed by established undisturbed workflows as productivity remains intact when these risks happen. According to workflow charts (Figures 7.5 and 7.6) it takes a bit more time to process the engineering and construction quantities in case of the model SD3. As a result, engineering should be completed in 23.8 months. The entire project should be done in 46.4 months. This means that engineering should take 12.3 per cent more time in comparison to the model SD1, whereas the entire project should take 6.4 per cent more time than in case of the model SD1.

The results for the model SD3 can be directly compared with the results for the L-SCRA Monte Carlo model. However, to make a meaningful comparison, the following points should be kept in mind. The model SD3 being the same as any other considered SD models

- can count only schedule uncertain events,
- cannot count general uncertainties,
- produces one-point value results that correspond to mean values in Monte Carlo models, and
- ignores JCL concept implications.

By these reasons, Figures 5.2 and 5.3 cannot be used for the comparison as they contain general uncertainties. Figure 5.8 that represents schedule risks only should be used instead. However, in place of P70 confidence level a mean-value confidence level P_{mean} should be used. Actually, as discussed in Chapter 5, the distribution of Figure 5.8 has a P_{mean} value of 1388 days that should be used for the direct comparison with the

SD3 model's results.[1] Hence, the contingency should be 6.0 per cent at the mean-value confidence level (1388 d vs.1309 d).

The model SD3 produces contingency 6.4 per cent which is in a good alignment with the P_{mean} schedule contingency 6.0 per cent produced by the L-SCRA model.

Results for models SD2 and SD3 differ very insignificantly (see Figures 7.3–7.6): minute perturbations to the project system are easily digested by undisrupted engineering and construction workflow rates that remain intact when risks occur. To sum up, reviewing the models SD2 and SD3 from risk interaction angle it could be noticed that:

- all risks act as standalone entities without any impacts on each other,
- when risks happen overall productivity (workflow) and quality stay unchanged, and
- two modules of the model—engineering and construction—are independent in the SD sense: a possibility for drawings with errors and omissions to enter construction is completely ruled out.

From the project risk management angle these three observations actually mean that in case of the models SD2 and SD3:

- the engineering quality control is absolutely ideal and does not allow any drawings with errors and omissions to leave engineering and enter construction,
- there is no knock-on and compounding interactions among risks or a project team is capable of decoupling any possible risk interactions, and
- a project system should be highly mature and adaptive that allows the isolation and absorption of all perturbations and the complete prevention their impacts on productivity.

These points represent quite rare practical cases for simple projects with highly mature project systems. Linear SD models are very feeble in terms of non-linearity as all non-linear risk interactions are ignored. This indicates that SD models are not terribly sensitive or useful when handling minute non-interacting perturbations that are part of the models SD2 and SD3.

Anticipating the discussion in Chapters 15 and 16, Chapter 16 introduces more realistic project SD models (the models SD4 and SD5) that take into account real-life risk interactions and hence are applicable to complex project systems (see Chapter 8). These SD models are well aligned with the non-linear Monte Carlo models N-SCRA-A, N-SCRA-B, and SCRA-C introduced in Chapter 15 by a simple reason: both SD and Monte Carlo models take into account internal risk amplifications stemming from

[1] This can be easily checked if running the real model available at the book companion website at www.oup.co.uk/companion/Raydugin.

project system issues as well as cross-risk interactions—knock-on risk interactions and compounding.

7.5 Chapter's Markup and Takeaway

- To avoid excessive complexity, SD models should be simple enough although still informative enough.
- Two rework cycles—engineering and construction—are introduced in the considered SD models to represent a project workflow. Only schedule aspects of SD modelling are included.
- The basic risk-free case (the model SD1) is needed to align the SD modelling with the project schedule.
- As a next step, the model SD2 describes two consecutive rework cycles—engineering and construction—that result from the poor work quality risks. Other project risks are left out.
- As an extension of the model SD2, the model SD3 takes into account the rest of the project risks to mirror the risk exposure represented by the Monte Carlo model L-SCRA.
- As SD models do not handle general uncertainties and spreads of outcomes, a one-point mean value of the model L-SCRA derived in absence of general uncertainties should be used for comparison with the SD3 model's results. In this case both models demonstrate similar results.
- The direct comparison of the project duration results for the models SD1, SD2, and SD3 reveals insignificant differences. This can be explained by ignoring of several factors of the project risk exposure. This includes the facts that risks are treated as non-interacting and that two rework cycles are completely separated. Simply speaking, a full-scale SD cannot kick off when risk interactions are non-existent or feeble.

PART TWO

PRINCIPLES OF RISK QUANTIFICATION IN COMPLEX PROJECTS

8

Introduction to Project Complexity

This chapter overviews possible approaches towards project complexity description. A literature review on the topic is undertaken in the chapter. Some most valuable ideas are extracted from the bulk of the literature and captured to be further developed in consecutive chapters. Keeping in mind the goals of this book, the discussion on project complexity zeroes in on project system issues and interactions of project risks.

Modern literature on project complexity could be divided into two distinctive bodies: complexity of projects and complexity in projects (Gull and Williams (2017)). The first comprises texts that describe complexity characteristics of various types of projects and proposes practical frameworks to manage complexity. The second is rather esoteric as it utilizes various ideas from science, mostly from physics, to better understand implications of project complexity.[1] It should be recognized that the relevance of some of the discussed complexity-in-projects ideas seems to be slightly shallow. In some instances, the relevance is apparent although their mathematically accurate representation in project management is not possible or easy.

The cornerstone works on the topic by Gull and Williams (2017) and Chapman (2017) are extensively cited in this chapter.

8.1 Complexity of Projects

This section provides a literature review on the topic followed by a list of most valuable findings that could be directly used in risk quantification in complex project systems.

8.1.1 Complexity of Projects: A Literature Review

Complexity of projects could be considered from various angles depending on a context of a particular challenge or study. However, there are comprehensive publications that are in a way context-free and that cover all or nearly all possible aspects of complexity in general and project complexity in particular (see, for example, Hass (2011, 2017)). Being extremely valuable descriptive pieces of research work that could serve as reference materials on various aspects of complexity regardless of a particular

[1] Due to features of author's background that includes both project risk management and physics both literature domains are overviewed in this chapter.

Modern Risk Quantification in Complex Projects: Non-linear Monte Carlo and System Dynamics Methodologies. Yuri G. Raydugin, Oxford University Press (2020).
DOI: 10.1093/oso/9780198844334.001.0001

context, they could be found too general for applications in a particular project context. This section narrows the general complexity discussion down to aspects that are directly relevant to risk quantification in complex projects.

In general terms, complexity refers to a cognitive difficulty to understand a particular phenomenon in a particular context or environment. In more specific terms, complexity usually points to intricate interactions between many parts. At the same time, there is no consensus on a precise meaning or definition of complexity: one may find more than 30 definitions of complexity for various types of systems in the literature (Gull and Williams (2017)). Complex systems are referred to as disorderly, instable, emergent, non-linear, recursive, uncertain, irregular, random, etc.

Long story short, to impose a focus and structure on the discussion, a straightforward project system classification should be adopted as a project system continuum that covers four categories (Gull and Williams (2017)):

- Simple systems: systems of linear functional projects.
- Complicated systems: fully knowable and predictable project systems with a high number of structural elements which can be reduced to a number of simple subsystems.
- Complex systems: not fully knowable but somewhat predictable project systems with a high number of structural elements which cannot be reduced to a number of simple systems.
- Chaotic systems: completely unknowable and unpredictable non-linear project systems.

A straightforward example of simple vs. complicated systems could relate to flight scheduling from/to an airport. For example, if it is about a John O'Groats airport in Wick, Scotland, scheduling of a few flights from/to other similar local airports does not look formidable. Scheduling is clearly a simple system. However, if Heathrow airport is considered,[2] scheduling has all the attributes of a complicated although not of a complex system: scheduling of each separate flight is still a relatively simple task.

The concept of reducibility that tells apart complicated and complex project systems is of paramount importance. Attempts to reduce truly complex systems to a number of simple subsystems destroys them and impedes their adequate understanding. Apparently, complex systems in some circumstances may become chaotic. The state of a system being between complex and chaotic is referred by Stacey et al. (2000) as the edge of chaos. However, structural elements of a system are not the causes for approaching the edge of chaos: corresponding causes are attributable to the dynamic interactions among the system's parts.

The consideration of systems as simple, complicated, complex, and chaotic is based on structural complexity that is characterized by the differentiation referring to the

[2] Heathrow airport has about 1,300 take-offs and landings a day which may occur as often as every 45 seconds during rush hours (pre-COVID19 and post-COVID19 statistics).

number of system's parts, and interdependencies pointing to interactions among the parts. The structural complexity is referred sometimes to as the sequential complexity, and the interdependence complexity—as the feedback complexity (Williams (2002)). In turn, the interdependence or feedback complexity may have three instances depending on a nature of interactions as follows.

- Pooled: a simplest interdependency when each differentiated structural element contributes a discrete input to a project and is not sequence bound.
- Sequential: an interdependency when element's contribution (output) becomes an input to other structural elements.
- Reciprocal: a most complex interdependency consisting of feedback loops and mutual impacts.

Taking a look at project risk exposure from an empirical angle, Merrow (2011) argues that bad things tend to happen in groups, not individually, because projects are so tightly woven. Even if one bad thing happens individually first, it tends to trigger a cascade of problematic effects due to the interactions in projects. It is pointed out by Gull and Williams (2017) that many small projects may display some characteristics of complexity that pertain to complex projects. This could be explained by a presence of a simplest type of interactions even in simple projects—pooled interactions—that stem from internal inefficiencies in a project system when the other two types of interactions—sequential and reciprocal—are negligible.

According to Chapman (2017), traditional project planning approach based on linear scheduling is more appropriate to simple projects only as it does not take into account feedback loops or mutual impacts related to pooled, sequential, and reciprocal interactions. Thus, the traditional project planning should be rethought and amended based on principles of system thinking.

One important angle on project complexity which should be remembered relates to uncertainty. Namely, some authors do argue that complexity stems from uncertainty as ambiguity or lack of clarity and unpredictability. Chapman (2017) points to importance of project risk management as a process to manage various aspects of project uncertainty. He also accentuates a role of system dynamics as a major methodology to model complex interactions in project systems including interactions of risks. However, as Chapman (2017) argues, a traditional risk identification process based on risk causes, risk events, and consequences is not adequate enough as it does not take into account relations among risks. He proposes a radical change to risk identification through visualizing feedback loops among risks using system dynamics software packages in place of traditional risk registers. This is a must to discover a combined effect of aggregated risks.

Similarly, risk treatment should address the feedback loops responsible for combined impacts of risks that increase overall project risk exposure. This approach is based on system dynamics' ability to describe project activities much better than conventional linear scheduling. System dynamics can also provide an enhanced view of risks including relationships between risks and implications of risk addressing actions implemented to address interactions in a conglomerate of risks.

Shenhar et al. (2002) divide uncertainty to internal and external, where internal uncertainty affects the process of product design (method uncertainty), while external uncertainty is limited to clarity of customer requirements (goal uncertainty).

In addition, all systems could be considered either closed or open (Chapman (2017)). Closed systems do not interact with their external environment and, hence, are not adaptive to change. Open systems do interact with the external environment which makes them adaptive to change. As a rule, projects are open systems: a project should be treated as a nested system located in a larger external environment system. As a result, a project is impacted from outside by its own organization, industry it operates in, a national economy, a government regulatory, and legal system, etc.

To keep things simple and practical, a mental note should be made on statements from the literature that link complex systems with their adaptability. If these buzzwords 'complex', 'systems', 'open', and 'adaptive' are considered in practical terms, what do they mean to a seasoned construction site manager or a foreman? Chapter 17 explains the logic behind and relevance of all these notions. Anticipating this discussion, in a project environment all these things mean a certain degree of flexibility and de-centralization of project planning and controls based on a short-term planning.

Another important aspect of project complexity discussed by Cicmil et al. (2009) relates to interactions of people in project environment that bring forward people-related uncertainty. In the context of this book, two aspects of people-related uncertainty are important:

- various realizations of psychological and organizational bias in risk management process, and
- skills of specialists involved in project delivery including leadership skills.

8.1.2 Complexity of Projects: Practically Valuable ideas

Referring to the previous section, the following ideas could be utilized for risk quantification in complex projects.

First, based on three types of interactions in a project system, it would be required to clearly define interactions of risks that apparently use the same channels of interactions as project parts. Risk can be treated as perturbations to the ideal part's interactions. This immediately requires the definition of an ideal part's interactions in a project system that could be established as an imaginary risk-free base case or as an interaction's baseline. A work of a well-maintained car engine or a high-quality mechanical watch or clock that have multiple moving and interacting parts could serve as examples of interaction's baselines.

There should be no deviations from the planned interaction's baseline in a well-oiled 'project engine' that works perfectly like a clock, although it is a trivial ideal case. The case when the project is a little jumpy and not well-oiled represents a much higher interest.

This case presumes that there should be inevitable perturbations in the project system in a form of deviations from the planned or ideal interaction's baseline. Some of these deviations could be treated as standalone risks, and the others as resulting from risk interactions. This should allow rewriting three types of the part's interactions and undertaking a transition to three risk interactions types as follows.

- From pooled part's interactions to internal risk amplifications: depending on internal inefficiencies in a project system (project system issues) a standalone risk might vary in terms of its severity from project to project.
- From sequential part's interactions to knock-on risk interactions: earlier occurred risks might impact down-the-line risks making them more severe.
- From reciprocal part's interactions to compounding of risks: if two or more risks happen simultaneously their combined impacts would be higher than if they occur independently.

It is acceptable to keep the same terminology for both part's interactions and risk's interactions as in both instances they have the same nature: part's interactions in an unideal project system do contain risks as deviations from part interaction's baselines, although it also stands to reason to tell them apart and to zero in solely on risk dynamics. The risk interactions are further discussed in Chapter 9.

On the one hand, the internal risk amplifications are applicable to standalone risks associated with particular isolated project activities. Presence of other risks is not required for the internal risk amplifications which can be treated as intra-risk or auto-risk interactions. On the other hand, as the internal risk amplification is still one of the three risk interaction types, it could be understood as an intra-risk interaction stemming from internal project system inefficiencies. These internal inefficiencies are defined as project system issues in this book. By these reasons, in some particular discussions the intra-risk interactions (internal amplifications of risks that require participation of single standalone risks) and two types of cross-risk (or inter-risk) interactions (knock-on and compounding ones that require more than one risk) are told apart. In some other more general cases, the three types of risk interactions are referred to as risk interactions all together.

Second, the described picture of risk interactions requires rethinking of risk identification and tools used for this. By this reason, a traditional bowtie diagram for a standalone risk without any interactions (Figure 2.2) should be rethought. Actually, three new bowtie diagrams are developed in Chapter 9 to comply with the nature of newly introduced three risk interactions in a project system.

Third, keeping in mind the three types of risk interactions and the idea of reducibility of project systems to elementary parts (see section 8.1.1), the four types of project systems—simple, complicated, complex, and chaotic—may be further specified as follows.

A simple system is a sum of a few elementary parts where some internal amplification of risks is possible. A complicated system contains many elementary parts where some risk interactions—mostly internal risk amplifications and a few cross-risk interactions—are possible. A complex system cannot be easily reduced to few elementary parts. Risk interactions belonging to various project parts are quite intensive and include both intra-risk and inter-risk interactions. For this reason both parts and associated risks largely lose their individual characteristics. No standalone risks should be considered. Instead risks affinities should be considered as dynamic patterns. In case of a chaotic system, individual part's and risk's characteristics are melted down due to highly intensive interactions of all three types. A chaotic project system itself melts away to a completely different state.

The last point about melting a project system away to form a completely new state belong to complexity-in-project discussion. This is about two-peak schedule and cost distributions and first-order phase transitions (see section 8.3 and Chapter 17). In other words, this is about how a complex project system might become chaotic.

Fourth, if a project system consists of interacting parts it could be a good idea to group them properly. A natural intent could be to tell apart all project parts that define a project as an entity to deliver (scope, locations, technologies, facilities, etc.) and project parts that define how the project should be delivered (project organization, applicable procedures, leadership, etc.). This idea is shaped as follows.

Two interacting project subsystems should be introduced—PSS and PDS.

Interestingly enough, words 'a project', 'a subject', and 'an object' stem from the same Latin root *jectum* that could be translated to English as 'a throw'. In combination with prefixes *pro-* (forth), *sub-* (under), and *ob-* (against) this root generates three important project management words. In the light of this discussion these could be interpreted as:

- 'a goal planning prior to an action (a pro-ject)',
- 'a recipient of a goal-oriented activity (a sub-ject)', and
- 'an activity against a goal (an ob-ject)'.

Despite originally a word 'pro-ject' meant planning only, it should be understood not only in a context of the FEL. Then it is fully complying with traditional definitions of a project cited in this chapter. However, the other two words—'a sub-ject' and 'an ob-ject'—seem to be helpful to better understand 'a pro-ject' as a system. In the context of this discussion, a PSS may be treated as a 'sub-ject' describing what should be delivered referring to the structural complexity. Whereas a PDS as an 'ob-ject' describes how it should be delivered referring to the delivery complexity. As discussed in Chapter 9, imbalances between these two subsystems of any origin are very important and they are referred to as PSS–PDS mismatches. In turn, types of PSS–PDS mismatches are closely associated with the types of risk interactions.

The ideas captured in this section allow us to come up with a new definition of complexity in projects (section 8.1.3) and put forward a new project complexity framework introduced in section 8.2.

8.1.3 Complexity of Projects: A New Definition

A new definition of complexity in projects is required to move on and develop a workable project complexity framework useful for project risk quantification. According to sections 8.1.1 and 8.1.2 the following aspects should be covered by the definition.

- Two aspect of complexity: structural (project parts) and interdependencies (part's interactions).
- Two high-level project subsystems: project structure and project delivery

As a result a following new complexity-of-projects definition can be made.

Complexity of projects is an evolving project-phase specific characteristic that reflects on project subsystems and their inherent reducibility to elementary interacting parts.

8.2 A New Project Complexity Framework

Various frameworks are discussed by Gull and Williams (2017) and Chapman (2017) to shape and visualize the project complexity concept. Based on their considerations but keeping in mind the context and goals of this book—project risk quantification—a new project complexity framework is introduced by Figure 8.1.

Two top-level complexity areas represented by Figure 8.1 are:

- the project structural complexity attributed to the PSS, and
- project delivery complexity attributed to the PDS.

8.2.1 A Fundamental Project Complexity Concept: A PSS–PDS Mismatch

Any instance of a mismatch due to any reason between these two subsystems (a PSS–PDS mismatch) is represented by the double-sided arrow on Figure 8.1.

The concept of the PSS–PDS mismatch is fundamental to understanding complexity of projects. Any misalignment or imbalance between the two subsystems can be understood as a source of dynamic behaviour of an entire project system. This section overviews the main facets of the PSS–PDS mismatches as well as their origin and implications.

Two fundamental sorts of the mismatches can be distinguished—static and dynamic.

A PSS–PDS mismatch stemming from long-term chronic project system issues can be labelled as static or chronic. The chronic system issues define a background of the entire project system. Apparently, the same project risk, if occurred, might have much higher severity in presence of chronic system issues. In accordance with section 8.1, this phenomenon should be called the internal risk amplification or intera-risk interaction due to the presence of the chronic system issues.

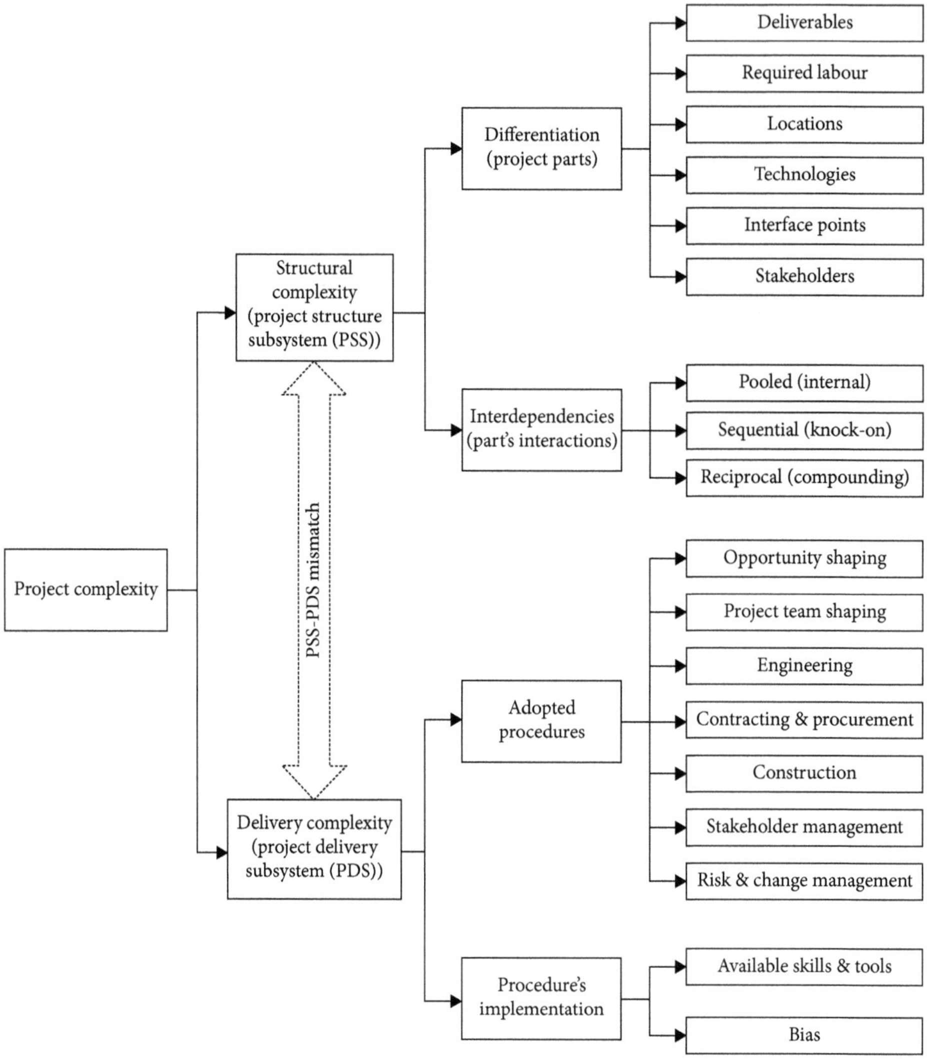

Figure 8.1 Project complexity framework

Simply speaking, a presence of a given static PSS–PDS mismatch should predetermine a severity (aka an expected value or consequence[3]) of a possible future risk event. In other words, the same risk if it occurs in two different projects that have different chronic system issues should have very different consequences $P \times I$. For instance, this defines a difference between two pipelines projects when one of them has a strong team of pipeline engineers and designers and the other does not. Both probabilities and impacts of some related project risks would be quite different.

This is easy to demonstrate using the immune-system example. If two individuals have two different probabilities to catch flu, and both do catch it, impacts on health in the two cases may be as different as life and death due to different statuses of the

[3] It is understood as a product of a probability P and an impact I of a risk on considered objectives.

individual's immune systems. This is not a call for annual flu vaccinations, just an attempt to accentuate the dual role of chronic project system issues and hence, static PSS–PDS mismatches in risk evaluations: they influence both probabilities and impacts[4] simultaneously.

This contradicts the traditional approach based on the bowtie diagram of Figure 2.2: regular causes of project risks influence (add to) probabilities of risk occurrences, whereas magnitudes of impacts are not defined by the causes at least explicitly. Apparently, if particular risk causes are actually chronic project system issues, they should have a say in defining magnitudes of risk impacts too.

It should be pointed out that the static PSS–PDS mismatch is not taken into account consistently in traditional Monte Carlo models as regular project causes and project system issues are not told apart. In other words, the true chronic project system issues even if they are identified as causes of risks (see Figure 2.2) are expected to increase probabilities only.

The internal amplification of a risk becomes visible and relevant only when corresponding standalone risk does happen. If that risk does not happen in a particular project at all, its internal amplification stays dormant. If it does happen the static PSS–PSD mismatch defines the scale of the dynamic risk-induced PSS–PSD mismatch. In the traditional Monte Carlo models, the static PSS–PDS mismatch is considered non-existent (full static PSS–PDS match).

The condition on the risk occurrence is important to accentuate. Sometimes project system issues are understood and treated as project risks in the same sense as uncertain events (see Hollamnn (2017a)). They are still risks in the broader sense as givens/issues, not as future risk events (see section 2.1). The structure of the bowtie diagram (see Figure 2.2) clarifies the role of issues as risk causes.

Regardless of intra-risk amplifications, when a standalone risk occurs it should induce a dynamic PSS–PSD mismatch. It can be imagined that an occurred risk results in a need for rework as well as in a shortage of required labour, equipment, or materials (work constraints). These induced issues can be understood as dynamic project system issues. The corresponding dynamic PSS–PDS mismatch can be tagged as risk-occurrence or risk-induced PSS–PDS mismatch, and the induced issues as dynamic risk-induced project system issues.

Obviously, this situation is in the primary focus of the traditional Monte Carlo models. (The cardinal difference with the traditional case is that the risk event can be already internally amplified due to the presence of chronic project system issues.) This case of the standalone although internally amplified risk does not imply any cross-risk interactions yet. The occurred risk impacts schedule and cost project objectives directly

[4] It should be an overshooting to start speculating on the 'risk consequence's uncertainty principle', although it is not possible to simultaneously define both a probability and an impact of a risk that has a chronic project system issue as a cause. Although a vague allusion to the quantum uncertainty principle of Heisenberg when it is not possible to simultaneously define particle's coordinates and impulse could be made (see for example Eisberg and Resnik (1985)). In any case, the non-linear N-SCRA Monte Carlo methodology developed in Chapter 15 does deal with risk consequences, not probabilities and impacts separately.

without bumping into other risks. The detailed mechanism of the direct impact based on the rework and work constraints is uncovered in Chapter 9.

So far, only standalone risks have been considered. What happens when occurred risks do bump into each other and start interacting? According to section 8.1, there should be knock-on and compounding risk interactions referred to as cross-risk (or inter-risk) interactions.

It would be reasonable to assume that cross-risk interactions induce an additional PSS–PDS mismatch that is obviously dynamic in its nature too. This mismatch can be called risk-interaction-induced PSS–PDS mismatch that, in turn, induces dynamic risk-interaction project system issues.

As a rule, the long-term static PSS–PDS mismatches cannot be promptly addressed and eradicated. They reflect on given long-term internal project inefficiencies or chronic project system issues.

Table 8.1 summarizes definitions introduced in this section. Corresponding terms are used interchangeably in this book. Sometimes the usage of particular terms depends on a context and a level of detail. For instance, if a dynamic PSS–PDS mismatch is discussed it covers both risk-occurrence and risk-interaction PSS–PDS mismatches, etc. unless only cross-risk interactions are discussed.

As a rule, occurred standalone risks are considered internally amplified before they take part in cross-risk interactions. The trivial type of dynamic risk interactions—risk correlations—is left out of the discussions for simplicity purposes although all Monte Carlo models introduced in this book do contain required correlations.

The discussion on various realizations of the PSS–PDS mismatches may be concluded by putting forward corresponding definitions.

> **A static PSS–PDS mismatch** is a state of a project system characterized by static (long-term) misalignment of its two subsystems due to possession of chronic issues that additionally obstruct project delivery through internal risk amplifications of risks.
>
> **A dynamic PSS–PDS mismatch** is a state of a project system characterized by dynamic misalignment of its two subsystems due to either occurrence of a standalone internally amplified risk or when occurred risks take part in cross-interactions.

Table 8.1 Main kinds of PSS–PDS mismatches, project system issues, and risk interactions

Nature of PSS–PDS mismatch	Origin of PSS–PDS mismatch	Sort of project system issues	Type of risk interaction	Impact mechanism
Static	Long-term (chronic/ static) system issues	Chronic	Intra (internal)	Internal risk amplification
Dynamic	Risk occurrence	Risk-induced	N/A (standalone)	Direct
	Risk interaction	Risk-interaction-induced	Inter (cross)	Knock-on Compounding

A complete PSS–PDS match is a risk-free and dynamics-free state of a project system characterized by a full alignment of its two subsystems.

A static PSS–PDS match is a state of a project system characterized by the absence of chronic issues although does not exclude the dynamic PSS–PDS mismatch when project risks occur and interact.

Actually, the case of the static PSS–PDS match (no chronic project system issues and intra-risk interactions) if cross-risk interactions are left out is a subject to traditional Monte Carlo L-SCRA models (see Chapter 5). Only standalone non-interacting risks are counted. It works perfectly for simple projects.

In case of complicated, complex, and chaotic projects both static and dynamic PSS–PDS mismatched should be taken into account consistently calling for a non-linear N-SCRA version of the Monte Carlo methodology (see Chapter 15).

8.2.2 Primary Complexity Aspects

As this book is devoted to quantification of risks as deviations from schedule and cost objectives, a topic on robustness of schedule and cost baselines immediately arises.

Goal setting is extremely important for system's definition at large and for risk quantification in particular. If project goals and baselines are not developed in accordance with applicable standards, contingency development becomes complicated and confusing. Actually, it is possible and easy to distinguish goal setting as a separate part of PDS although for simplicity purposes it is not done. It is just considered one of the procedures adopted by a project under opportunity shaping category (see Figure 8.1).

Technical aspects of scope, base schedule and estimate development are outside the scope of this book.[5] At the same time bias-related factors stemming from stakeholder's misalignment, hidden agendas or late scope changes are part of the discussion and actually can be quantified (see Chapter 15). They are under the procedure implementation category of Figure 8.1. This category includes any relevant scheduling and estimating procedures too. Some of the most notorious bias factors may be associated with Hiding Hand mentioned in the introduction when technical aspects of goal and baseline development are ignored or distorted on purpose. Usually this results in lower aggressive duration and budget baselines to keep a project attractive. If a propensity to reduce project baselines is pinned down as one of chronic project system issues, this should lead to internal amplification of corresponding project risks. As discussed in Chapters 13–15, this in turn should compensate the excessive aggressiveness of project baselines when these risks occur.

[5] Technical aspects of project schedule development could be found, for example, in the Schedule assessment guide GAO-16-89G of US Government Accountability Office (GAO (2015)). Development of project base estimates is described in AACE recommended practices 34R-05 (AACE (2014)), 35R-09 (AACE (2015b)), etc.

If goal-setting and stakeholder-related factors are not treated as standalone independent factors, the primary complexity aspects of Figure 8.1 relevant to project risk quantification are:

- differentiation,
- interdependencies,
- adopted procedures, and
- procedure's implementation.

These four primary categories can be treated as generic and applicable to any project system. However, the right part of Figure 8.1 that specifies these four cornerstone categories should reflect on a context of an upcoming project. Some corresponding categories are still generic enough, although some should be more project specific. It should be kept in mind that proposed detailed categories of Figure 8.1 are developed for a typical oil and gas project. Projects in other industries might have different categories to better reflect on their contexts.

Differentiation reflects on the number of project parts (number of subprojects, packages, involved engineering and other disciplines, required technical and other skills, locations, hard and soft interface points, etc.). Instances of new technology applications are counted too. Project owners and regulators are considered project parts unless they are included to another project subsystem describing a structure of the project external environment.[6]

Apparently, Figure 8.1 is not supposed to list all possible project parts, which is simply impossible due to specific contexts of particular projects. For instance, stakeholders may include local communities, mass media, industry associations, environmental non-government organizations, etc. The next level of details relevant to a particular project may be reflected in the unshown layer of Figure 8.1. This approach is similar to development of risk breakdown structures (RBS) discussed in Chapter 3 where top RBS levels are usually quite standard whereas bottom levels reflect on uniqueness of project context differentiators. From this viewpoint Figure 8.1 may be referred to as a complexity breakdown structure (CBS) that is obliged to fully reflect on the project context of a given project.

As discussed in section 8.1.1, true system dynamics kicks off only when project parts start interacting. Figure 8.1 includes three types of interactions that may occur in a project system among parts and among occurred risks. They are listed together having the same nature in both instances.

Another important complexity aspect relates to project delivery (a project delivery subsystem (PDS)). It covers both procedures selected for project delivery and their actual implementation.

[6] See section 9.2.4 on a possibility to distinguish a third project subsystem—external environment subsystem (EES).

The adopted procedures mentioned by Figure 8.1 should include all relevant procedures actually used by a project organization to deliver a project. As project procedures should cover all aspects of project development and execution from stakeholder management, engineering, procurement, and construction, to safety, to commissioning and startup, a list of relevant procedures should be tailor-made.

Notably, selection of a risk quantification method should be part of the picture under risk management category: as discussed in Chapter 3, a selection of a risk quantification method that is not aligned with project context or general risk quantification principles leads to low accuracy and/or precision of developed contingency (see Figure 3.4).

The implementation aspects of Figure 8.1 include availability of required skills and tools including leadership skills. It also counts various realizations of bias. The implementation aspects are relevant not only to the owner's project teams but also to contractors and sub-contractors involved in project development and execution.

A general approach towards reviewing PSS, PDS, and the PSS–PDS mismatches is that PDS should mirror PSS as well as possible to avoid or minimize any possible PSS–PDS mismatches, both static and dynamic. Due to dynamic nature of some mismatches, PDS should be flexible to make the entire project system adaptive.

The discussion on a project system, its subsystems, and interactions in them including risk interactions is continued in Chapter 9.

8.3 Complexity in Projects

The topic on complexity of projects discussed in sections 8.1 and 8.2 may be followed by a more esoteric and less practical topic on complexity in projects. Although it does not provide direct practical value, it could be informative enough for deeper understanding of dynamic processes in real project systems. This in turn could yield valuable insights on risk quantification and addressing. Additional topics on complexity in projects are introduced in Chapter 17. These new topics can be considered spin-off results of this book.

Most insights on complexity in projects come from physics where studies on non-linearity, emergence, self-organization, adaptive systems, bifurcations, phase transitions, system thinking, feedback loops, etc. have been undertaken for decades. It should be kept in mind that some borrowed complexity topics can serve only as qualitative near-scientific analogies that are not fully justified mathematically. Moreover, sometimes these topics should be taken with a grain of salt as layman-type speculations. In any case they all can easily catch the reader's imagination in an attempt to better understand and interpret complexity in projects.

A book by Strogatz (2015) is recommended for looking up various mathematical aspects of non-linear dynamics. A high-level although very informative review of the complexity themes borrowed from physics is done by Chapman (2017). Three valuable observations on non-linearity and system thinking could be drawn from these two works to support the discussion in this book.

- Stacey et al. (2000) argues that a project system becomes self-influencing as non-linear interactions amplify both positive and negative feedbacks leading to non-linear responses into a chain of circular causality.
- Lorenz (1963) points out that even minute differences in initial conditions can lead to significant and unpredictable consequences in non-linear systems such as complex projects (a butterfly effect).
- Lorenz (1963) also mentions 'strange attractors' to predict long-term behaviour of a project system—long-term states where a system could end up depends on initial conditions.

Two points above attributed to Lorenz (1963) have a robust mathematical explanation. 'A butterfly effect' may be discovered when a minute difference in initial conditions of a non-linear differential equation (actually this is a reference to the equation called after Lorenz (1963)) might lead to completely different solutions over time. A dynamic pattern that represents all possible states of a system (a trajectory) around and on the way to the two 'strange attractors' visually resembles butterfly wings (Strogatz (2015)). Another less scientific metaphora is that a flap of butterfly's wings might be sufficient to make a difference and result in a tsunami somewhere else, although it does not contradict too much the scientific explanation and general principles of the chaos theory.

At a first glance these three observations could be seen as purely esoteric. In reality they may provide food for thought and a relevant analogy on initial level of project scope development, proponent-driven scope changes, or some other occurred project risks that could be perceived as initial small perturbations to a project system that eventually could lead to the chaos and project failure.

Merrow (2011) puts forward a Jamima principle[7] to point out that 'large capital projects are by their nature either quite good or quite bad' reminding of the two strange attractors. He also argues that large projects are fragile and terribly sensitive to seemingly small mistakes which hints at the relevance of the strange attractors and the butterfly effect analogy. Merrow (2011) observes that 'most smaller projects follow the single mode character with a skew toward poorer outcomes'. This is a single-modal distribution comparable with a bell curve. This observation tells apart larger supposedly more complex projects and smaller more simple ones. The more complex projects supposedly follow some laws of non-linear dynamics which does bring forward an analogy related to two strange attractors. As Merrow (2011) argues, this 'strange pattern of results' should be linked with a one of the most important characteristics of megaprojects: their fragility. They do not tend to simply degrade towards poor outcomes; they tend to collapse instead.

[7] Merrow (2011) refers to a well-known children's rhyme by Henry Longfellow about a little girl Jamima that '. . . when she was good, she was very, very good, and when she was bad, she was horrid!' This hints at an additional contribution of regression-based methodology to descriptive statistics of capital projects and complexity in projects. Although there are other lines in the rhyme not cited by Merrow (2011) that also accentuate an importance of proper project governance and controls that read '. . . she stood on her head, on her little trundle-bed with nobody by for to hinder'. This can be declared a Jamima's project controls principle.

Merrow's observation that more complex projects are more fragile does not contradict the idea that perturbations (aka risks) in a complex project system interact due to intra-risk and inter-risk interactions. Instead of working out as standalone risks as in the case of simple projects, each occurred risk injects ripples that are further amplified due to presence of feedback loops. Examples of such loops—risk interaction's mapping inside the affinity of project risks—are demonstrated in Chapter 14 where dynamic patterns of risks are introduced. Risk interactions mean a possibility of 'a chain reaction' or 'avalanche' of consequences that might lead to the chaotic project system behaviour. In simple project systems it is not possible by definition: a simple project does not have a medium for risk propagation due to feeble interactions among its parts.

Merrow's statement that complex projects should have bi-modal duration and cost distributions should be further investigated. As discussed in Chapter 17, complex projects may have bi-modal (two-hump) distributions in some cases but not always. Namely, this may happen if there is a single dominant risk or a small group of risks in the dynamic risk pattern that rises as a result of risk interactions. The single risk should have a dominantly high sensitivity to be responsible for a single failure mode that is associated with the second distribution's peak which leads to a project collapse. On the contrary, when all risks are comparable in terms of sensitivity, each of them may represent its own failure mode resulting in multiple failure modes and degrading of a project towards poor outcomes.

A sports analogy could be drawn to illustrate the situation. Well-trained marathon runners normally do not have single failure modes although they start losing endurance gradually under intense physical loads and psychological pressure (multiple failure modes). On the contrary, less advanced athletes usually do have a single failure mode and may encounter an abrupt failure of one of their body's important organs (knee joints, heart, lungs, etc.).

This discussion is continued in Chapter 17 where an analogy from physics is drawn describing a particle in a two-well potential. One-hump and two-hump schedule distributions are developed and investigated to define criteria when complex projects should have the bi-polar distributions and when not.

8.4 Chapter's Markup and Takeaway

- The modern literature on project complexity had two distinctive domains: complexity of projects and complexity in projects.
- Four sorts of projects are distinguished depending on their level of complexity—simple, complicated, complex, and chaotic. This classification is based on an idea of reducibility of a project system to elementary parts, if it is possible at all.
- A simple system is a sum of a few weakly interacting elementary parts and corresponding standalone risks.
- A complicated system contains many weakly-interacting elementary parts and risks.

- A complex system cannot be easily reduced to a few elementary parts due to intensive part and risk interactions that largely lose their individual characteristics and should be considered as dynamic patterns.
- In case of a chaotic system, individual part and risk characteristics are melted down due to highly intensive interactions; this is a mechanism how a complex project melts away and ends up in a different (chaotic) state.
- There is no single dominating definition of complexity although one is required to zero in on risk quantification in complex projects.
- A newly proposed definition says that complexity of projects is an evolving project-phase specific characteristic that reflects on project subsystems and their inherent reducibility to elementary interacting parts.
- The proposed definition can serve as a foundation of the newly proposed project complexity framework which covers four primary complexity aspect—differentiation (parts), interdependencies (interactions), adopted procedures, and procedure's implementation. The first two aspects belong to PSS, whereas the other two are covered by PDS.
- A fundamental concept of a PSS–PDS mismatch explains nature and origin of dynamic risk interactions in a project system. There are one static and two dynamic types of the mismatches.
- A static PSS–PDS mismatch stems from presence of chronic project system issues.
- The first dynamic mismatch is induced by standalone non-interacting risks. It is subject to the traditional Monte Carlo SCRA modelling. In reality these standalone risks can be amplified if there are chronic project system issues.
- The second dynamic mismatch is additionally induced by interacting risks due to cross-risk interactions—knock-on interactions and risk compounding.
- The case when none of the PSS–PDS mismatches are possible (a complete PSS–PDS match) should be considered as a risk-free complexity baseline.
- Complexity-in-projects topic draws some ideas from physics such as non-linearity, strange attractors, butterfly effect, bifurcations, bi-modal probabilistic distributions, phase transitions, etc. Some of them seem to be marginally relevant, although all of them may catch imagination and help better understand project system's behaviour.
- It is argued that complex projects should have bi-polar (two-hump) schedule and cost distributions. This argument requires further investigation.

9

Interactions in a Project System

Project complexity concept discussed in Chapter 8 encourages the adoption of a new dynamic non-linear project system paradigm. First, it should be based on dynamic interactions of project system's parts. Second, it should be based on understanding that risks cannot be considered standalone any longer due to their interactions. These topics are discussed in this chapter. New definitions of projects and project risks are provided that are applicable to complicated, complex, and chaotic projects. Corresponding modified bowtie diagrams are discussed to describe, identify, and address risk interactions.

9.1 Traditional Definitions of Project and Project Risk

A project complexity and system thinking concepts discussed in Chapter 8 require a definition of a project through its interacting parts. Standard definitions of project as 'a temporary endeavour undertaken to create a unique product or service' (PMI (2017)) would be considered right although not terribly helpful or relevant in terms of project complexity. Or rather it may be relevant to simple projects only. By this reason section 9.2 is devoted to exploring ways to modify a project definition for complicated, complex, and chaotic projects. As creating of a new product or service is normally done in phases, it would be reasonable to replace 'an endeavour' by 'a process'. As a project in different phases is represented through intermediate deliverables (e.g. drawings in engineering and actual deliverables in execution), it would be required to accentuate the evolving nature of project deliverables in the new definition too. Moreover, capital projects do not work on products or services only—they mostly work on physical facilities. It should be good to add a word 'facility' to the definition which is done in section 9.2.

Risk is usually defined as a future event that, if it occurs, might have impact on project objectives (PMI (2009)). A similar although more concise definition treats risk as possible impact on project objectives (Raydugin, 2013). However, both definitions are based on an assumption that risks are considered standalone one-off events that may occur in complete isolation. Again, these definitions could be applicable to simple projects only where risks don't interact being standalone. It is of paramount importance to consider affinities of interacting risks or dynamic risk patterns to define overall project risk exposure in complex projects in place of considering standalone risks. This is done in section 9.3.

Modern Risk Quantification in Complex Projects: Non-linear Monte Carlo and System Dynamics Methodologies. Yuri G. Raydugin, Oxford University Press (2020). © Yuri G. Raydugin.
DOI: 10.1093/oso/9780198844334.001.0001

9.2 A Project System as Two Interacting Subsystems

Before risk interactions in project systems are considered, main project subsystems and interactions between them should be reviewed.

According to Figure 8.1, any project as a system should have two fundamental interacting subsystems. One describes a project structure and interactions in it, and the other refers to a project delivery method that is supposed to match the challenge at hand. Definitions of PSS and PDS are developed in Chapter 8.

The next legitimate question is if the representation of a project system as two interacting subsystems is sufficient enough. A first sign of doubt about this is instigated when thinking about a role and a position of a project owner and other important external stakeholders in a project system.

9.2.1 Where a Project Owner Belongs To?

It is required to delineate a role of a project owner: does it belong to PSS or to PDS? As normally a project owner is a part of operations of a project owner organization, both answers might be considered. Although for the purpose of this discussion a project owner should be treated as a part of PSS (not a part of PDS) along with other external stakeholders. They do not take part in the project delivery directly although may add additional structural complexity to a project.

As this opinion could be easily challenged, the assignment of project owners to PSS could be a temporary solution. Keeping in mind possible roles of the other important external stakeholders and due to the ambiguity of the owner's role, a third project subsystem could be contemplated as an alternative.

9.2.2 Is It Enough to Consider Two Subsystems?

Introduction of interactions among risks in a project system due to various PSS–PDS mismatches implies that all significant interactions pertain to these two subsystems by default. This assumption could be easily challenged if one keeps in mind project's external environment. A hint on this is provided in section 8.1 when internal and external project environments are discussed. Then a third project subsystem could be introduced–external environment subsystem (EES). Presumably all relevant and significant factors in project external environment that do not belong to and are not included to PSS and PDS can be collected in EES. Factors related to project owners, external stakeholders, permitting process and institutional environment, marketing and commercial factors, political and cultural factors, security situation, level of corruption, etc. all could be aggregated in EES. This would require a more detailed considerations of possible imbalances and mismatches pertaining to the three subsystems:

PSS–PDS, PDS–EES, and PSS–EES. In turn this should give rise to a rich spectrum of risk interactions.

Despite an obvious benefit of more detailed study of non-linear interactions among risks, the idea to introduce the third subsystem—EES—is dropped at least for the time being for the sake of simplicity. This means that most important factors of the external environment should be taken into account in the two remaining subsystems: PSS and PDS. However, this topic is open for further investigation. The addition of EES might further enrich the project system dynamics concept. Although it is not the path selected in this book for the sake of simplicity.

9.2.3 A New Definition of Project

To sum up the project system discussion of section 9.2 and presuming that a project system consists of just two subsystems—PSS and PSD—it would be beneficial to come up with a new definition of the project, keeping in mind complexity and system thinking. It may replace previously used definitions and it could read as follows.

> **A project** is a dynamic system accompanied by a phased process to deliver a new product, service or facility that is characterized by evolving phase-specific structure (the project structure subsystem (PSS)) matched by a required delivery organization (the project delivery subsystem (PDS)).

Previously used project definitions can still be applicable to projects of low complexity (aka to simple projects). In other words, the newly-proposed definition is a must in case of complicated, complex, and chaotic projects.[1]

9.3 From Standalone Risks to Dynamic Risk Patterns: A New Definition of Risk

According to the discussion in Chapter 8, representation of a project system as the interacting subsystems implies interactions of risks. Project risks might involve three types of interactions: internal amplifications, knock-on interactions, and compounding. The former are also referred to as intra-risk interactions, the other two as cross-risk or inter-risk interactions.

A current best practice to define project risks is based on the bowtie diagram and the three-part risk naming (see Chapter 2): several causes might lead to a risk event followed by impacts on project objectives. In the traditional paradigm based on standalone risks, causes contribute to risk probabilities only and cannot influence impacts.

[1] Very often in this book including the book title all three categories of non-simple project—complicated, complex, and chaotic—are covered by a single generic category—complex—for short.

Traditional project risk registers do not contain risk interactions handling largely standalone risks.

Apparently, a more sophisticated and adequate approach should be adopted when project risks can be treated as an affinity of interdependent and interacting elements. In this case they form dynamic risk patterns specific to and supported by a given project system. Before this approach is further explored a nature of standalone risks in complex projects should be better understood.

9.3.1 Standalone Non-interacting Risks in Complex Projects

The traditional bowtie diagram (Figure 2.2) should be rethought in the light of the system thinking. It would be worthwhile to better understand fundamental mechanisms for a standalone non-interacting risk to impact schedule and cost objectives. This is about the risk event—project objectives link of the traditional bowtie diagram of Figure 2.2. The causes–risk event links of bowtie diagram may be temporary ignored.

A closer look should be taken at the system dynamics rework cycle introduced in Chapter 7 to obtain some valuable insights.

Project system perturbations (aka project risks) rarely result in project delays or extra costs directly as the bowtie diagram of Figure 2.2 suggests. As a rule, an occurred risk results in a sort of rework which may take four distinctive forms as follows:

- additional work,
- adjustment of previously finished work,
- scrapping previously finished work, and
- scrapping unfinished work.

Apparently, additional work and adjustment of previously finished work requires extra time and funds. Both kinds of scrapping may require some time although they certainly imply corresponding sunk costs related to wasted worked man-hours, materials, and equipment. Also, all four rework sorts may result in lower productivity of works.

Regarding both types of scrapping costs, their direct impacts on cost objectives (sunk costs) and schedule-driven costs should be told apart. Sunk costs are part of fixed costs, whereas schedule-driven costs are variable costs.

However, the immediate rework as well as continued normal work are rarely possible right off the bat as an occurred risk often results in so-called work constraints due to an emerged lack of required:

- information (data sheets, drawings, interface points, etc.)
- labour,
- materials,
- equipment, and
- permits.

Project delays and extra costs stemming from work constraints can be explained by a need to eliminate listed factors first before the rework starts or normal work continues. Upon removal of the work constraints the rework itself will require additional time and money to fulfil.

Apparently, the induced work constraints are of a different nature in engineering, procurement, and construction, although they all usually result in lower work productivity.

This discussion leads immediately to the broader and fundamental topic on work availability. For example, SD models of Chapter 7 show two major stocks: 'Work Remaining (E)' and 'Work Remaining (C)'. When 'work flow (E)' and 'work flow (C)' are constant and do not depend on any other system factors such as occurred risks, this implies that both engineering and construction works are fully available at any moment in time. In other words, there are no work constraints listed above at all. Corresponding engineering and construction work flows that stay intact despite some occurred risks should keep depleting these stocks methodically. The problem with this viewpoint and with the linear SD models of Chapter 7 is that work constraints do emerge and productivity does drop inevitably when risks occur and reworks are required.

It can be imagined that some risks might not lead to the rework at all. Although they would lead to creation of work constraints only. In this case they impact normal workflows directly.

Both emerged reworks and the work constraints can be temporary addressed through undertaking out-of-sequence works to quell risk consequences at least partially. Project activities could be reshuffled and workers temporary reassigned. In a way this could be considered a mitigation action if a project team is flexible enough. However, out-of-sequence work itself usually results in:

- schedule pressure,
- lower quality,
- overtime work,
- workers' burnout,
- hiring more workers,
- etc.

These factors are also known as culprits responsible for reduction of productivity of works.

The observations and speculations regarding rework, sunk costs, work constraints, and out-of-sequence work provided above should help evaluate impacts of risks on schedule and cost objectives in a more structured and thoughtful way.

Figure 9.1 summarizes the discussion above. It is far from being complete as some logical links and feedback loops are not shown for simplicity purposes.[2] However,

[2] Ideally this part of the bowtie diagram should be represented as a standalone micro-system dynamics model for a given risk event. This is not done here as it is considered an evasion.

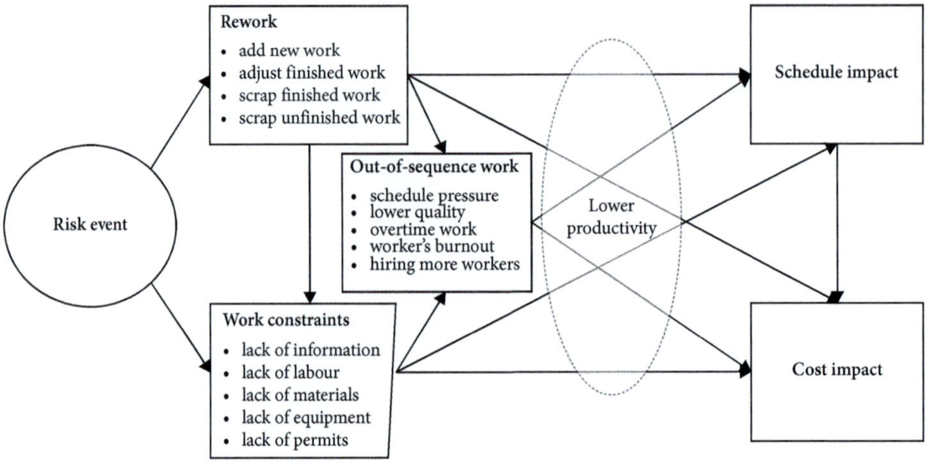

Figure 9.1 A mechanism of risk impacts on schedule and cost objectives (the right part of the bowtie diagram) in the case of a standalone risk

Figure 9.1 accentuates the fact that a reduction of productivity accompanies all impact factors. This chart should amend the traditional bowtie diagram (see Figure 2.2) for standalone risks in complex projects (its right part describing the risk impacts links) even if intra-risk and cross-risk interactions in a project system are not considered. The risk interactions and ways to include them to bowtie diagrams are discussed in the following section. However, the impact mechanism represented by Figure 9.1 should be imbedded to the bowtie diagrams for the interacting risks too.

9.3.2 Dynamic Patterns of Interacting Risks in Complex Projects

The static and dynamic PSS–PDS mismatches, their origin and implications are introduced in Chapter 8. They presume a possibility for risks to participle in multiple interactions. This points to a need to refrain from viewing risks as standalone events as they should be considered a close affinity. The risk affinity in a particular project system can be expressed through dynamic risk patterns. Types of dynamic patterns realizable in a given project system would be fully stipulated by the configuration of a project system itself and interdependencies in it (see Figure 8.1). The reason for this is simple: risks as perturbations resulting in the static and dynamic PSS–PDS mismatches propagate and interact through the same channels as project parts use to interact with each other (see Chapter 8).

Following this approach, one could guess that changes to a project system itself should amend supported dynamic risk patterns. Changing of schedule logic or durations could be examples of this. For example, even changing of timing of two schedule activities may induce or prevent a possibility of risk compounding, etc.

A fresh look at risk identification and addressing from the system-thinking angle that includes risk interactions is required. Section 9.4 introduces modified bowtie diagrams that show single instances of risk interactions. These bowtie diagrams can be used for identification of possible risk interaction instances. Dynamic risk patterns that comprise and visualize all risk interactions in a given project system are introduced in Chapter 14. Meanwhile it is worth coming up with a definition of risk in complex project systems in the light of the discussion above. It could read as follows.

> **Project risk** is a possible aggregated impact on project objectives by an affinity of interacting risks that form dynamic risk patterns supported by a project system.

The same as in the case of project definition of section 9.2, this definition of project risk is a must in case of complicated, complex, and chaotic projects where risk interactions are significant by definition. When intra-risk and inter-risk interactions are negligible as in the case of simple projects, a standard standalone-risk definition is still valid.

9.4 Risk Identification in Complex Projects: Modified Bowtie Diagrams

As pointed out in section 9.3, the complexity concept and system thinking can provide powerful insights on feedback loops and non-linear interactions in a project system including interactions of risks. As soon as a risk occurs it may bump into several other occurred risks resulting not just in a simple additive superposition of impacts of standalone risks but in much more complicated and enlarged non-linear consequences.

Three types of risk interactions introduced in Chapter 8 are worth specifying further. This logically calls for the design of three types of modified bowtie diagrams. They are introduced by Figures 9.2–9.4. It should be kept in mind that all risk event–objectives links should be represented and understood in a more sophisticated form of Figure 9.1. This is not done in the modified bowtie diagrams for simplicity purposes. Besides discussed risk interactions all three diagrams are a reminder of schedule-driven costs by schedule-to-cost arrows.

The modified bowtie diagrams show only impacted objectives and links directly relevant to discussed risk quantification. For example, the quality objective is shown only for the knock-on bowtie diagram where the role of occurred quality risk (impact on quality objective) is to provide an induced dynamic systematic cause to the down-the-line quality risk.

Despite dotted-line arrows pointing to impacts of chronic or dynamic project system issues (systematic causes) on project objectives they actually symbolize cases when non-linear impacts on project consequences are expected (see section 9.4.3). This means that the product $P \times I$ should be amended, not just probability P or impact I separately. The situation is specific in the case of compounding as it is believed that probabilities of risks are not amended by risk compounding.

9.4.1 A Bowtie Diagram for Internal Risk Amplification

The first type of bowtie diagram describes the internal risk amplification when a given standalone risk is amplified in presence of chronic project system issues resulting in static PSS–PDS mismatch. This type of bowtie diagram was discussed first by Raydugin (2018a), (2018b) and represented by Figure 9.2. This bowtie diagram requires regular risk causes and causes associated with chronic project system issues being systematic risk causes to be told apart.

Hurricanes or sand storms that took place previously in the project location area could be treated as regular risk causes. Another example could be a need for adopting a new technology by a project. In the immune-system example discussed in Chapter 8, an on-going flu or corona virus pandemic could serve as regular risk cause too. However, poorly developed and implemented construction planning procedures should be considered chronic project system issues being systematic risk causes. Needless to say, in the case of the pandemic example, a weak immune system is clearly a systematic cause. The weaker the immune system, the more severe consequences of the disease contracting.

One project system issue A is shown by the bowtie diagram of Figure 9.2. As any risk cause, it gives rise to higher risk probability. Although in addition it influences impacts on schedule and cost objectives which directly depends on importance and power of the chronic system issue. In case of a full static PDS–PSS match, this issue does not exist and so makes zero additional impact. In practice, the power of a particular project system issue as a systematic cause can be assessed by corresponding non-linearity parameter (see section 9.4.3).

9.4.2 A Bowtie Diagram for Knock-on Interactions

The knock-on bowtie diagram is introduced by Figure 9.3 for a case of two risks. Each of these risks could have non-linearities related to internal amplification as discussed in section 9.4.1 (see Figure 9.2) although those are not shown. The up-the-line risk, if it occurs makes a knock-on impact on the down-the-line risk if the

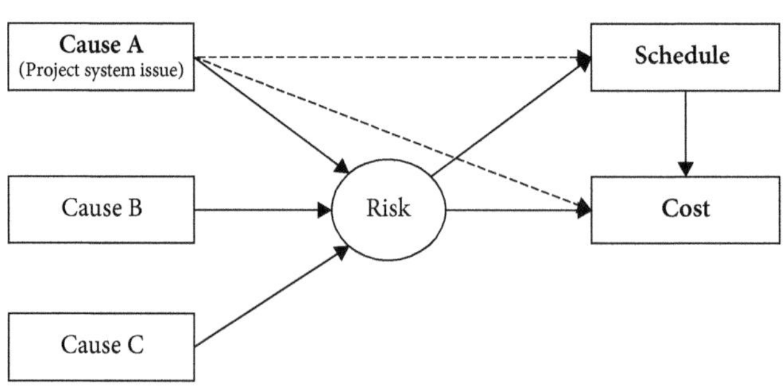

Figure 9.2 Bowtie diagram for the internal risk amplification

latter also occurs. Figure 9.3 demonstrates a typical situation when the impact on the quality objective inflicted by the first risk turns to an additionally induced dynamic project system issue for the second risk: it serves as a dynamic systematic cause of the down-the-line risk. The presence of the dynamic cause results in an additional impact on schedule and cost objectives by the second risk which is shown by dotted-line arrows.

It is important to keep in mind that two risks taking part in a knock-on interaction usually do not have overlapping windows of occurrences. On the contrary, they usually occur in completely different project phases. It might take a substantial time to discover that a previously occurred quality risk has a knock-on impact on the down-the-line quality risk. A typical example of this is a case when engineering errors and omissions are not timely discovered and corresponding drawings enter construction. The knock-on impact on construction quality risk could be discovered in several months or even in a few years and as late as in commissioning and startup.

It could be noticed that the part of the bowtie diagram showing the second risk is the same as in the case of the bowtie diagram for the internal risk amplification (Figure 9.2). This points to a common mechanism of influence of both chronic and dynamic project system issues on risk consequences when a project system issue (any origin) is in presence.

Getting back to the immune-system example, the knock-on impact of the flu contracting risk are post-flu complications.

Following this logic, a corresponding knock-on non-linearity parameter should be introduced (see section 9.4.1).

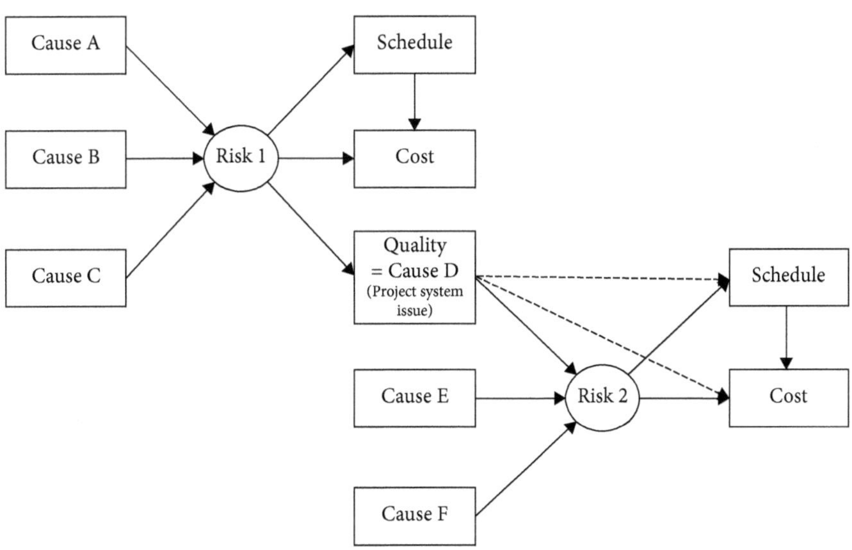

Figure 9.3 Bowtie diagram for dynamic knock-on risk interaction

9.4.3 Bowtie Diagram for Risk Compounding

Figure 9.4 introduces a bowtie diagram for two compounding risks. Both standalone risks could have their own internal amplification non-linearities stemming from corresponding static PDS–PSS mismatches (Figure 9.2) that are not shown. The two risks may interact resulting in compounding of schedule and cost impacts, if both occur simultaneously. However, the possibility of their simultaneous occurrence should be allowed by a dynamic risk pattern supported by a project system. These two risks should be allowed to have overlapping windows of occurrences. The project system issue C symbolizes a possibility of simultaneous occurrences of the two risks. It could be mentioned that the cause C is not linked with the two risk events. This presumes that the cause C does not influence probabilities of the risk occurrences being responsible only for inflicting compounded risk impacts.

The resulting impacts on schedule and cost objectives are expected to be higher than sums of individual impacts. In practice, more than two project risks may become compounded depending on a supported dynamic risk pattern.

The immune-system example suggests that a risk of contracting of any other infection along with the flu or corona virus could exacerbate impacts of both.

9.4.4 Using Amended Bowtie Diagrams for Risk Interaction's Quantification

In case of the traditional bowtie diagram (see Figure 2.2) the role of any risk cause in risk quantification is limited to contribution to overall probability of risk occurrence. As discussed in Chapter 2, in the traditional case the probability of the risk shown by

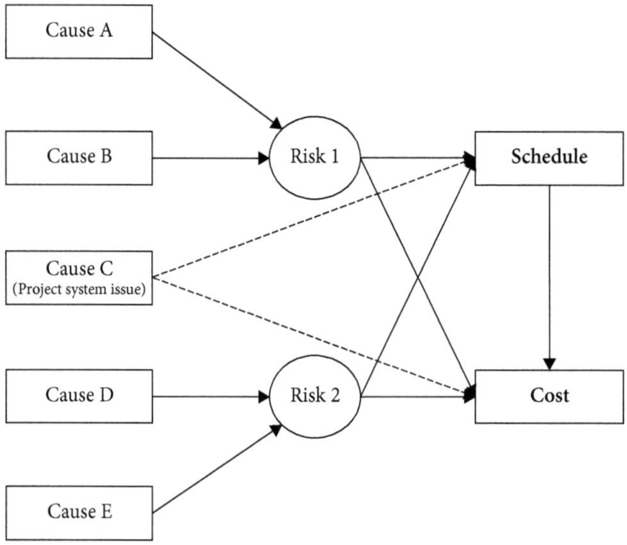

Figure 9.4 Bowtie diagram for dynamic compounding risk interaction

Figure 2.2 is assessed as $P = P_A + P_B + P_C$, where P_A, P_B, P_C are probabilities associated with causes A, B, and C. Regular and systematic risk causes are not distinguished.

If the cause A is a project system issue, it plays also a role in defining overall risk consequences not just contributing to the occurrence's probability.

Using an example related to intra-risk amplification (Figure 9.2), it could be imagined that two different project teams work on two similar projects in similar locations and have the same relevant chronic project system issue A. Those issues could result in a particular risk event. The first project may have a more mature project system with a near-ideal static PSS–PDS match and the very feeble chronic system issue, whereas the second project may have a less mature project system with a substantial static PSS–PDS mismatch and the very powerful system issue. In simple terms, the first project should have the same chronic issue A less sensitive than the second project. Certainly, the occurrence probabilities of this risk should be different in the two projects. If nevertheless the discussed risk does occur in both projects, it is logical to expect that the second project must encounter a more severe impact on the project objectives.

Figure 9.2 introduces this possibility through directly connecting the cause A with schedule and cost objectives. All in all, it is suitable to speculate about increase of the overall risk consequences and not about increase of its probability or impacts separately.

This approach is formalized through introduction of non-linearity parameters in Chapter 13. In case of internal risk amplification due to the presence of a project system issue A shown by Figure 9.2, internally amplified risk event R can be represented as:

$$R \times (1 + \beta \times R) = P \times I \times (1 + \beta \times P \times I), \tag{9.1}$$

where β plays a role of an intra-risk non-linearity parameter. Expression (9.1) represents the internal amplification of risk R due to the presence of the systemic issue A. It should be pointed out that the non-linearity parameter β does modify the risk consequence (a product $P \times I$), not the probability P or the impact I separately resulting in the quadratic non-linearity R^2. Apparently, when there are no relevant project system issues ($\beta = 0$), the risk R has the consequence $P \times I$, as in the traditional linear case.

This approach is fully applicable to knock-on interactions as an occurred up-the-line risk provides an additional dynamic project system issue (see Figure 9.3). However, the resulting non-linear multiplier should show both risks—the up-the-line risk R_1 that creates the dynamic system issue and the knocked-on risk R_2:

$$R_2 \times (1 + \beta_{21} \times R_1), \tag{9.2}$$

where β_{21} is a knock-on non-linearity parameter that couples risks R_1 and R_2.

Quantification of compounding of two risks R_1 and R_2 can be formalized based on the bowtie diagram of Figure 9.4 as follows:

$$R_1 + R_2 + \beta_{12} \times R_1 \times R_2 = R_1 \times (1 + \frac{1}{2} \times \beta_{12} \times R_2) + R_2 \times (1 + \frac{1}{2} \times \beta_{12} \times R_1), \tag{9.3}$$

Expression (9.3) splits the total mutual coupling impact $\beta_{12} \times R_1 \times R_2$ of the two risks to two equal parts.

These generic formulas 9.1–9.3 for the three risk interactions are further discussed in Chapter 13. Needless to say, the non-linearity parameters used in expressions 9.1–9.3 are not the same having different natures and origins. This aspect is further explained in Chapter 13.

9.4.5 Using the Amended Bowtie Diagrams for Risk Interaction's Addressing

If risk interactions–intra-risk and inter-risk interactions—are identified as an additional component of project risk exposure they should be addressed to obtain reductions.

Risk addressing in a complex system should be rethought. It should be aimed at suppressing risk interactions where possible besides regular addressing of standalone risks. In simple terms, risk interaction's addressing in complex projects should include addressing actions to deprive a complex project of some of its complexity attributes related to risk interactions. These are attempts to push a complex project to behave as a simple project where possible through suppressing of its dynamic risk pattern.

Addressing of internal risk amplifications directly relates to addressing of static PSS–PDS mismatches stemming from chronic project system issues. This is a most challenging activity. A variety of steps from development of the project team's skills to improvement of all relevant project development procedures could be considered. Some of them are overviewed in Chapter 15.

One of the defining factors to stop knock-on risk interactions in a project system would be enforcement of PDS by quality assurance and quality control (QA/QC): major deliverables should be thoroughly checked to prevent possible deficiencies from moving to next project phases. This is the same as addressing a broader spectrum of chronic system issues but related to QA/QC only. As a case in point, most advanced engineering companies spend an excessive amount of time on final engineering model reviews prior to issuing engineering packages for bidding or construction. This is a way to suppress possible knock-on risk interactions. For instance, undiscovered engineering risks should not be allowed to knock on procurement or construction risks.

Compounding risk interactions may be suppressed by proper planning of project activities when simultaneous occurrence of several risks is prevented where possible: their windows of occurrences should not overlap. Where required and possible, corresponding potentially impacted normal tasks should be separated in time and space.

9.5 Chapter's Markup and Takeaway

- Traditional definitions of projects and project risks ignore the dynamic nature of complex projects. New definitions are required.
- A possibility to distinguish a third project subsystem—ESS—is contemplated to include project owners and other external stakeholders. If only two subsystems are kept, the following definition of complex projects may be put forward.

- A project is a dynamic system accompanied by a phased process to deliver a new product, service or facility that is characterized by evolving phase-specific structure (the PSS) matched by a required delivery organization (the PDS).
- A paradigm shift from considering standalone risks to recognizing affinities of interacting risks is required. The affinities of interacting risks create dynamic risk patterns supported by a project system. Any change in a project system can result in amendment of the supported dynamic risk pattern.
- Even standalone risks in complex projects should be understood differently to explicitly take into account rework, work constraints, out-of-sequence works, and productivity reduction due to occurred standalone risks.
- In the case of complex projects where risk interactions are important the following definition of project risk may be put forward.
- Project risk is a possible aggregated impact on project objectives of an affinity of interacting risks that form dynamic risk patterns supported by a project system.
- Risk identification in complex projects should include identification of risk interactions. Risk registers should reflect on identified risk interactions.
- Instead of the traditional bowtie diagram that links risk causes with a risk event and further with impacts, modified bowtie diagrams are developed for the three types of risk interactions.
- Regular risk causes and systematic risk causes stemming from either chronic or dynamic project system issues should be told apart. Systematic risk causes have a capacity to influence impacts on project objectives in a non-linear way.
- Systematic risk causes stemming from chronic project system issues (the static PSS–PDS mismatches) are responsible for non-linear intra-risk amplification. Systematic causes associated with induced dynamic project system issues (the dynamic PSS–PDS mismatches) define non-liner impacts on project objectives due to knock-on and compounding risk interactions, correspondingly.
- Each instance of risk interactions can be mathematically described by a corresponding non-linearity parameter that brings forward a square non-linearity to impacts on project schedule and cost objectives.

10

A Project Structure Subsystem (PSS)

Figure 8.1 introduces two interacting subsystems: a PSS and a PDS. This chapter provides a more detailed although still relatively generic description of the PSS including its parts and interactions among them. It should be recognized that a description of a structure subsystem for each project should be highly tailor-made. Only two high-level categories of the Figure 8.1—differentiation and interdependencies—can be the same for the majority of projects. The proposed description of PSS could be relevant to some typical projects in the oil and gas industry.

10.1 Differentiation (Project Parts)

Following the discussion started in Chapter 8, a high-level overview of project parts is undertaken in this section. As argued in Chapter 8, PSS is evolving in the course of project development. The discussion in this chapter is not aimed at describing final deliverables—products, services, or facilities. It zeroes in on evolving interim project parts that eventually result in final deliverables.

Project parts as defined in this chapter are highly interdependent and rarely can be reviewed separately. For instance, any change in project deliverables could amend labour requirements, interface points, and possibly technologies and locations. Moreover, some additional stakeholders might emerge in some cases, etc. At a large scale, corresponding amendments to PDS could be required to restore the PSS–PDS match in case of these changes. This refers to the inherent reducibility mentioned in the definition of projects. The more complex a project is, the more important interactions between parts and risks become, the less reducible it is to elementary parts. Complicated projects can be reduced to elementary parts much easier whereas it is virtually impossible in chaotic projects.

As most of the topics in this chapter could be quite lengthy, a primary focus of the discussion is on risk quantification in complex projects. Topics that are not directly related to this are omitted.

10.1.1 Project Deliverables

A standard work breakdown structure (WBS) approach towards project planning presumes a hierarchical decomposition of all to-be-executed works. It should include all

Modern Risk Quantification in Complex Projects: Non-linear Monte Carlo and System Dynamics Methodologies. Yuri G. Raydugin, Oxford University Press (2020). © Yuri G. Raydugin.
DOI: 10.1093/oso/9780198844334.001.0001

works and deliverables required to complete a project. Typical WBS contains elements organized in hierarchy layers as follows:

- a project,
- project phases,
- project deliverables, and
- work packages.

A WBS depth (number of layers) and a width (number of WBS elements in the layers) depends on project scope, size, budget, complexity, interdependencies, contracting strategy, locations, etc. For instance, some deliverables or even work packages of large and complex projects are often bigger than entire small simple projects in the same industry.

On the one hand, WBS should cover an entire scope of a project which is called 'a 100 per cent rule'. This means the idea that the whole is greater than a sum of its parts is not applicable to WBS development. However, a case when the whole is smaller than a sum of its parts does not work either. The first case would mean missed components of scope. The other points to a duplication of works among WBS components. Both situations could result in various project risks including risks of late scope changes.

As WBS is used in project planning and scheduling, its components should be logically developed based on work areas, geographical locations, production units, project phases, activity lists, milestones, etc. or their combinations. Additional important considerations relate to contracting strategy and interface management. A project owner should evaluate alternative scenarios on the number of contractors involved in working on a project and the number of interfaces managed by contractors. A higher number of small packages and a higher number of involved contractors result in higher number of interfaces managed by different organizations. This increases a possibility of risks related to interface management failure.

10.1.2 Required Labour

WBS should point to all involved disciplines and specialties from engineering disciplines to construction labour specialties. This means that to deliver a project in accordance with a developed WBS a pool of required labour resources should be timely available and efficiently allocated to each WBS component being reflected in loaded schedules.

If labour resources are unlimited, the resource levelling should be undertaken to avoid excessive peaks and valleys of resource's utilization. This can usually be done without changing of project durations. In a more realistic case of limited resources with constraints a series method is used for resource allocation when activities and overall project duration may be altered. An activity starts as soon as both its predecessors are completed and required resources are available. An activity should not be interrupted once work is started. Lack of effective resource allocation results in work constraints which is a primary reason for schedule delays and extra costs. Only a few things in projects could be less efficient than a construction crew idling due to low work availability when required drawing, labour, materials, equipment, or work permits are not timely available (see Figure 9.1). This points to a need for flexible short-term work availability planning to make a complex project an adaptive system (see Chapter 17).

To undertake resource planning overall resource requirements should be well understood. This needs information about planned productivity too. Lack of required man-power or available required head count with lower skills are typical causes of lower productivity followed by schedule delays and extra costs.

No doubt timely availability of engineers or construction workers is important for execution of project works on time. However, two groups of workers have paramount importance: project leadership at large and construction supervision in particular. A fundamental reason for this was pointed out in Chapter 8. Complex projects are considered fragile and non-linear systems and may fail due to perturbations that might be considered minute and unimportant initially being in reality just the thin end of the wedge. This accentuates a need for the close work availability planning, performance monitoring, and everyday project management at both project and work package levels: if projects are complex systems, they should be adaptive and flexible to be successful. Otherwise they will be rigid and, as a result, very fragile.

The word 'leadership' in this context should not be taken lightly as another buzzword. If possible, this should be an intelligent tool to ensure that a project is on the right track despite multiple inevitable perturbations: there is no auto-pilot regime in projects, only manual piloting, especially in complex projects. A work of a good project leader resembles a wack-a-moll game to handle—to predict and to react to—abrupt perturbations, deviations, disruptions, disarrays, disturbances, and cataclysms. Project risk quantification results directly depend among other things on how efficient the project leadership is in this project wack-a-moll game. Actually, this should be factored in Monte Carlo modelling through one of the project system issues.

A list of involved project disciplines and required skills should be considered a next level of detail for the skills category.

10.1.3 Locations

This aspect has two major implications. First, physical project facilities may be developed in one or several locations. For example, in case of oil or gas pipelines, the facilities are quasi-one-dimensional being spread for many miles. Each location has its own characteristics related to remoteness, geotechnical conditions, weather, and availability of local contractors, suppliers, utilities, etc. In earlier project phases as FEL1 and FEL2A business characteristics of locations are important including institutional environment, regulatory regime, local content requirements, social and cultural considerations, level of corruption, etc.

One striking example could be put forward to illustrate this. A megaproject where the budget was counted in a few dozen billion dollars to construct a refinery was initiated in a semi-remote area. That area was characterized by regular sandstorms in spring and summer time. A word 'regular' meant at least one major and a few small sandstorms a week. This dictated design adjustments to take into account harsh weather conditions in both construction and operations. The story was not over when this was done. When the project entered FEL3 phase it was discovered that local utility companies would not be able to provide required utilities. This led to a very major change to the project scope due to a need to add a huge utility plant. This example could be considered rather extreme. Although it illustrates location-related pitfalls in business case decision-making in FEL1.

Even if extreme perturbations like this are not relevant to a particular project, the location part of the PSS should be taken into account seriously and in an adequate way to avoid some specific project risks.

Second, engineering and procurement teams working on a project may be located around the globe in both developed high-cost and developing low-cost countries. Recently this became one of mandatory competitive advantage requirement for EPC companies. Namely, if so-called engineering blended hourly rate is above a certain threshold it is almost impossible for an EPC company to stay competitive.[1] However, this approach brings forward additional risk topics to discuss related to interface management, quality, safety, etc.

Sometimes international EPC companies win EPC contracts but assign a vast majority of engineering works to their low-cost country offices. This seems to provide an opportunity for corner cutting and profitability boosting. Quite often quality of engineering works is high enough. However, there are multiple known cases in the oil and gas industry and beyond when a lack of advanced engineering experience led to errors and omissions and other risks.[2]

This topic is closely related to a propensity of some project owners to award EPC contracts to lowest bidders. This approach should be considered a major project system issue as this could become a systematic cause of multiple project risks.

A detailed list of project locations is required as a next level of detail for the location category.

10.1.4 Technologies

NASA uses technology readiness levels (TRL) to evaluate technology maturation (Mankins (1995)). A lowest TRL level corresponds to observation of basic physical

[1] If an engineering blended hourly rate for a capital project is above, say, \$55–65\$ an hour these days, a corresponding EPC bid might likely be considered uncompetitive.

[2] For example, discussing recent examples like this in the airspace industry that are linked with two deadly plane crashes is outside the scope of this book. Too often such terrible blunders stem from severe financial pressure imposed by corporations on technical project teams regardless of industry types.

principles that could be realized as a new technology in a distant future. The top level corresponds to an actual system that was tested in real-life mission conditions. The next to the top TRL level corresponds to a situation when an actual system was qualified through test and demonstration, but was not used in a real mission yet, etc. The US Department of Energy (DoE (2008)) and some other US government organizations such as Department of Defence (DoD (2011)) use similar technology readiness assessment (TRA) methodologies. This approach was summarized recently by the US Government Accountability Office (GAO (2016)).

Merrow (2011) uses a simplified set of technology maturation levels that is directly applicable to oil and gas industry as follows.

- Conventional: off-the-shelf technology, widely used previously.
- First-time integration: a conventional technology used somewhere else but adopted for the first time by an organization.
- Minor modifications to a conventional technology: unchanged core technology but with some improvements.
- Substantially new core technology.

Whatever technology maturation assessment methodology is used by a project, the core issue with any application of a new technology could be unknown, wrong, or incomplete data sheets used in FEL2B to shape a project scope. This normally leads to project changes and rework, not to mention delays and extra costs. A technology novelty is simply capable of making a PSS–PDS match impossible even in cases of most advanced and mature project teams.

That is why there is an unwritten rule in the industry that unless a new technology reaches the level of a successful pilot project when a project enters FEL2A, it should not be adopted by the project to avoid possible multiple project risks. It is important to have a new technology proven at least as a pilot by FEL2A when selection of project alternatives is normally undertaken.

One extreme example of a failed new technology was about a megaproject in rural Northern Alberta, Canada, initiated by a major European oil and gas company about twelve years ago. It was to adopt a new promising *in-situ* heavy oil upgrading technology. However, it was discovered during early construction works undertaken before FID that the proposed new technology was actually not workable (Raydugin (2013)). The project was back to square one to prove the technology's viability. Its budget was downsized from about twelve billion dollars to twenty-four millions. It was eventually closed soon after. Sensitive sunk costs were incurred by the company as several partially constructed objects were abandoned. These included a camp, an airfield, and a road, a road to nowhere.

A technology report on all new technologies adopted by a project including their level of readiness is required to clarify the technology category.

10.1.5 Interface Points

Any project has hundreds of interface points of two types.

- Hard interfaces: tangible physical interfaces among pieces of project equipment. Main hard interfaces can be defined, for example, by a process flow diagram (PFD), piping and instrumentation diagram (P&ID) and corresponding drawings.
- Soft interfaces: intangible data input-output interfaces that support transfer of data among project disciplines, project teams based in different location, contractors, suppliers, vendors, etc. and reflect on workflow steps to support project development.

Apparently, both types of interfaces could be both internal and external to a project. For example, external hard interfaces define connections with equipment located outside of the project scope[3] and may include pipelines, aerial electric lines, utilities, roads, etc. External soft interfaces link project disciplines with external parties involved in project development.

Interfaces are of interest to risk quantification in two respects. First, missed interfaces of both types may lead to bad discoveries and late changes. Some missed interfaces could be discovered as late as during commissioning and startup. Second, interfaces should be considered from system dynamics angle as a medium for propagation of perturbations. For instance, phenomena such as ripple effects or snowballing are often referred to when monitoring project changes. According to Cooper and Lee (2009), parts of a project that are not explicitly impacted by changes may be actually impacted too. Any other occurred risks especially quality-related risks have a capacity to propagate through interfaces in a similar way.

A most ridiculous instance of interface management failure was observed by the author in a greenfield refinery project that has a huge water intake canal in its scope. The construction of the canal was quite successful until it was stopped: a 180 ft aerial electric transmission tower was 'discovered' right on the canal's way.

A list of all relevant interfaces is required to shed additional light on the interface points category. This list should be a part of a project interface management plan.

10.1.6 Stakeholders

A practical definition of project stakeholders can be based on their intent and power to either claim part of expected project value or impact it. A list of project stakeholders could be quite long and may include a local government, regulatory bodies,

[3] Sometimes equipment located outside a project scope is called 'outside-battery limit (OSBL)' equipment as opposed to equipment belonging to a project scope ('inside-battery limit (ISBL)'). These terms come from petrochemical industry where any equipment that supports a given petrochemical process is called ISBL equipment traditionally.

local communities, mass media, employees, competition, joint venture or consortium partners, non-government organizations (NGO), a project proponent, etc. As discussed in Chapter 9, project owners should be understood as project stakeholders too.

Unless all stakeholders are identified and their claims are all settled by the end of FEL1, any emerging new stakeholders or new claims of previously identified stakeholders in FEL2 may lead to significant project changes and should be considered project system disturbances. Some changes are expected in FEL2 when a government or regulators consider project permits. Environmental impact assessment (EIA) reports with recommendations attached to project permits are a typical example of this. In contrast to somewhat expected EIA-related changes, another type of change could be more devastating to a project. These could come from a project owner itself after FEL2 when any scope changes should be banned. A situation may become exacerbated if a consortium of companies is considered a collective project owner. Any unsettled misalignments among consortium partners after FEL2 including unresolved business case issues after FEL1 are the major culprits responsible for late and substantial project changes. As previously discussed, perturbations like this may propagate and interact with other risks in a project system eventually leading to the project failure.

All relevant project stakeholders should be listed in a project stakeholder management plan that includes their relevance and power (stakeholder mapping) as well as approaches to manage them and their expectations. Procedures related to stakeholder management are overviewed in 11.1.6.

10.2 Interdependencies (Parts' Interactions)

A detailed enough discussion on parts' interactions is provided in Chapter 8. In place of general interactions of project parts—pooled, sequential, and reciprocal—their perturbation-related or risk-related analogues are used in this book: internal amplification, knock-on interactions, and compounding.

The previous paragraph has a paramount importance to understanding of project risk exposure. Risk identification, addressing, and assessments in complex projects should differ from corresponding PRM steps in simple projects. Needless to say, both intra-risk and cross-risk interactions are completely missed in the traditional risk identification and quantification approach. Even most advanced traditional integrated schedule and cost risk analyses (see Chapter 5) largely miss all three types of interactions.

A paper by Hulett and Whitehead (2016) represents an attempt to factor project system issues into Monte Carlo modelling, although it is done based on linear paradigm using so-called risk factors (linear multipliers). These system issues are treated as risk events themselves which is not exactly right in the light of the bowtie diagram discussion of Chapter 9. In addition, none of the cross-risk interactions are taken into account.

When can this approach be good enough? It is good enough when these intra-risk and cross-risk interactions are insignificant, suppressed or may be ignored, which could be the case of simple and some complicated projects. In the general case of complex projects, a different methodology that takes all risk interactions into account consistently is required.

Identification of all risk interactions discussed in Chapter 9 means revealing dynamic risk patterns supported by a project system. An example of this is provided in Chapter 14 when dynamic risk patterns are identified and visualized using so-called risk interaction mapping (see Figures 14.1 and 14.2). It shows knock-on and compounding integrations for risks included in the L-SCRA business case (see Chapter 5). Internal risk amplification is not part of risk interaction mapping though and should be considered separately. The risk interaction mapping is used to develop a non-linear N-SCRA business case of project Zemblanity (see Chapter 15) to quantify risk interactions.

Methods to suppress (address) non-linear risk interactions are discussed in Chapter 14. Addressing of non-linear interactions should be done along with traditional addressing of standalone risks.

10.3 Chapter's Markup and Takeaway

- A project structure subsystem (PSS) of a real project may have a few layers. A top-level layer is generic and includes two categories: structure parts (differentiation) and parts' interactions (interdependencies). These two categories are applicable to any types of projects.
- The next more-project specific layer defines major components of the differentiation and interdependencies categories.
- The differentiation category may include project deliverables, sorts of required labour, locations, technologies, interface points, and stakeholders.
- The interdependency category comprises three types of parts' interactions (pooled, sequential, and reciprocal) and three corresponding risk interaction's types (internal risk amplification, knock-on interactions, and compounding).
- An additional layer could be added to further specify discussed items as related to a context of a particular project.
- The deliverables category is closely related to project objectives and work breakdown structure (WBS).
- The required labour should be aligned with WBS categories as well as with a loaded project schedule. An important role of work availability including timely resources availability is accentuated.
- A role of project leadership is pointed out to effectively manage a project in a flexible way: a complex project should be an adaptive system to be successful. This directly relates to the work availability planning.

- The location aspects differ in different project phases. Business and commercial aspects are important in FEL1, whereas aspects related to engineering and construction take over in later project phases.
- Any new technology that is planned to be adopted by a project should be mature enough (a proven pilot project) by the FEL2A phase.
- Interface points—both hard and soft—provide a means for project parts' interactions and risks' propagations, although they themselves can be sourses of risks.
- Stakeholders have a power either to claim a part of project value or to influence its overall size. They could be a source of project risks especially ones related to project changes. Project owners are the top sources of late project changes after FEL2.
- Parts' and risks' interactions define overall level of project system complexity and non-linearity. Affinities of interacting risks create dynamic risk pattern supported by a project system. Project risks especially quality-related risks may propagate through interface points.
- As risk interactions contribute to overall project risk exposure they should be identified, addressed, and assessed accordingly.

11

A Project Delivery Subsystem (PDS)

The PSS is overviewed in Chapter 10. This chapter provides a descriptive overview of the other fundamental project subsystem—a PDS. Various industry institutions—the IPA Institute, Construction Industry Institute (CII), Association for Advancement of Cost Engineering (AACE), etc.—have developed various project-readiness-type indices that are supposed to characterize various aspects of capital project delivery. There is no intent to overview them because according to Figure 8.1 only the PSS–PDS mismatches are of primary interest: the absolute PSS and PDS characteristics are secondary in a way. Simply speaking, a particular PDS may be praised as highly efficient in a case of a relatively small and simple project, whereas the same PDS might fail dismally in a case of a more complex project. An example from the English Football League should further illustrate this. Even if a team wins the second tier English Football Championship this particular year it would be preposterous to expect that it would be the champion of the Premier League next year.

In addition, the above-mentioned indices have limited applicability to capital projects only and cannot be used in other industries anyway.

Ideally, features of PDS and its maturity should mirror PSS and its complexity to avoid at least static PSS–PDS mismatches. Dynamic risk-induced or risk-interaction-induced PSS–PDS mismatches should be also kept in mind. Although the static PSS–PDS mismatches are in the primary focus in this chapter to review possible chronic project system issues as systematic causes of project risks (see Figure 9.2).

According to Figure 8.1, the two fundamental PDS categories–adopted procedures and their implementation–are applicable to any types of projects. The next layer that specifies these two categories should be more project-specific. Some project procedures and their implementation aspects are considered in this chapter as mostly applicable to some typical oil and gas projects. Any particular project should have its own set of relevant project development procedures that reflects on its context.

11.1 Adopted Procedures

Following the discussion started in Chapter 8, a high-level overview of procedures adopted by a project is done in this section. Apparently, it is not possible to cover here all relevant procedures even for a single project: the number of procedures used during a project life cycle is usually counted in dozens if not hundreds, so very few of the most obvious and typical procedures are mentioned in this section. Another reason

Modern Risk Quantification in Complex Projects: Non-linear Monte Carlo and System Dynamics Methodologies. Yuri G. Raydugin, Oxford University Press (2020). © Yuri G. Raydugin.
DOI: 10.1093/oso/9780198844334.001.0001

for this is the project context. Even two very similar projects developed by two different organizations will certainly have a different set of adopted procedures. Some procedures would differ in terms of comprehensiveness and quality, some would be simply missing. When projects in different industries are compared, the sets of adopted procedures should differ substantially. For example, the engineering procedures in oil and gas and in information technology (IT) industries would have very few factual similarities.

In some cases, particular unwritten practices are not included in the company procedure list at all although actively used. This means that a word 'procedure' should be understood in a broader sense of the actual practice.

Every adopted procedure or groups of procedures should be evaluated qualitatively first. For simplicity purposes, quality of all types of procedures and their groups could be measured qualitatively using three simple parameters as follows.

- Excellent: world-class advanced procedures developed by highly professional and experienced specialists that cover all relevant areas of expertise and fully reflect on contexts of upcoming projects.
- Acceptable: relatively robust procedures developed by experienced enough professionals that cover most relevant areas of expertise although might not be fully aligned with contexts of upcoming projects.
- Poor: inconsistent procedures with major gaps either developed by inexperienced workers or borrowed from other industry organizations without paying due attention to contexts of upcoming projects.

It might be a case that just one particular adopted procedure out of a given group is especially poor. This could be a reason why entire procedure's group should be considered poor due to its weakest link. For example, all engineering procedures adopted by a project might be excellent but one related, say, to process engineering. In this case the entire engineering procedures group could be considered poor as it is engaged in input-output relations with the other engineering procedures and defines their efficiency playing the role of the black sheep. At the same time, absolute characteristics of procedures have a secondary importance. A conclusion about procedure's quality should be drawn against a context of a particular project: a procedure could be seen as poor for a complex project, but might work out well enough in a simple project.

Main groups of procedures—opportunity shaping, project team design, engineering, contracting, construction, stakeholder management, and change and risk management—are reviewed one by one below. Certainly, the list of procedures reviewed in this section is far from being exhaustive even for oil and gas projects, not to mention projects in other industries. Safety, environmental protection, scheduling, estimating, project controls, document controls, etc. procedures could be added to the list. They are understood here as being hidden under other categories as engineering, construction, etc. As an alternative, depending on a project context, some specific procedure categories could be shown as important standalone categories.

This section could be concluded by a real-life story describing a specific project planning procedure or rather practice—using sticky notes on a board for project planning purposes. This is a good illustration on procedure's quality and consistency of its implementation.

This real story took place in a Calgary office of a major European oil and gas company about a dozen years ago. Being led by an old-planning-school project manager, a project team put hundreds of sticky notes with upcoming project tasks on a huge board in a conference room. Tasks were regularly added, rearranged, reshuffled, and withdrawn as a result of heated debates on upcoming actions or when new ideas were put forward. However, it was observed on several occasions for quite a long period of time that agreed action plans became somewhat disorderly and inconsistent. These observations resulted in more and more heated debates, finger pointing, and delays. Accidentally, it was discovered that a diligent night-shift cleaner put sticky notes back on the board in a perfectly random manner when they came off every now and then.

11.1.1 Opportunity Shaping

Top industry organizations have very high mortality rates of projects in early phases and very low failure rate in execution. Less successful organizations have a reverse statistic. They try to provide life support to feeble projects. Moreover, sometimes in place of front-end loading (FEL) they seem to actually preach back-end loading (BEL): important project development phases, steps, and decision gates are ignored or skipped leading to multiple changes in later phases.

As Merrow (2011) argues, the main purpose of opportunity-shaping process is finding reasons for killing projects, not reasons to pursue them. According to Merrow (2011), the opportunity-shaping process has five distinctive steps as follows.

- Understanding a context of a coming project: features of physical location including environmental issues, history of other projects in the area and current competitive projects, political and regulatory environment, local content requirement and local labour and contractor availability, as well as social, cultural, and religious considerations.
- Evaluating a value of a coming project: full cost-benefit analysis against the project context and expected stakeholders.
- Assessing competitive advantage of a coming project: development of robust business objectives accompanied by consolidating and aligning of business and technical parts of a project team.
- Managing stakeholders: expected stakeholders and their power and intent to claim their share of project value.
- Identifying project partners: organizations that may have vested interests in a project and could provide valuable expertise and resources to support a project.

A number of business, economic, accounting, engineering, etc. procedures are required to accomplish the opportunity-shaping mission in FEL1. Without going into further details, all these procedures are referred to as opportunity-shaping ones.

Ideally, opportunity-shaping should be finalized by the end of FEL2 when a project scope is agreed upon by all main stakeholders. The stakeholders should all become happy about shares of project value allocated to them including a project owner. No further changes are allowed afterwards. If it is not achievable, a current version of a project should be rejected: a project should be abandoned or go to square one to brush up on the business case.

It is widely believed that final investment decision (FID) after FEL3 is the event that defines the future of a project. Actually, this defines funding of the project only. A real go-ahead decision should be done after FEL2 when a project scope and project value allocation is agreed upon by stakeholders. FEL3 phase is just used to rationalize, justify, and formalize this decision in a form of funding.

If the scope is changed after the end of FEL2 by any reasons, this should be considered a serious perturbation to a project system which real impact depends on a nature of the change (timing, size, affected work breakdown structure (WBS) components, etc.). The later the changes happen, the higher their impacts. If several changes happen in a row, they could become compounded.

11.1.2 Project Team Design

Project team design/shaping might not be based on particular procedures and could be represented by a particular practical approach adopted by an owner organization. In any case, an effective development of a project requires timely deployment of a project team in accordance with the project context (clearly formulated business objectives, a robust workflow, deliverables, planned labour requirements, locations, etc.). This immediately puts forward two relevant topics as follows (Merrow (2011)).

- Project team size and integration: in accordance with a project context all required specialists are available as project team members being representatives of either owner or contractor organizations.
- Project leadership: a capable and experienced project leadership should be assigned to the entire duration of a project on a continuity basis who are savvy in efficient upward, outward, downward, and inward communications to promptly resolve and address upcoming issues and risks.

One of the key challenges of project leadership in the case of a complex project is to ensure that a project system is flexible and adaptive enough. Another challenge related mostly to early project phases is to consolidate and balance business and technical part of a project team to avoid a situation when either business or technical aspects of project development are played down or even ignored.

Apparently, expected characteristics of a project team evolve in the course of project development. For this reason a project team's size and leadership traits in different project phases are different.

Any shortcomings in project team design/shaping can be understood as project system issues and should result in static PSS–PSS mismatches. These may significantly amplify regular project risks. For instance, there may be a static PSS–PDS mismatch related to inability to timely manage stakeholder's expectations. As a project system issue and a systematic cause, it should amplify project risks associated with change management (see Figure 9.2). Moreover, a project owner or other key stakeholders could stand behind this systematic cause.

11.1.3 Engineering

As engineering procedures are in the very core of project activities in FEL2, FEL3, and detail design, their quality has paramount importance. Engineering is quite important in FEL1 too but as a function supporting business decisions. A set of relevant engineering procedures directly depends on a context of an upcoming project. Their detail review might be near-endless. Instead of undertaking a comprehensive review of the kind, it would be worthwhile to identify top engineering goals in all of these phases and then overview required groups of engineering procedures against these goals. It is done below for a case of an oil and gas processing project.

In FEL1 high-level evaluations of locations, technologies, marketable products, etc. related to a project business case should be undertaken by using corresponding engineering procedures resulting in corresponding reports and recommendations.

In FEL2 there are two primary engineering goals. First of all, a selection of a preferable project alternative should be completed in FEL2A. Second, development of data sheets for a selected project alternative is the engineering goal to support development of a project scope in FEL2B. A process flow diagram (PFD) along with a facility plot plan can be considered primary outcomes of engineering in FEL2B. Various engineering studies from environmental impact assessment (EIA) to security assessment could be undertaken to support scope development. A number of preliminary engineering studies should be initiated as well that will be completed in later phases.

Engineering studies initiated in FEL2 should be further worked on in FEL3: from interface management plan, to commissioning and startup, to operational readiness plans. They should all be completed in the detailed engineering design phase. A piping and instrumentation diagram (P&ID) and a project execution plan for detail design, procurement, and construction along with issued-for-bid (IFB) packages, could be understood as primary engineering deliverables. They are used for development of a project schedule and a base estimate considered a part of a final investment decision (FID) package.

It is well-recognized that quality of front-end loading (FEL) that is a joint business-engineering activity is directly correlated with chances of project success (Merrow

(2011)). Several industry organizations developed special indices to measure FEL quality. Two of them—IPA FEL index (Merrow (2011)) and CII project definition rating index (PDRI) (CII (1996))—can be used as examples. They represent a wealth of accumulated descriptive statistics information on previously completed capital projects and prescribe what deliverables should be completed by the end of FEL. The value of these indices is to accentuate the importance of quality of involved engineering procedures and their outcomes. By their nature these capital project FEL indices are scoring-type indices. Other industries involved in different types of projects should have similar indices. A detailed review of these or similar indices is outside of the scope of this book.

Upon positive FID when a project enters detail engineering phase, all required drawings should be produced and issued-for-construction (IFC). All previously initiated engineering studies should be completed by the end of detail engineering phase.

Engineering activities in various industries differ from oil and gas engineering significantly. For example, an IT project may include:

- planning step to define functionality requirements,
- an architecture development step to select a most efficient solution,
- a system design step to describe all involved modules such as databases and their interactions with hardware,
- a development step to write a code or building system components,
- implementation, integration, and testing step to ensure that a newly developed IT system meets business goals,
- operation and maintenance step when the system is fine-tuned.

Apparently the IT project steps may be easily matched with FEL and execute phases of a typical oil and gas project (see Figure 1.1). A point here is that engineering activities of projects, used procedures, and major deliverables may differ significantly in full accordance with their contexts.

Any weaknesses or project system issues in relevant engineering procedures would lead to the static PSS–PDS mismatches. In turn this will lead to internal amplifications of regular engineering-related project risks. However, one type of engineering risks stands out from the list of other engineering risks. Namely, possible errors and omissions due to poor engineering quality assurance and control procedures have two different implications.

One scenario is when errors and omissions are discovered with a delay in the same engineering phase. This triggers the engineering rework cycle discussed in Chapter 7. The similar rework cycle could be triggered in construction if poor quality in construction units is discovered with a delay. However, if errors and omissions are allowed to leave detail design phase and enter procurement or construction this may trigger a different rework cycle that combines both rework cycles together. This scenario is not considered in Chapter 7 at all but is considered in Chapter 16. In other words, engineering errors and omission have a capacity to knock on quality-related risks in procurement and construction. Moreover, their delayed impact may be discovered as late as in commissioning and startup. This situation is referred to as a 'banquet of consequences' in the introduction.

As pointed out in Chapter 9, any perturbations in engineering can lead to rework and/or engineering work constraints. As a rule, the work constraints themselves may lead to out-of-sequence engineering works leading to schedule delays and extra costs, not to mention delays to fulfil regular work and rework. All three factors—rework, work constraints, and out-of-sequence work—usually result in lower work productivity (see Figure 9.1).

The following situations describe origin of some engineering work constraints leading to rework and out-of-sequence work.

- Due to project system issues or occurred risks (e.g. poor basic engineering data, lack of data about existing facilities or from vendors, delay with geo-tech, environmental assessment, etc. studies) planned engineering works should be made based on assumptions or out of planned sequence which in turn could lead to rework.
- Due to discovery of engineering errors and omissions or poor quality in previous engineering activities a rework is required before engineering work proceeds as planned.
- Due to a major change order leading to added scope of work, scrapped finished work, elimination of unfinished work, and adjustment of finished work, the planned work cannot be fulfilled as planned.
- Impacts of perturbations previously occurred in earlier engineering activities may be discovered in later engineering activities leading to a substantial rework.
- Etc.

To conclude the discussion on engineering procedures it is recommended to contemplate the following points.

- Adopting of robust engineering procedures is required to prevent amplification of regular engineering risks (the internal risk amplification).
- Rethinking of traditional engineering procedures and approaches is required to ensure that a project team is capable to handle work process constraints and prevent proliferation of knock-on and compounding risk interactions.
- Paying special attention to QA/QC procedures that prevent propagation of quality issues to next project phases is mandatory.
- Paying special attention to engineering management and project control approaches that ensure resilience and flexibility in handling project system perturbations and their combinations is mandatory.

One of the best although unwritten engineering practices is to allocate an abundance of time and skills to model reviews prior to issuing drawings for bidding in FEL3 (IFB packages) and for construction in detail design (IFC packages). Instead of being considered a luxury, this should be treated as a necessary barrier to prevent propagation of quality issues down-the-line and mitigate knock-on risk interactions in later phases.

Implications of poorly developed engineering procedures are discussed in this section so far. Excellently developed engineering procedures that are poorly implemented may have a devastating effect too. Poorly developed and poorly implemented engineering procedures would highly likely be a recipe for a project disaster.

A discussion related to construction procedures (11.1.5) strongly resembles the discussion in this section. This is done on purpose as project system issues in engineering and construction as related to risk interactions are very similar from the system dynamics viewpoint.

11.1.4 Contracting and Procurement

Three main types of EPC contracts are used (with variations) in capital projects—lump-sum, reimbursable, and mixed. The last type usually supposes a reimbursable type of contract in engineering and procurement, and lump-sum in construction.

As Merrow (2011) states, no particular type of contract guarantees success of a project. This does not mean that a project owner organization should not care about a contracting strategy.

In any case, a project owner should have a contracting strategy that provides guidelines on selection of contract types and justifies a selection of a particular contract type. Several interdependent considerations that might increase efficiency of contracting strategy are as follows.

- Number of project packages vs. sizes of packages vs. number of contractors: as discussed in Chapter 8, this defines the number of managed interfaces between different organisztions and relations (including contractual ones) between them.
- Level of project risk exposure: due to an illusion of effective risk transfer through lump-sum contracts in projects with high risk exposure, high-risk-exposure projects may require reimbursable contracts, whereas low-risk exposure projects might work well under lump-sum ones.
- Owner's project management experience and capacity: if a project owner has a strong in-house project management capacity it may handle a mixed type contract; otherwise a lump-sum contract might be a better fit.
- Owner's vs. contractor's EPC market: if the EPC market is overheated it could be quite difficult to attract qualified EPC contractors. As a result, reimbursable or mixed types of contract may work out as being more attractive. Otherwise lump-sum or mixed contracts could be a choice.
- Low bidder approach: an owner might have a written or unwritten rule to select lowest bidders always. This could be a major reason for low quality of works, missed components of project scope, assignment of less experienced labour to a project, disputes and claims, not to mention safety issues.

Although all points above are important for selecting a right contracting strategy, the last point requires special attention. In the majority of cases the illusive low-bidder cost savings result in major project system issues that feed project risks resulting in project failures.

The same way as in the case of engineering procedures, contracting and procurement procedures should be efficient enough to avoid any internal risk amplifications stemming from the presence of corresponding project system issues and instances of static PSS–PDS mismatches. This covers all procurement steps from purchase order (PO) placement to logistics. The role of QA/QC procedures in procurement is as important as their role in engineering. It has two instances. First, it relates to prequalification of bidders to avoid as many procurement risks as possible prior to contract awarding.[1] The second should handle quality of provided materials and services. For example, various QA/QC steps should be undertaken to check quality of purchased materials including factory and site acceptance tests (FAT and SAT, correspondingly). From the system dynamics viewpoint, the purpose of QA/QC activities in procurement is to stop knock-on impacts on quality risks in construction.

11.1.5 Construction

Construction planning, contractor's coordination, interface management, project controls including schedule and cost control, safety management, work permitting, material's management, quality control, etc. are main focus areas in construction that require corresponding procedures. If commissioning and startup is considered part of construction, corresponding planning, integration, testing as well as project close-out procedures should be added to the list.

Besides traditional construction procedures mentioned above, more advanced and sophisticated approaches related to work constraints and work availability are required.[2] Very often these are not formalized as procedures being fully dependent on experience of construction management.

The following typical situations could describe work constraints and reworks leading to out-of-sequence work, schedule delays, and extra costs in construction.

- Due to occurred risks or project system issues (e.g. delayed equipment delivery, delay of up-the-line works belonging to another work package or project, poor construction planning, a weather event, etc.), planned construction works should be put off or made out of planned sequence.

[1] The selection of bidders can utilize the same integrated risk-based and economic-based alternative's selection methodology with minute modifications as is used for selection of project alternatives in FEL2A (see 3.3).

[2] Work constraints in construction can be also tagged as construction site constraints.

- Due to discovery of engineering errors and omissions or poor quality in previous construction activities a rework is required before construction work proceeds as planned.
- Due to predecessor work delays construction crews are idling and require re-deployment.
- Due to a lack of specific required construction labour (e.g. welders) a planned work cannot be fulfilled, so some other works are done instead.
- Due to a major change order leading to additional scope of work, scrapped finished work, elimination of unfinished work, and adjustment of a finished work, planned construction work cannot be fulfilled.
- Impacts of perturbations previously occurring in engineering and/or construction may be discovered in commissioning and startup leading to substantial late rework.

Some examples above refer to previously discussed knock-on risk interactions (see Chapter 8). This accentuates the importance of quality assurance and control (QA/QC) procedures in their execution. If quality issues are discovered and fixed promptly without exiting a particular project phase this would lead to some schedule delays and extra costs only without inducing any knock-on effects on down-the-line activities except delays explained by normal schedule logic.[3]

To conclude the discussion on construction procedures, the following points are recommended:

- adopting of robust construction procedures aimed at preventing amplification of regular construction risks through addressing chronic project system issues,
- rethinking of traditional construction procedures and approaches to ensure that a project team is capable to handle work constraints to ensure work availability and prevent proliferation of knock-on and compounding risk interactions,
- paying special attention to QA/QC procedures that prevent propagation of quality issues to commissioning and startup,
- paying special attention to construction management and project control approaches that ensure resilience and flexibility in handling project system perturbations and their combinations.

Implementation of construction procedures is discussed in section 11.2 as excellently developed construction procedures may be poorly implemented. Poorly developed and poorly implemented construction procedures would likely lead to a project disaster.

[3] In reality this often leads to some additional work constraints down-the-line anyway.

11.1.6 Stakeholder Management

If stakeholders are defined as parties that have an intent and power to claim part of project value, stakeholder management should be understood as one of the tools to preserve project value in a balanced way. As stakeholders are understood as project parts (see 8.1.6), robust procedures should be adopted to identify stakeholders, communicate with them, assess their value claims, and manage their expectations. Specialized techniques based on stakeholder mapping could be used to fulfil this task.

Whereas a certain degree of unpredictability related to external stakeholders (government, regulators, local communities, mass media, non-governmental organizations (NGO), etc.) is not considered unusual, a real dichotomy stems from a project owner's role. On the one hand, project owners should be interested in keeping as much project value as possible. On the other hand, they often destroy it through putting forward late project changes with an intent to increase the value. The road to project failure is paved with good change intentions.

Apparently, impacts of the changes depend on their magnitude and timing, although this is not the end of the story. As discussed on several occasions previously, their impacts also depend on their propagating in a project system and combining with other risks which is sometimes called a ripple effect. A ripple effect may be understood as a propagation of impacts of risks that are subject to internal amplification, knock-on interactions, and compounding through a project system in accordance with dynamic risk patterns pertaining to a given project system. In case of the disputes and claims, implicit impacts of changing project parts on the parts that are not supposed to change at all are in the focus of the debates (Cooper and Lee (2009)).

In many organizations project owners consider themselves entitled for any project changes at any time. Even if project changes are properly carried out from the change management procedures viewpoint, some hidden impacts of changing project parts on seemingly unchanging parts are usually overlooked. This destroys project value.

At the same time, project owners could be under severe pressure to undertake project changes. Some reasons for changes include:

- results of ongoing reservoir simulations or exploration drilling in case of oil and gas upstream projects,
- changing market demand in case of petrochemical projects,
- competitive projects,
- new regulations,
- low readiness of new technologies and a need to replace them by proven ones,
- re-estimation of utility requirements or shortage of utility's supply available in a given project location,
- reassessments of project economics,
- etc.

A review of this list points to a conclusion that most of these issues should be resolved in FEL1 (business case) and FEL2A (selection of alternatives). In any case, corporate stakeholder management procedures should address rules and criteria when major owner-driven changes are allowed to avoid project value destruction. These procedures should be developed in conjunction with change and risk management procedures.

11.1.7 Change and Risk Management Procedures

There is a belief that if a robust change management procedure is applied in the right way, there should not be any project risks associated with a change because project baselines will be simply re-defined. First of all, this assumption implies that an owner organization accepts longer duration and higher capital budgets through redefining project schedule and cost baselines and as a result, a poorer project economics possibly accompanied by jeopardizing some strategic considerations such as early market penetration, competitive advantage, etc.

Higher project durations and costs are expected not only when a project scope is increased. The same happens when a project scope is decreased as rework, scrapped completed work, additional work, etc. would be required even when project scope shrinks.

Apparently, project changes make sense if the additional value outweighs extra duration and cost losses associated with the changes. Second, this assumption means that all impacts stemming from changes are taken into account including changes to seemingly unchanging parts.[4] This requires taking into account any realizations of the previously-discussed ripple effect (Cooper and Lee (2009)). In other words, change management procedures among other things should establish rules when changes are acceptable and well-defined.

A traditional risk management process overviewed in Chapters 2 and 3 is based on reviewing standalone risks, although the context of this discussion points to an overdue need to amend standard risk management procedures. In the light of this discussion the following points should be kept in mind, as modern project risk management should:

- be aimed at identification of any project system issues (additional systematic risk causes) giving rise to the static PSS–PDS mismatches and internal risk amplifications using the amended bowtie diagram of Figure 9.2,

[4] The primary focus of the entire discussion is this book is on project durations, capital expenditures, and to a certain degree on project quality and scope. A more consistent approach towards change management should cover all other relevant project objectives including operating budgets, cash flow, return on investment, safety, reputation, environment, etc.

- be aimed at identification of channels—dynamic risk patterns—that should allow propagate knock-on risk interactions based on the modified bowtie diagram of Figure 9.3,
- be aimed at identification of cases when project risks are compounded using the bowtie diagram of Figure 9.4,
- pay special attention to consistent identification of all risks related to engineering and construction work constraints, re-work, and out-of-sequence work based on Figure 9.1,
- adopt amended schedule and cost risk analysis (SCRA) quantification methodology that can intake internal risk amplifications, knock-on interactions, and compounding to produce adequate quantification results which is done in Chapters 13–15.

11.2 Procedure's Implementation

The remaining area of Figure 8.1 that requires attention describes procedure's implementation and various realizations of psychological and organizational bias discussed in section 2.3. However, as opposed to bias referring to systematic errors in risk identification, assessment, and addressing, comparable realizations of bias applicable to procedure's implementation can be contemplated.

A key point of this section is that even excellently developed project procedures that are poorly implemented will contribute to the static PSS–PDS mismatches (internal risk amplifications) and propagation of knock-on and compounding risk interactions. Similar to evaluation of the procedure's quality, procedure's implementation is evaluated using the following three simple qualitative categories.

- Excellent: consistent and effective application of adopted procedures supported by all required skills;
- Acceptable: application of adopted procedures characterized by some gaps, non-critical lack of some required skills, and some realizations of bias;
- Poor: poor application with multiple gaps associated with critical lack of required skills and exuberant manifestations of bias.

A lack of skilled workers is directly linked with implementation bias as lack of skills may result in misunderstanding of implemented procedures. However, some realizations of bias (conscious or hidden agenda bias) might result in poor procedure implementation undertaken on purpose. Most notorious examples of this are discussed in the introduction and referred to as the Hiding Hand, strategic misrepresentation bias, or optimism bias. Cases of wrongly implemented estimating, scheduling, and risk quantification procedures result in lower than would be reasonable project contingencies to keep projects attractive to stakeholders (see e.g. Telegram (2017) or National Post (2018)). One of the realizations of this is previously mentioned 10 per cent syndrome pinned down by Baccarini (2005). In a nutshell, the Hiding Hand, strategic misrepresentation bias, 10 per cent syndrome, and optimism bias are different facets of psychological and organizational bias that is tagged as a hidden agenda bias (Raydugin (2013)).

Although development of schedule and cost baselines is outside of the scope of this book, notorious realizations of bias related to inadequate baseline development could be quantified somehow. This could be done through taking into account inadequate implementation of corresponding procedures. An obstacle here is purely political: to honestly identify such realizations of bias in the first place.

The same as in the case of procedure's quality, only procedures that are characterized by poor or acceptable implementation are of interest: excellently developed and implemented procedures should not cause a project any additional troubles such as extra systematic causes of project risks.

The same way as in section 11.1, in place of individual procedure implementation reviews, a higher-level review of implementation of their groups could be undertaken. Even a single poorly implemented procedure may drag the entire group down playing the role of the black sheep.

11.3 Chapter's Markup and Takeaway

- The PSS–PDS mismatches may stem not only from issues in PSS but from issues in PDS too. Two main categories that characterize PDS are adopted procedures and their implementation. Well-developed project procedures could be poorly implemented and vice versa.
- Issues with quality or consistency of implementation of adopted procedures are chronic project system issues. By this reason the static PSS–PDS mismatches stemming from these project system issues should be in the focus of the PDS evaluation. However, some dynamic risk-induced or risk-interaction-induced project system issues should not be written off completely too.
- The high-level categories related to procedure's quality and implementation are applicable to any types of projects. Although groups of procedures adopted by a particular project and reviewed under these categories should be tailor-made to reflect on the project context.
- In a case of a typical oil and gas project the following groups of procedures should be reviewed: opportunity shaping, project team design, engineering, contracting and procurement, construction, stakeholder management, change and risk management.
- Consistency of procedure's implementation can be assessed using three categories: excellent, acceptable, and poor. Lack of required skills and various realizations of bias play a critical role in procedure's implementation.

12
Project System Maturity Evaluation

Chapters 8 and 9 introduce project system concept by telling apart two interacting project subsystems: the PSS and the PDS. Chapter 10 is devoted to overview of structural complexity associated with PSS through reviewing of project components (differentiation or project statics) and interactions among them (interdependencies or project dynamics). Chapter 11 describes delivery complexity pertaining to PDS through overviewing quality of adopted project procedures and consistency of their implementation.

This chapter continues this discussion and introduces a workable qualitative framework to evaluate the two project subsystems simultaneously which allows evaluating the maturity of the entire project system. The static PSS–PDS mismatches that stem from the presence of chronic project system issues and result in internal risk amplifications are in the primary focus of the maturity evaluation. The evaluations of the two subsystems are merged to produce a qualitative project system maturity evaluation framework. It is important to keep in mind that the level of project system maturity may vary and could be different in various parts of a project. Hence, ideally the maturity evaluation should be undertaken at a project deliverables or disciplines levels. This maturity evaluation framework is adopted as a basis for quantitative evaluation of relevant non-linearity parameters used in the non-linear Monte Carlo models (see Chapter 13–15).

The three evaluation frameworks developed in this chapter distantly resemble the risk assessment matrix (RAM) for qualitative risk assessments discussed in Chapter 3. Some features of the RAM such as three-colour code could be applicable to the frameworks developed in this chapter. It is not done for simplicity purposes.

12.1 Evaluation of the PSS

As argued in Chapters 8–11, project structural complexity has two dimensions:

- differentiation (statics): all relevant parts that constitute a project, and
- interdependencies (dynamics): interactions among the parts.

These dimensions are used to build a qualitative structural complexity evaluation framework which is introduced by Figure 12.1.

Modern Risk Quantification in Complex Projects: Non-linear Monte Carlo and System Dynamics Methodologies. Yuri G. Raydugin, Oxford University Press (2020). © Yuri G. Raydugin.
DOI: 10.1093/oso/9780198844334.001.0001

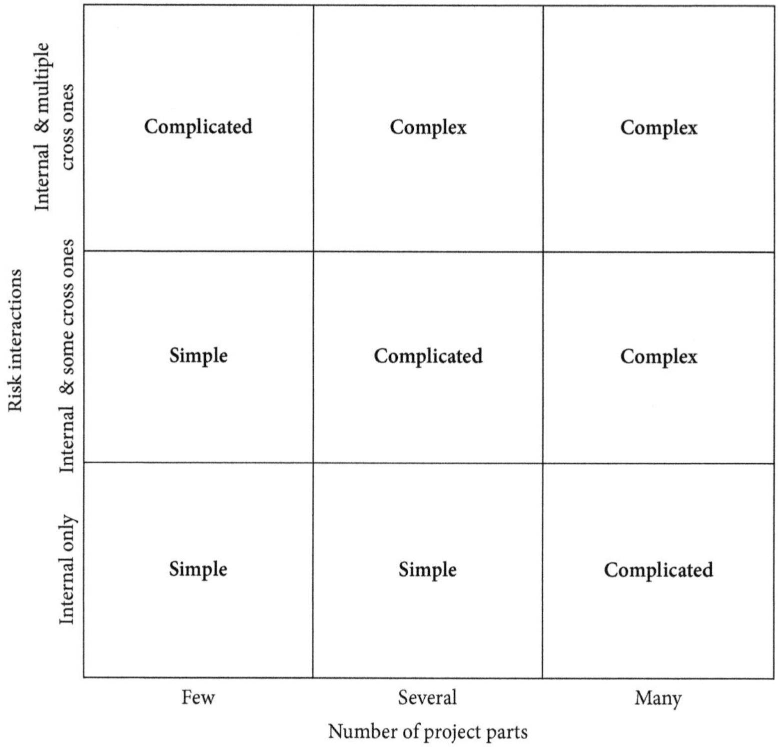

Figure 12.1 Project structure subsystem (PSS) complexity evaluation framework

Three qualitative categories are used to evaluate numbers of project parts:

- few,
- several, and
- many.

These simple qualitative categories that may look subjective count project parts introduced by Figure 8.1. Their company-specific definitions should be developed based on structural complexity of usual projects that an organization develops.

Apparently, all projects should have deliverables related to project parts, although their number could vary. Also, there should be at least a few interface points and stakeholders. Normally at least one location is considered a project part, although in cases of some IT projects devoted to development of virtual information systems this might not be required. Apparently, the technology category might be dropped if only routine technical solutions are used, etc.

As discussed in Chapter 8, it is possible to introduce an interaction's baseline in a risk-free case describing interactions in an ideal PSS in absence of any project risks. Occurred risks being perturbations to the ideal part's interactions propagate through the same channels as part's interactions. Propagating and interacting risks form dynamic risk patterns that can be supported in a given project system. If PSS changes in the course of the project development and execution by any reason, the dynamic risk patterns should be adjusted accordingly.

As only perturbations to ideal part's interactions are of interest, the following three categories are introduced to evaluate the nature and intensity of risk interactions in place of three part's interactions as follows.

- Internal (intra-risk) interactions only: any static PSS–PDS mismatches in a project system (aka chronic project system issues) lead to amplifications of standalone project risks that have the chronic project system issues as systemic causes (Figure 9.2).
- Internal and some cross-risk interactions: besides internal amplifications of project risks some of their cross-interactions are possible (see Figures 9.3 and 9.4).
- Internal and multiple cross-risk interactions: besides internal amplifications of project risks multiple cross-interactions are expected.

If a hierarchy of the three risk interactions is considered, internal risk amplifications should be on top representing chronic project system issues resulting in the static PSS–PDS mismatches. These exist in all three types of projects including even simple projects. Knock-on risk interactions follow as they actually represent fundamental project system's characteristics related to quality of engineering and construction works and the rework cycles. Compounding of interacting risks could be severe if risks occur simultaneously. This is a more random type of interaction that further exacerbates project risk exposure.

Figure 12.1 merges the two PSS characteristics—number of interacting parts (statics) and the nature and intensity of risk interactions (dynamics). Strictly speaking, Figure 12.1 should include part's interactions types—pooled, sequential, and reciprocal—as these are more general and actually include risk's interactions as perturbations to normal part's interactions, although ideal risk-free part's interactions do not represent any interest in the light of risk interaction discussion. For this reason only interacting perturbations to the ideal part's interactions (the part's interaction's baseline)—the risk interactions—are reflected by Figure 12.1.

The selection of these three categories is based on the understanding that intra-risk interactions exist even in simple project systems when risks are still standalone (see Chapter 9). However, the true complexity kicks off only when cross-interactions among risks—knock-on and compounding—become intensive enough. According to this, the results of the PSS evaluation can be categorized using the following three complexity types:

- simple,
- complicated, and
- complex.

No surprise that these are the three project complexity types that are already discussed in Chapter 8. The fourth type—chaotic—is not introduced explicitly as it is believed that a complex project could become chaotic when situation deteriorates enough due to overwhelming risk interactions of all three types. In other worlds, chaotic projects are understood as a limiting case of complex projects.[1] If required, adding chaotic projects as a standalone PSS evaluation category to the framework of Figure 12.1 does not represent any difficulties. Moreover, it might be thought logical to keep it as a standalone category or a dumping ground for failing projects. An additional discussion on project transfers to chaos is provided in Chapter 17.

Despite complexity of a project system represented by both interacting subsystems, it is actually represented in non-linear Monte Carlo risk quantification models only through three types of risk interactions (see Chapter 13). They in turn reflect on static and dynamic PSS–PDS mismatches realized in the project system. This implies that even simple projects may support intra-risk interactions. That is why even simple projects can demonstrate non-linearity as discussed in Chapter 8. The point is that these static intra-risk interactions are not taken into account in the traditional linear Monte Carlo modelling. Only generic dynamic risk-induced PSS–PDS mismatches stemming from occurrences of standalone unamplified risks are factored in the traditional modelling (see Table 8.1) where all projects are considered not only simple but also risk-interaction-free.

It is recommended to formulate short statements for the categories used in Figure 12.1. This should be done based on the context of typical projects that an organization normally works on.

It is well known that many organizations already have methodologies to evaluate project complexities. Along with project sizes these are used to distinguish project types in the organization's project portfolios and stipulate project development requirements for each project type. These methodologies could be adopted as starting points for the PSS evaluations. The framework discussed in this section takes into account the intensity and diversity of risk interactions.

Figure 12.1 might be represented in three colours if required—green, yellow, and red—to resemble risk assessment matrix (RAM) discussed in Chapter 3. It might help to visualize the three PSS complexity categories although it does not bring forward any additional information.

12.2 Evaluation of the PDS

Chapter 8 points out that project delivery has two dimensions:

- quality of adopted procedures, and
- consistency of procedure's implementation.

[1] Hypothetically a situation cannot be excluded when complicated or even simple projects jump to chaos due to very strong intra-risk interactions.

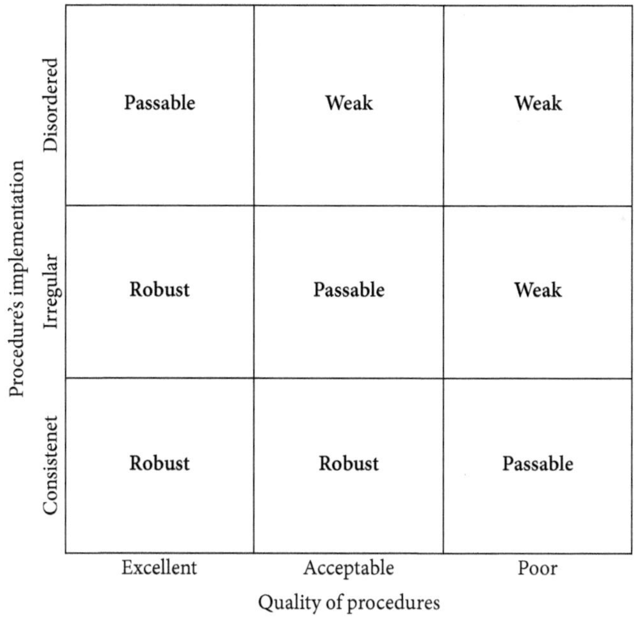

Figure 12.2 Project delivery subsystem (PDS) robustness evaluation

These two dimensions may be used to construct a qualitative framework for evaluation of the PDS robustness which is introduced by Figure 12.2.

Similar to the PSS complexity evaluation undertaken in Section 12.1, the following three qualitative categories may be utilized for the evaluation of quality of procedures adopted by a project:

- excellent,
- acceptable, and
- poor.

Detailed evaluation of particular procedures is outside the scope of this book. However, definitions for excellent, acceptable, and poor procedures could be easily developed based on organizational context of undertaken projects.

Evaluation of procedure's implementation may be based on the following three categories:

- consistent,
- irregular, and
- disorderly.

Once again, these three categories may be further specified as applicable to a particular procedure, to a discipline, to a given project or even to an entire project owner organization.

Combining evaluations of the procedure's quality and implementation results in three qualitative PDS robustness categories as follows:

- robust,
- passable, and
- weak.

A short statement would be required to qualitatively define the categories used for PDS robustness evaluation. This should be done on the basis of organizational contexts of projects normally developed by an organization.

It is a well-known fact that many EPC organizations undertake (or should undertake) procedure reviews and audits every now and then. Often it is undertaken when a new large project is awarded. Their results could be used as starting points for the procedure's quality evaluation.

The same as in the case of the PSS evaluation in section 12.1, a three-colour code could be used to tell apart the three PDS evaluation categories which is not done here to keep things simple.

12.3 The Project System Maturity Evaluation Framework

Two evaluation frameworks introduced in sections 12.1 and 12.2 represent an attempt to characterize the two subsystems—PSS and PDS—independently. In order to carry out an evaluation of an entire project system these two standalone frameworks should be merged. The primary reason for the merger is that it is important they are not standalone PSS and PDS evaluations but the PSS–PDS mismatch evaluations.

Figure 12.3 merges PSS complexity and PDS robustness evaluation's frameworks to assess the overall maturity of a project system.

Resulting categories of the two independent PSS and PDS evaluations are represented as the two dimensions of the project system maturity evaluation framework (see Figures 12.1–12.3). The resulting categories of the project system maturity evaluation are:

- mature,
- ordinary, and
- immature.

A three-colour code can be used to distinguish these three categories. However, as in the case of the other two frameworks it is not done for the sake of simplicity.

Similar to the RAM methodology, short statements should be developed to clarify the resulting project system maturity categories, although this should be done in a full alignment with categories adopted for PSS (section 12.1) and PDS (section 12.2) evaluations.

Project structure subsystem (PSS)	Robust	Passable	Weak
Complex	Ordinary	Immature	Immature
Complicated	Mature	Ordinary	Immature
Simple	Mature	Mature	Ordinary

Project delivery subsystem (PDS)

Figure 12.3 Project system maturity evaluation framework

The purpose of the combined maturity evaluation is to verify that an upcoming project represented by PSS can be successfully delivered by available PDS in principle. If a project system that combines PSS with PDS is mature, a belief that there should be no significant static PSS–PDS mismatches (aka chronic project system issues) is favoured. In terms of applicable Monte Carlo methodology, integrated linear schedule and cost risk analysis (L-SCRA) should be accurate enough as it should not have any significant systematic error (see Figure 3.4). Otherwise, a modified Monte Carlo methodology is required.

In practical terms, the maturity evaluation described by Figure 12.3 as well as the separate PSS and PDS evaluations correspond to the project system level at large. It presumes that the same level of the static PSS–PDS mismatch exists across the entire project system. In practice, various parts of the project system could have different local levels of the PSS–PDS mismatches. For this reason the PSS–PDS mismatches and project system maturity evaluation should be done locally at least at the discipline level: some particular core disciplines should certainly be evaluated individually. For example, PDS for a pipeline project might have a passable civil engineering and weak hydraulic simulation disciplines. A level of the static PSS–PDS mismatches and, as a result, of a project system maturity would differ for these two disciplines. This approach is applied in Chapters 14 and 15.

Figure 12.3 sheds quite a pessimistic light on a possibility for complex projects to have a mature project system. Namely, even in a case of a robust PDS only an ordinary maturity level of a complex project system is achievable or worse. This is in complete alignment with the iron law of megaprojects put forward by Flyvbjerg (2017). On the other hand, this accentuates one more time a need for quantification of not only standalone project risks but their internal amplifications and cross-risk interactions too: project complexity does call for extra duration and cost contingencies!

Chaotic project systems are not considered by Figure 12.3, although Figure 12.3 points to the situation when a PDS applied to a complex project is weak (see the top right cell of Figure 12.3): this is a most likely entry point to the edge of chaos and beyond.

12.4 Chapter's Markup and Takeaway

- Two separate qualitative PSS and PDS evaluations should be undertaken to result in the combined evaluation of a project system's maturity.
- The PSS complexity evaluation framework combines the differentiation characteristics (number of project parts) with the interaction characteristics (types and intensity of risk interactions).
- The three differentiation categories are based on number of project parts: few, several, and many.
- The three interaction characteristics refer to three cases when only intra-risk interactions exist, when intra-risk and some cross-risk interactions are realized, and when multiple intra-risk and cross-risk interactions are presented.
- Combining differentiation and interaction characteristics results in the PSS complexity evaluation framework that covers three PSS types: simple, complicated, complex. This is in line with previously discussed types of projects. The chaotic type is not represented being treated as a limiting case of the complex project category when a project is passing the edge of chaos.
- The PDS robustness evaluation can be undertaken based on two characteristics: quality of adopted procedures and consistency of their implementation.
- Quality of the procedures is described by three categories: poor, acceptable, and excellent.
- Procedure's implementation is having three categories: disordered, irregular, and consistent.
- When these two PDS characteristics are merged they result in the PDS robustness evaluation framework that covers three PDS robustness types: weak, passable, and robust.
- Merging the two PSS and PDS evaluation frameworks produces the project system maturity evaluation framework that distinguishes the three maturity types: immature, ordinary, and mature.
- All evaluation characteristics, categories, and types used in the three evaluation frameworks require a further detailed specification before they are used for evaluations of particular project systems. They should be clearly defined based on typical contexts of projects in a project portfolio.
- When fully developed, the three qualitative evaluation frameworks may serve as a starting point for quantitative evaluations of risk interactions.

13

Non-linear Multipliers for Risk Interactions

Chapter 8 introduces the non-linear interactions of parts in project systems in general and the non-linear interactions of risks in particular. Three types of risk interactions—internal risk amplifications, knock-on interactions, and compounding—are introduced by modified bowtie diagrams in Chapter 9 (see Figures 9.2–9.4). It is explained that a core reason for internal risk amplifications in project systems are static PSS–PDS mismatches stemming from chronic project system issues. They can be found even in simple projects. Both static and risk-interaction-induced PSS–PDS mismatches associated with chronic and dynamic project system issues, correspondingly, become important in complicated and complex projects. The risk-induced PSS–PDS mismatches represent standalone occurred risks that are in a primary focus of the traditional Monte Carlo methodologies (the linear case), however they could be preliminary amplified (the non-linear case).

Projects are treated as system dynamic models in Chapter 7 showing that even when risk interactions are next to non-exiting an overall behaviour of project systems is substantially non-linear anyway at least as related to time series of main project system parameters (stocks).

This chapter summarizes a discussion on main types of interactions among risks and builds a mathematical foundation for the non-linear Monte Carlo modelling. For simplicity and demonstration purposes it is done for a case of just two risk events R_1 and R_2 which could be easily expanded to a general case of multiple risks.

In mathematical terms all three types of risk interactions are expressed through corresponding non-linearity parameters and risk multipliers which is already briefly discussed in Chapter 9. These mathematically describe bowtie diagrams for the risk interactions (Figures 9.2–9.4). If no risk interactions are expected in a project system or if all risk interactions are ignored, all risk multipliers are equal to one by design. This corresponds to a limiting case of the traditional linear L-SCRA modelling where four types of projects (simple, complicated, complex, and chaotic) and three types of risk interactions (internal amplification, knock-on, and compounding) are not recognized and not distinguished.

13.1 A Monte Carlo Modelling without Correlations

Applying a mathematical expectation (aka expected value) concept, any risk R can be represented as a product of its probability of occurrence P and its impact I as $R = P \times I$. This is equally relevant to both schedule and cost impacts.

Modern Risk Quantification in Complex Projects: Non-linear Monte Carlo and System Dynamics Methodologies. Yuri G. Raydugin, Oxford University Press (2020). © Yuri G. Raydugin.
DOI: 10.1093/oso/9780198844334.001.0001

In the case of the Monte Carlo modelling without correlations, a combined consequence of two risks may be expressed as:

$$R_l + R_2 = P_1 \times I_1 + P_2 \times I_2, \tag{13.1}$$

Impacts I_1 and I_2 are understood as distributions of either durations or costs. Quite often three-point distributions (e.g. a triangular distribution) are used to evaluate the impacts unless reliable historic data are available that tell otherwise.

Normally a range of probabilities $(P^{min}; P^{max})$ is defined for a given risk R during risk assessment step. However, the probability ranges are contemplated for discussion purposes only: all Monte Carlo software packages usually recognize average probabilities $P = (P^{\min} + P^{\max})/2$. This corresponds to a mean value of a uniform distribution $(P^{\min}; P^{\max})$.

A standard deviation of the resulting joint distribution σ_{12} is defined through standard deviations of individual distributions σ_1 and σ_2 as:

$$\sigma_{12}^2 = \sigma_1^2 + \sigma_2^2 \tag{13.2}$$

Formulas (13.1) and (13.2) represent a clearly linear case.

13.2 A Traditional Monte Carlo Modelling with Correlations

Actually, some embryonic non-linear interactions among risks are taken into account even in the traditional models through correlations. Correlations add realism to Monte Carlo models through suppressing iterations with unrealistic combinations of sampling parameters (Raydugin (2013)). Correlations can be considered attempts to make sampling used in Monte Carlo modelling biased on purpose. Actually, correlations can be qualified as instances of judgement sampling discussed in section 1.4. The subjective judgements of samplers are realized through imposing generic correlations rules imbedded in the Monte Carlo model. They are required to prune less realistic and irrelevant combinations of modelling parameters and, hence, play a positive role. These less realistic combinations should be naturally suppressed in real-life samples. This resembles distantly the supervised training of artificial neural networks (ANN) models discussed in Chapter 4.

For example, it is quite natural to expect that the higher impacts of errors and omissions in engineering should go hand-in-hand with the higher impact of defects in construction. A practical hint is that if some correlations are contemplated in a traditional Monte Carlo model, they might be discerned as precursors of cross-risk interactions in the non-linear models.

A correlation between two risks R_1 and R_2 may be represented as $K_{12} \times R_1 \times R_2$, where K_{12} is a correlation coefficient (Kreyszig (1999)). As both risks can be represented as two distributions with standard deviations σ_1 and σ_2 a joint distribution should have a standard deviation:

$$\sigma_{12} = \left(\sigma_1^2 + \sigma_2^2 + 2 \times K_{12} \times \sigma_1 \times \sigma_2\right)^{1/2} \tag{13.3}$$

In simple terms, as standard deviations can serve as measures of distribution's widths, in a case of positive correlations ($0 < K_{12} \leq 1$) the joined distribution will be wider than in the case of two uncorrelated distributions when $K_{12} = 0$. Wider distributions mean higher contingencies for given confidence levels (Raydugin (2013)). The case $K_{12} = 0$ is considered in section 13.1. Negative correlations or anticorrelations ($-1 \leq K_{12} < 0$) result in a narrower joint distribution.

When two identical risks are considered ($\sigma_1 = \sigma_2 = \sigma$), the resulting standard deviation is defined as:

$$\sigma_{12} = \sigma \times \sqrt{2 \times \left(1 + K_{12}\right)}. \tag{13.4}$$

In case of the full correlation of two risks ($K_{12} \to 1$). the resulting standard deviation $\sigma_{12} \to 2 \times \sigma$.

Despite the standard deviations of the resulting joint distribution in presence of correlations (expression (13.3)) and in their absence (expression (13.2)) differ, a consequence of the joint distribution can be represented by the same expression (13.1). This means that correlations change widths of joint distributions without changing their mean values.[1]

A practical importance of this conclusion is that when mean values are selected to define the contingency criteria, the defined contingencies in both cases will be identical. Although when contingency criteria differ from mean values (for example, P70), defined contingencies will somewhat differ. An overall contribution of correlations to project contingencies usually measured by few percentage points. This does not mean that correlations should be completely ignored. Although it should not be expected that correlations drastically increase contingencies. As correlations increase standard deviations of resulting distributions and do not change the mean values, they increase the random errors keeping systematic errors intact (see Figure 3.4). A bowtie diagram for two correlated risks is shown by Figure 2.3, where two risks have at least one common or similar cause.

13.3 Linear vs. Non-linear Regression Models

Regression-based models discussed in Chapter 4 that are traditionally used in project management are clearly linear (see expression (4.1)).

In statistics, the linear regression is just a limiting case of its more general non-linear version. Instead of linear predictors (functions describing relations of independent and dependent parameters in regression analyses), non-linear predictors should

[1] The behaviour of distributions with and without correlations comply with a Central Limit Theorem (CLT) (see for example Kreyszig (1999)).

be used by default. For this reason, some serious researchers include squared predictors in regression analysis models where there is no a priori reason to assume linearity (Doane et al. (2010)).

Following this logic, there should be a need to justify linear regression analysis models, as non-linear models are more general. Using squared predictors is recognized as a most straightforward way to introduce non-linearity to regression analysis models, although an exact form of non-linear predictors directly depends on a nature and context of a studied system. As a case in point, non-linear regression is widely used in various branches of science from ornithology (Genidy (2017)), to agriculture (Whetton et al. (2017)), to physics (Stepanov and Sheindlin (2017)) where non-linear functions used for regression are often more sophisticated than just a square function.[2]

Non-linear regression is also discussed occasionally in project management. For example, non-linear behaviour of per cent cost growth (after contingency) versus level of project definition for projects with new technologies is derived as outcome of Rand corporation regression analysis model discussed by Merrow et al. (1981). The same study points to a weaker non-linear dependence of total project costs on level of project definition even in cases of projects without new technologies. The clearly non-linear (quadratic) regression dependency of labour productivity on timing and magnitudes of changes is demonstrated in works devoted to evaluation of cumulative impact of changes on labour productivity (Ibbs and McEnery (2008)).

Probabilistic (Monte Carlo) risk models should be able to accommodate non-linear types of predictors and must be capable of reflecting on the empirically observed non-linear behaviour in project systems to produce more accurate results. This does not contradict the non-linear nature of interactions in complex project systems discussed in Chapter 8.

13.4 Risk Interactions in Non-linear Monte Carlo Models

The purpose of this section is to mathematically describe the three types of risk interactions in order to incorporate them to Monte Carlo models. This is done through introducing three types of non-linearity parameters. Apparently, the non-linearity parameters should be different in schedule and cost risk analysis cases. They also should be different in these cases for simulations before and after addressing. (Addressing in this instance means addressing of static and dynamic project system issues responsible for risk interactions.)

[2] More complex distributions such as exponential, logarithmic, etc. could be used as predictors for project risk quantification too. Although they can be represented as Taylor series that have quadratic (sometimes cubic) terms as first non-linear terms (Kreyszik (1999)). This makes use of squared predictors most simple and practical when expected non-linearity is relatively weak. Technically there is no obstacle to use more sophisticated non-linear predictors where really required which is not done in this book.

The following convention is adopted in this book to label the non-linearity parameters. Non-linearities in SRA Monte Carlo models are measured by project system non-linearity parameters γ (before addressing) and ρ (after addressing). In case of CRA models non-linearity parameters β (before addressing) and α (after addressing) are used. As only after-risk-addressing cases are discussed in this book, only after-addressing parameters ρ and α are actually used in non-linear Monte Carlo models of Chapter 15. However, parameter β is utilized to develop general formulas for the three types of risk interactions in this chapter.

13.4.1 Internal Non-linear Amplifications of Standalone Risks

Any static or dynamic PDS–PSS mismatch can be compared with a balance weighing scale where factors on both sides are equally important. Neither an absolute level of project complexity represented by PSS (see Figure 12.1) nor project delivery system robustness (see Figure 12.2) evaluated independently are relevant. Only the PDS–PSS mismatch reflecting on the level of project system maturity (see Figure 12.3) is important.

In case of the internal risk amplifications (intra-risk interactions) only chronic project system issues define the PSS–PDS mismatch which is static. The immune-system example is discussed in Chapters 8 and 9 to illustrate internal risk amplification of the flu-contracting risk. Another illustration from football could be called for to illustrate the static mismatch associated with chronic system issues. If a footballer has a mild foot or ankle injury but can still play, his or her performance will be at a lower level due to the background cause. A risk of not scoring will be higher. He or she still might give resultative passes or even score although due to the partial mismatch (or rather the misfit in this example) he or she would unlikely secure a most valuable player (MVP) award. The degree of the mismatch directly depends on the severity of the injury and could be considered a background factor of all the player's actions, although it is not a single factor, just one of the systematic risk causes.

In the non-linear quadratic regression analysis terms, the expression (13.1) for overall consequence of two identified risks (a joint distribution) with internal amplification should take the form:[3]

$$R_1 \times (1 + \beta_{11}^{\text{int}} \times R_1) + R_2 \times (1 + \beta_{22}^{\text{int}} \times R_2), \tag{13.5}$$

where β_{11}^{int} and β_{22}^{int} represent project system non-linearity parameters for internally amplified standalone risks R_1 and R_2 Apparently, formula (13.5) does not include cross-interactions between the two risks. These two risks may be represented in a modified form as:

[3] As internal-amplification or intra-risk non-linearity parameters are always symmetrical describing self-interaction of a risk with itself the superscript 'int' is not quite informative and it is not actually required: $\beta_{jj}^{int} = \beta_{jj}$. It is kept here for consistency only.

$$\tilde{R}_1 = R_1 \times \left(1 + \beta_{11}^{int} \times R_1\right), \tag{13.6}$$

$$\tilde{R}_2 = R_2 \times \left(1 + \beta_{22}^{int} \times R_2\right). \tag{13.7}$$

Ideally a detailed evaluation of the internal (or intra-risk) non-linearity parameters for each individual risk included to a Monte Carlo model is required. In practice, this means evaluation for each chronic project system issue: several risks may have common chronic project system issues as causes in their bowtie diagrams. By this reason the evaluation could be done for groups of risks related to particular disciplines. For instance, an issue with poor QA/QC in engineering could be common for several standalone engineering risks. All of them should have the same chronic project system issues as systematic causes resulting in the same internal amplification non-linearity parameter. As an additional simplification step, it is possible to do a high-level evaluation assuming the same single non-linearity parameter for all risks, issues, and disciplines. This rule-of-thumb evaluation would correspond to a generic project system level. Although it is much easier to undertake, usually such an evaluation cannot be specific enough.

The structure of expressions (13.6) and (13.7) allows the natural introduction of internal amplification multipliers based on the non-linearity parameters as $1 + \beta_{11}^{int} \times R_1$ and $1 + \beta_{22}^{int} \times R_2$ for standalone risks. Both will be equal to one in the absence of internal amplifications.

A reverse engineering approach could be adopted in a form of what-if scenarios for planning purposes. Namely, the following questions should be asked. What should be a set of project system characteristics at risk, issue, discipline or project levels to deliver a project with zero overspending and schedule slippage? What should those project system characteristics be to ensure that both overspending and slippage would not exceed 5 per cent, 10 per cent, 25 per cent, 70 per cent, etc.? Answers to these questions can be transferred to internal non-linearity parameters in order to build required non-linear Monte Carlo models. The comprehensive answers to these questions could be obtained only when cross-risk interactions are taken into account too. The reverse engineering of this kind is further discussed in Chapter 17.

13.4.2 Knock-on Impacts from Up-the-line Risks

The system dynamics models SD2 and SD3 of Chapter 7 include two separate project rework cycles related to engineering and construction, although rework induced by errors and omissions in engineering that can sneak in to construction undetected is not considered yet in Chapter 7. What happens if some engineering drawings with errors and omissions do enter construction undetected? This case is investigated in Chapter 16 when engineering errors and omissions (a risk R03) do sneak in to construction undetected and exacerbate a construction quality risk (a risk R04). It could be expected that commissioning and startup risk (a risk R06) should also be aggravated. This type of interaction is identical to previously discussed knock-on risk interactions.

In general, this knock-on effect can be observed between engineering in FEL2, FEL3, and detail engineering, between engineering and procurement, between procurement and construction, or between construction and startup. In terms of system dynamics modelling this depends on the granularity of developed SD models. The key point here is that a discovery of any errors and omissions or poor-quality implications is delayed for a significant period of time (from several months to a few years). Moreover, the discovery takes place in distinctively different project phases down-the-line.

Apparently, the knock-on interactions are a one-way street. Earlier occurred risks may impact with a delay to the later risks if both occur.

In terms of the bowtie diagram of Figure 9.3, the two interacting risks become coupled in a non-linear way due to the additional dynamic systemic cause provided by the up-the-line risk. In addition, it should be kept in mind that risks take part in the knock-on interactions after being internally amplified (expressions (13.6) and (13.7)) where applicable. On the contrary, in a traditional linear Monte Carlo modelling two risks of this kind are either considered standalone or are correlated at best.

In alignment with the knock-on bowtie diagram of Figure 9.3, an expression for an additional non-linear knock-on consequence for risk R_2 from earlier-occurred risk R_1 is as follows:

$$R_2 + \beta_{21}^{knock} \times R_2 \times R_1 = R_2 \times (1 + \beta_{21}^{knock} \times R_1), \tag{13.8}$$

where β_{21}^{knock} is a knock-on non-linearity parameter that couples the two risks. Expression (13.8) demonstrates that the knocked risk R_2 is modified by a non-linear knock-on multiplier $(1 + \beta_{21}^{knock} \times R_1)$. According to expressions (13.6) and (13.7) both risk R_1 and R_2 should be already modified by their internal amplification multipliers if applicable. For this reason expression (13.8) should be replaced by the following expression reflecting on intra-risk interactions (see expressions (13.6) and (13.7)):

$$\widetilde{R}_2 + \beta_{21}^{knock} \times \widetilde{R}_2 \times \widetilde{R}_1 = \widetilde{R}_2 \times (1 + \beta_{21}^{knock} \times \widetilde{R}_1), \tag{13.9}$$

If a risk R_2 is internally amplified and takes part in a knock-on interaction, overall non-linearity of expression (13.9) is defined by expression $\sim R_2 \times (\beta_{22}^{int} \times R_2) \times (\beta_{21}^{knock} \times R_1) \times (\beta_{11}^{int} \times R_1)$ which is actually bi-quadratic. If this risk takes part in additional interactions, an even higher degree of non-linearity can be expected.

There is no mutual impact of risks during the knock-on interactions. Only down-the-line risks are impacted. The expression (13.9) can be used when modelling instances of non-linear knock-on interactions in Monte Carlo models.

13.4.3 Risk Compounding

More than one risk could occur in a project at a given point in time. Their combined impact on project duration or cost would be higher than a sum of separate impacts. This is referred to as risk compounding and represented by the modified compounding bowtie diagram of Figure 9.4. The same as in the instance of the knock-on interactions, risks become compounded after being internally amplified where applicable. They also may take part in knock-on interactions before or after compounding. Each instance of interactions is described by a corresponding non-linear multiplier.

The word 'simultaneously' above should be clarified. If a risk R_1 occurs in a moment of time t_1 it brings forward an abrupt risk-induced perturbation to a project system including rework, work constraints, and reduced productivity (see Chapter 9). It could require a period of time Δt_1 for the project system to recover. Quite often no full recovery is possible as inflicted delays or extra costs are accrued. However, after a period of time Δt_1 the project system arrives at a new equilibrium in terms of productivity. The productivity can be eventually restored in most cases. However, if another risk R_2 occurs in a moment of time t_2 before the project system gets to the new equilibrium ($t_2 - t_1 < \Delta t_1$), the two risks would interact resulting in their compounding. A summary consequence of two risks would be higher than just a sum of two standalone occurred risks. Mathematically the compounding of two risks can be introduced as an additional dynamic risk-interaction-induced PSS–PDS mismatch which could be expressed as follows:

$$R_1 + R_2 + \beta_{12}^{comp} \times R_1 \times R_2 = R_1 \times (1 + 1/2 \times \beta_{12}^{comp} \times R_2) + R_2 \times (1 + 1/2 \times \beta_{21}^{comp} \times R_1), \quad (13.10)$$

where β_{12}^{comp} and β_{21}^{comp} are compounding non-linearity parameters. An assumption on the mutual nature of compounding interactions should be done in expression (13.10) so that $\beta_{12}^{comp} = \beta_{21}^{comp}$. Apparently, unlike the intra-risk symmetrical non-linearity parameters β_{11}^{int} and β_{22}^{int} describing internal risk amplifications of single risks, the compounding cross-risk interaction is described by non-linearity cross-parameters β_{12}^{comp} and β_{21}^{comp} that couple two risks R_1 and R_2.

Expression 13.10 justifies the origin of factor 1/2 in both multipliers: the overall compounding impact is shared by the two mutually impacted risks. This factor does not exist in case of knock-on interactions as down-the-line risks receive entire impacts. The expression (13.10) is split into two parts. One amends risk R_1 by adding a compounding multiplier $1 + 1/2 \times \beta_{12}^{comp} \times R_2$. The other amends risk R_2 in a similar way by adding the other compounding multiplier $1 + 1/2 \times \beta_{21}^{comp} \times R_1$. If compounding does not take place, both multipliers are equal to one.

This simple representation of compounding has to be further modified if both risks R_1 and R_2 are internally amplified due to the static PSS–PDS mismatches (see expressions (13.6) and (13.7)): they enter the compounding interaction being already amplified. Combining expressions (13.6) and (13.7) for internal non-linearity and (13.10) for compounding, results in the following expression for compounding consequences:

$$\widetilde{R}_1 \times (1 + 1/2 \times \beta_{12}^{comp} \times \widetilde{R}_2) + \widetilde{R}_2 \times (1 + 1/2 \times \beta_{21}^{comp} \times \widetilde{R}_1). \quad (13.11)$$

Expression (13.11) that is also bi-quadratic should be used when modelling instances of non-linear compounding interactions in Monte Carlo models. If a particular risk takes part in other cross-risk interactions, more multipliers should be assigned that increase the overall level of project system non-linearity.

13.4.4 A Dual Nature of Non-linear Multipliers

Initially identified risks pertaining to the traditional linear case are subject to the usual Monte Carlo sampling (see Chapter 5). Their impacts can be interpreted as results of risk-induced PSS–PDS mismatches when all risk interactions are excluded. They may happen in some iterations and may not in others depending of probabilities of their occurrences.

Expressions (13.5)–(13.11) point to an important role of the non-linear multipliers leading to higher consequences or higher severities of project risks. These multipliers are characterized by a dual nature: they do not modify probabilities or impacts separately. This topic is discussed in detail in Chapter 8 where the 'risk consequences uncertainty principle' is mentioned.

The discussed multipliers that modify the original risks are represented through relevant non-linearity parameters, given linear-case probabilities, and given linear-case impacts (see formulas 13.5–13.11). Their participation in Monte Carlo iterations is defined by amending the products of linear-case probabilities and impacts (aka consequences). That is why the risk multipliers have a nature similar to general uncertainties—they scale up linear-case risk consequences.[4] Getting back to bowtie diagrams for the risk interactions (see Chapter 9), non-linearity parameters that define the multipliers are attached to project system issues serving as either static or dynamic systematic risk causes.

The discussed dual nature of the multipliers seems to answer to complaints by Merrow (2011) on low accuracy of Monte Carlo methods due to difficulties in obtaining the right distributions in Monte Carlo models. Application of non-linear multipliers describing features of risk interactions in a project system adds required realism to Monte Carlo modelling. In terms of Figure 3.4, the discussed multipliers redefine both mean values and standard deviations of resulting Monte Carlo distributions and, hence, decrease the systematic errors although they increase the random errors. The features of project risk exposure in complex projects based on dynamic risk patterns supported by project systems are credited for both factors.

In the case of linear Monte Carlo modelling, corresponding multipliers are called risk factors (Hulett (2019)). These multipliers $m = (1+\delta)$ are linear by their nature as parameters δ are fixed constants ($\delta = const$) which are independent of any risk

[4] In a more complicated modelling alternative, the multipliers may play a role similar to the role of uncertain events if they take part in Monte Carlo iterations according to defined probabilities. However, a full correlation of all corresponding probabilities (the correlation of occurrences) in a model would be required in all relevant multipliers. For simplicity this method is not used in this book.

interactions and their intensities. They just scale up risk consequences in a linear way based on some expert judgements and possibly available empirical data. Quite often they are set up as triangular distributions to amend ranges of general uncertainties for groups of normal tasks or for individual uncertain events.

It could be justifiable to call the non-linear multipliers discussed in this chapter the non-linear risk factors $M = (1 + \Delta)$, but they are not fixed as they do depend on nature as well as on intensity of risk interactions in a non-linear way as $\Delta \sim \beta_{ij} \times P_i \times I_j$. Moreover, several non-linear multipliers are increasing overall non-linearity of a project system whereas the linear multipliers cannot do so.

If a dynamic risk pattern supported by a project system changes due to, say, risk addressing, non-linear risk multipliers should be adjusted accordingly. If the dynamic risk pattern vanished (all $\beta_{ij} = 0$), this would be the limiting case of the linear Monte Carlo modelling.

13.4.5 Non-linear General Uncertainties

At this point in time a reader should ask a question about applicability of non-linear interactions to general uncertainties. Technically there is no obstacle to applying non-linear multipliers to general uncertainties at all. The question arises, is there any reason to do this?

There are general uncertainties as future events and there are general uncertainties that are givens.

First, if discussed are future events with expected probability of occurrence $P = 1$ they should be treated as any other regular uncertain events (aka future project risks). Future general uncertainties should certainly be a subject to the non-linear interaction modelling.

Second, regarding existing general uncertainties, they should not be considered from the angle of non-linear risk interactions being givens comparable with issues in their nature although with unknown impacts. They are a primary subject to activity and line-item Monte Carlo ranging methodologies discussed in Chapter 3. It is argued that given general uncertainties at normal-task and line-item levels should be limited to uncertainties related to a current level of project development. They should never include or become a substitute to future uncertain risk events.

Should general uncertainties be higher in cases of complex projects and immature project teams? Yes, although given general uncertainties do not require any predictions or amendments being actuals for corresponding project phase, not future factors: they are rather a subject to reviews and audits, not to forecasting and modelling. If there is a compelling reason to increase general uncertainty ranges drastically, than a new standalone project risk should be identified instead (see section 3.4.3). This corresponds to branching out from activity ranging or line-item ranging to the Monte Carlo modelling that does consider uncertain events.

A well-known guideline is that if a reason emerges to significantly expand a maximum value I_{max} of an uncertainty range above a most likely I_{ML} value, this should result in identification of a new relevant uncertain event instead (see expression (3.1)).

As discussed on several occasions (see Merrow (2011)), activity and line-item Monte Carlo range estimating consistently result in about the same project contingencies of about 10 per cent unless input distributions are artificially and substantially skewed. This consistency produced by activity and line-item range estimating is explained in Chapter 3: higher upside deviations from a most likely (ML) value would be simply outweighed by higher downside deviations in a more or less consistent way. This characterizes the Monte Carlo ranging methodology as one of the 10 per cent syndrome methodologies (see Baccarini (2005)).

Downside risk events are cardinally different as they are not parts of project schedule or cost base estimates: they are pure additions that may occur with some probabilities. Both higher and lower impact values of risk events are additions to base line values and do not outweigh each other. In case of upside risks both higher and lower impact values are subtracted from base estimate values.[5]

Following this logic, there is no need to attach any non-linearity to existing general uncertainties although corresponding ranges could be higher. Once again, these ranges should be defined as a result of reviews or audits as givens. Their contribution to overall project contingencies is usually quite well predetermined by a few contingency percentage points and they do not need any additional attention.

Chapter 17 provides a detailed study devoted to investigation of a mechanism of general uncertainty's origination. A general uncertainty around a particular normal task or cost line item can be interpreted as summary impacts of multiple low-severity risks (tier-two risks) that did not make it to become included in the project risk register (tier-one risks). The identification of new risk events when there are reasons to inflate general uncertainty ranges simply means moving one of tier-two risks to a project risk register declaring it the tier-one risk.

13.4.6 Non-linear Multipliers in Case of Upside Risks (Opportunities)

The upside risks (aka opportunities) are not discussed in this book as the focus is on quantification of downside risks (aka threats). Moreover, only downside intra-risk and inter-risk interactions that have clearly negative connotation have been discussed so far. What if there are opportunities as well as some givens in a project system that bring forward an upside effect on project duration and cost?

First, favourable consequences of regular opportunities that exist in traditional linear Monte Carlo models should be reduced due to intra-risk and cross-risk interactions of threats. Second, the situation should be different if there are favourable

[5] Risk events that has both positive and negative impacts could be imagined (for instance, an economic risk when the price of oil could be outside of an expected range) although are not used terribly often in project risk management.

givens in a project system (project system anti-issues) that amplify opportunities and positive cross-risk interactions of opportunities. In this case opportunities should bring forward even better upside consequences. All in all, the following four cases should be considered for threats and opportunities.

In the case of a threat exposed to downside intra-risk and cross-risk interactions, the non-linear multipliers take a form $M_{threat} = (1+\Delta_{downside}) > 1$, where the parameters $0 < \Delta_{downside}$ represent relevant downside intra-risk and cross-risk interactions. This case is already discussed in section 13.4.5.

In the case of opportunities exposed to downside intra-risk and cross-risk interactions, the non-linear multipliers should take a form $M_{opportunity} = (1-\Delta_{downside}) < 1$. This reduces upside opportunity's consequences.[6]

In the case of threats exposed to upside project system issues (positive anti-issues) and positive cross-risk interactions, the non-linear multipliers should take a form $M_{threat} = (1-\Delta_{upside}) < 1$, where the parameters $0 < \Delta_{upside}$ represent non-linear effects of relevant upside system issues and cross-risk interactions on a given threat. Apparently, this should reduce downside threat's consequences.

Finally, in the case of opportunities exposed to upside project system issues and positive cross-risk interactions, the non-linear multipliers should take a form $M_{opportunity} = (1+\Delta_{upside}) > 1$. This further increases the upside opportunity's consequences.

Certainly, these additional cases require some attention. They could be practically considered when required. This could be easily done when following the same protocol for development of non-linear inputs to Monte Carlo models that is introduced in Chapters 14 and 15, although for simplicity purposes, these cases are not further considered in this book.

13.4.7 Units of Measure of Non-linearity Parameters

According to expression (13.5)–(13.11) all types of non-linearity parameters introduced in this chapter have a unit of measure $1/R = 1/(P \times I)$. In case of cost risks the unit of measure is 1/£ where £ refers to British pound which could be replaced by any other currency used for project budgeting. In case of schedule risks, it is $1/d$, where d refers to days.

13.5 Chapter's Markup and Takeaway

- A most simple linear Monte Carlo modelling case takes into account additive risk consequences without risk correlations.
- The currently widely-used traditional Monte Carlo modelling besides additive risk consequences takes into account correlations among risks. The correlations

[6] Consequences of opportunities have always negative values to reduce project durations and costs.

can be treated as embryonic non-linear risk interactions. Correlations increase standard deviations of resulting distributions without changing mean values.

- The Monte Carlo modelling should be adequate in cases when regression analyses produce non-linear results (non-linear dependencies of dependent and independent parameters). The non-linear regression analysis should always be used by default unless there are compelling reasons to reduce it to the limiting linear regression case.
- A most simple type of non-linearity that can be imbedded in regression analysis and in the corresponding Monte Carlo models is quadratic. More sophisticated types of non-linearity could be adopted when required. This is usually the case when intrinsic non-linearities in investigated systems are particularly strong.
- Intra-risk interactions (internal risk amplifications) can be mathematically represented by symmetrical non-linearity parameters β_{ii} due to auto-risk interactions of single risks R_i. Corresponding risk multipliers take a form $1+\beta_{ii} \times R_i$.
- Knock-on risk integrations being a type of cross-risk interactions can be represented by cross-type knock-on non-linearity parameters β_{ij} and non-linear multipliers $1+\beta_{ij} \times R_j$ where R_j are impacting (up-the-line) risks.
- Compounding interactions can also be expressed by cross-type compounding non-linearity parameters β_{ij} and non-linear multipliers $1+1/2\times\beta_{ij} \times R_j$, where R_j represent impacting risks.
- Risk compounding may occur when at least two risks happen simultaneously or near-simultaneously.
- Non-linearity parameters are attached to static or dynamic project system issues that serve as systematic risk causes in corresponding bowtie diagrams.
- Non-linear multipliers have dual nature as they modify risk consequences, not probabilities or impacts separately. This gives a reason to speculate about risk consequences uncertainty principle.
- As non-linear multipliers allow taking into account project risk exposure more adequately, they reduce systematic errors of non-linear Monte Carlo modelling although naturally increase random errors.
- Existing general uncertainties are not subject to modifications using non-linear multipliers. Being givens, they are subject to audit and evaluations. If there is a compelling reason to inflate general uncertainty ranges, a possibility of identification of a new risk event should be considered instead. This means moving one of the tier-two risks covered by general uncertainties to a project risk register declaring it a new tier-one risk.
- General uncertainties as future uncertain events with probability $P=1$ should be treated as any other future events. They should be subject to the non-linear interactions and modifications.
- Both threats and opportunities can be a subject to the modifications by non-linear multipliers. Along with project system issues that have negative connotation, positive project system issues may be contemplated (positive anti-issues).

- As opposed to constant linear risk factors used in the traditional Monte Carlo models, non-linear multipliers depend on risk interactions in a non-linear way being dynamic-risk-pattern specific. If a dynamic risk pattern supported by a project system is changing by any reason including addressing, corresponding non-linear multipliers should be fine-tuned accordingly.

14
Non-linear Monte Carlo Modelling Requirements

Project risks R01–R06 included to the model L-SCRA of Chapter 5 are developed for a case when no risk interactions in a project system are recognized or distinguished. The same risks are used as inputs to the system dynamic models SD2 and SD3 of Chapter 7. The model SD3 mirrors the model L-SCRA of Chapter 5 and yields comparable results. Following the initial linear Monte Carlo and system dynamics modelling attempts, Chapters 8 –13 lay a foundation for taking project risk interactions into account in Monte Carlo models that are referred to as the non-linear SCRA (N-SCRA) models in this book (see Chapter 15). System dynamics models that are aligned with the non-linear Monte Carlo models are run in Chapter 16.

To undertake the transition from L-SCRA to N-SCRA modelling a set of clear rules is required. As a matter of fact, most of them are already introduced in the previous chapters. The purpose of this chapter is to consolidate them and to come up with a list of technical requirements to develop and run N-SCRA models that take risk interactions into account. These requirements include identification, addressing, and quantification of project risk interactions and development amended (non-linear) risk registers that include these factors.

Results of this chapter are applied to a second project Zemblanity business case in Chapter 15 to practically demonstrate the paradigm shift from L-SCRA to N-SCRA modelling. Some conclusions of this chapter should be considered along with modelling steps of Chapter 15. This is required to avoid a perception that the cart is being put before the horse: a full relevance of some topics of this chapter should be clear when reading Chapter 15 where they are used as practical modelling guidelines.

14.1 Cross-risk Interaction Mapping: Dynamic Risk Patterns

As discussed in Chapter 9, internal risk amplifications are stemming from chronic project system issues. Strictly speaking intra-risk amplifications do not belong to the affinity of the interacting risks as they belong to standalone risks. But when standalone risks do take part in cross-risk interactions they enter these interactions already internally amplified.

Knock-on and compounding risk interactions may result in induced dynamic project system issues reflected by knock-on and compounding bowtie diagrams of Figures 9.3 and 9.4.

Modern Risk Quantification in Complex Projects: Non-linear Monte Carlo and System Dynamics Methodologies. Yuri G. Raydugin, Oxford University Press (2020).

DOI: 10.1093/oso/9780198844334.001.0001

Whereas the static PSS–PDS mismatch can be qualitatively evaluated using project system maturity evaluation framework (Figure 12.3), some additional qualitative tools are required to evaluate cross-risk interactions.

Based on the mathematical approach towards quantification of project risks (see Chapter 13) and using system dynamics experience (see Chapter 7), two risk interaction maps for risks R01–R06 are developed (Figures 14.1 and 14.2) for schedule and cost impacts. They reflect on stable dynamic patterns of risks supported by the system of project Zemblanity.

The development of the traditional Monte Carlo schedule risk models is based on mapping of standalone risks to impacted normal tasks. The entire schedule logic should be kept (see Figure 5.1) and utilized as a foundation for the risk interaction mapping. Then all instances of cross-risk interactions are analysed and added to visualize the dynamic risk pattern supported by the project system. One difficulty with this approach is that traditionally risks are mapped to the ends of impacted normal tasks. However, in high-level proxy schedules (see Chapter 5) normal tasks may have quite long durations exceeding several months. If occurred risks induce temporary impacts on project productivity (for instance, for a few weeks) compounding of two risks might not happen if one of them could occur at the beginning of the normal task and the other by its end. Timing of risk impacts can be taken into account in system dynamics models more precisely (see Chapter 7) to include or exclude compounding. An example aimed at comparing of two system dynamic models—with and without compounding—is provided in Chapter 16. Hence, when developing the risk interaction mapping, windows of risk occurrences should be defined to pin down all compounding possibilities.

Figure 14.1 defines knock-on and compounding risk interactions for schedule impacts of risks R01–R06 initially included to the L-SCRA model of Chapter 5. The foundation of the schedule risk mapping is the project Zemblanity schedule (Figure 5.1) with standalone risks mapped to impacted normal tasks. Figure 14.2 shows the interactions for the same risks as related to cost impacts of project Zemblanity (see Chapter 5). A higher level of map's sophistication in case of schedule risks is explained by the fact that schedule risks are schedule-logic specific.

Figures 14.1 and 14.2 point to a clear need to expand project risk identification beyond traditional identification of standalone risks (see Chapter 2) and their mapping to impacted normal tasks (see Chapter 5). In the case of complex projects all significant instances of cross-risk interactions should be identified to discover supported dynamic risk patterns. This should be accompanied by defining possible windows of risk occurrences to point out possibilities of risk compounding.

The newly-understood risk identification can be supported by using bowtie diagrams for interacting risks of Chapter 9 (see Figures 9.3 and 9.4).

The bowtie diagram of Figure 9.2 should be used for identification of chronic system issues that are not reflected on although presumed by the dynamic risk patterns of Figures 14.1 and 14.2.

In unison with Table 8.1, the discussion above points to a need to tell apart four types of project system issues as follows. Distinguished should be:

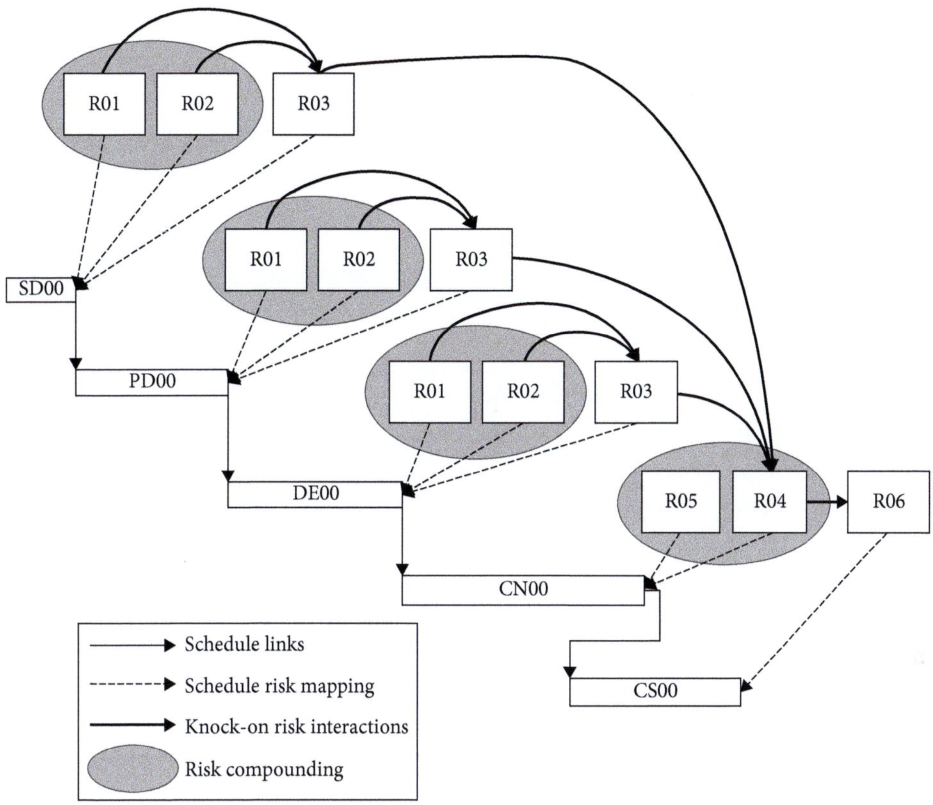

Figure 14.1 A dynamic pattern of schedule risks: cross-risk interaction mapping for project Zemblanity

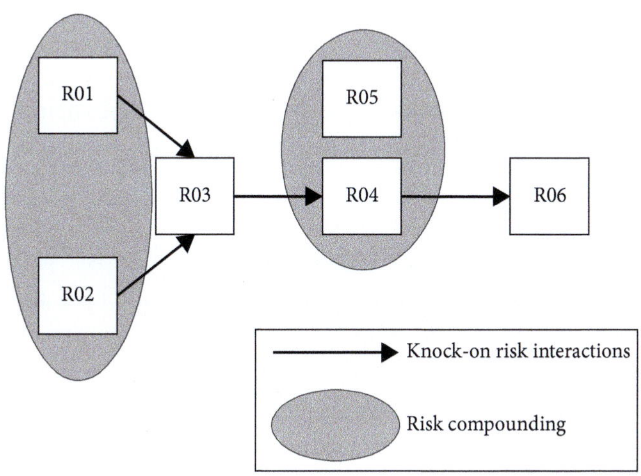

Figure 14.2 A dynamic pattern of cost risks: cross-risk interaction mapping for project Zemblanity

- chronic project system issues associated with the static PSS–PDS mismatch (Figure 9.2) leading to internal risk amplifications,
- dynamic risk-interaction-induced project system issues associated with the dynamic PSS–PDS mismatches due to knock-on interactions (Figure 9.3),
- dynamic risk-interaction-induced project issues associated with the dynamic PSS–PDS mismatches due to compounding (Figure 9.4), and
- dynamic risk-induced project system issues due to occurrence of standalone non-interacting project risks that are associated with the risk-induced dynamic PSS–PDS mismatches (Figures 2.2 and 9.1).

According to Chapter 13, each instance of risk interactions associated with corresponding types of system issues could be evaluated by corresponding non-linearity parameters. In turn, these non-linearity parameters define three corresponding non-linear multipliers (see expressions (13.5), (13.8), and (13.10)). For example, specific non-linear risk multipliers for models N-SCRA-A and N-SCRA-B discussed in Chapter 15 are developed in Appendix D to include all instances of risk interactions. This is done through amendment of risk registers traditionally used for L-SCRA modelling (see Chapter 5). Risk interaction mapping (Figures 14.1 and 14.2) should be incorporated to the non-linear project risk register. The amended risk register for project Zemblanity that includes all relevant instances of risk interactions and their non-linear multipliers is developed in Chapter 15. All risk consequences included to the linear L-SCRA model should be recalculated to take in all required non-linear multipliers. These modified risk consequences are used as inputs to the previously developed linear L-SCRA model of Chapter 5. Although after taking in the modified non-linear consequences this previously linear model reincarnates as a non-linear N-SCRA model.

14.2 Risk Interactions: An Overall Calibration and Individual Evaluations

Upon risk identification that includes all risk interactions and developing of risk interaction mapping, a next step of a risk management process is risk assessment (see Figure 2.1). It should factor in all relevant non-liner multipliers attributed to each instance of risk interactions. Apparently, if a risk takes part in several interactions it should accrue several non-linear multipliers.

Each instance of identified risk interactions could be of various magnitudes depending on relevant static or dynamic project system issues. This brings forward a challenge to properly calibrate risk interactions which has two stages.

First, there should be generic guidelines and criteria that help tell apart weak, medium, and strong interactions associated with the entire dynamic risk pattern. This corresponds to the project-level assessment. This topic is discussed in section 14.2.1.

Second, as soon as these guidelines and criteria are established it would be required to get down to individual risk interactions to assess their power one by one. This is a subject for discussion in section 14.2.2.

14.2.1 An Overall Risk Interaction Calibration

Several generic categories should be introduced to define when risk interactions of all project risks—an overall risk interaction impact delivered by the entire dynamic risk pattern—could be qualified as weak, medium or strong. When there are no risk interactions at all—that is, all non-linear risk multipliers are equal to one by default—an overall impact of standalone risks on schedule and cost objectives should be the same as in the limiting case of the model L-SCRA (see Chapter 5).

Apparently, a magnitude of an overall risk interaction impact as an addition to the linear case should belong to a continuum from zero to extremely high values. To introduce areas of this continuum (e.g. weak, medium, and strong additions to a linear case) some characteristic one-point threshold values or dividers should be established.

Generally speaking, establishing of calibration dividers is directly associated with risk tolerance approach adopted by a particular project owner organization. As an alternative, the selected calibration dividers can be aligned with recognized industry criteria for project failures. For instance, a project-failure definition by the IPA Institute (Merrow (2011)) could be adopted: a project with 25 per cent or more schedule slippage and/or 25 per cent or more overspending can be qualified as a failure.

Table 14.1 proposes criteria and thresholds for non-linear Monte Carlo model's calibration. It introduces three categories of summary risk interactions—weak, medium, and strong—through imposing 10 per cent and 25 per cent calibration thresholds. It could be agreed and established that these categories mirror project system maturity evaluation categories (mature, ordinary, immature) of Figure 12.3. All instances of all three sorts of risk interactions that form dynamic risk patterns in a given project system should contribute to overall project duration or cost.

The 10 per cent and 25 per cent thresholds can be replaced by any other values to better comply with corporate risk tolerance guidelines.

By design, risk interaction summary impacts are counted as additional schedule slippages and overspendings above the ones of the linear L-SCRA modelling case of Chapter 5 that plays a role of the benchmark.

Table 14.1 Three categories of risk interactions and project system maturity

Risk interactions	Project system maturity	Schedule slippage, %	Overspending, %
Weak	Mature	0–10%	0–10%
Medium	Ordinary	10%–25%	10%–25%
Strong	Immature	>25%	>25%

It is acceptable to use more than three categories of overall risk interactions, if desired. As a case in point, five categories—very weak, weak, medium, strong, and very strong—are used initially to calibrate internal cost risk amplifications (Raydugin (2018a), (2018b)), although this would require redefining the PSS, PDS, and project system evaluation frameworks of Chapter 12.

Table 14.1 can be used not only for evaluation of the overall impact of all risk interactions at the project level used in this section for calibration purposes. It can be used 'locally' for evaluation of a particular project system area (for example, a discipline): particular project system issues and associated individual risk interactions can be also evaluated (see section 14.2.2).

The risk interaction calibration is undertaken by recalculation of risk consequences using relevant non-linear multipliers followed by running of existing (linear) Monte Carlo models with these recalculated risk consequences as inputs.

First, amended risk registers for schedule and cost impacts that include all relevant risk interactions should be run several times for various but in each run equal values of non-linearity parameters: each run is done under a condition that all non-linearity parameters for all interactions are equal. This provides evaluation for all corresponding non-linear risk multipliers (see expressions (13.5), (13.8), and (13.10)). The calculated multipliers amend consequences of initially identified risks R01–R06.

Second, these amended expected values are used as inputs to the previously developed linear L-SCRA models. After accommodating the non-linear risk multipliers, the L-SCRA model can be pronounced the non-linear N-SCRA model.

This calibration method is based on the recognition of stable dynamic patterns of risks discussed in section 14.2. Although as all non-linearity parameters are set the same for each instance of risk interactions in every calibration run, a capacity of a particular risk to impact the overall project outcome if fully stipulated by its initial assessment in the linear case (the model L-SCRA of Chapter 5) and its role and power in the dynamic risk interaction pattern.

The last point refers to number, nature, and strength of risk interactions a particular risk takes part in. In other words, it depends of the overall sensitivity of a particular risk that takes into account both factors.

If a standalone risk has low initial severity and takes part in just a few risk interactions, its resulting sensitivity would be much lower than the one for a risk of higher initial severity that takes part in multiple strong risk interactions. Higher initial (linear) risk consequences will be amplified much more if a risk takes part in multiple strong interactions despite the same non-linearity parameter being used for all risk interactions.

For example, risk R05 has quite low schedule and cost consequence assessments (see Table A.4 of Appendix A) and does not play much role in the dynamic risk pattern (see Figures 14.1 and 14.2). The other risks have higher assessments and play a much more active role in the dynamic pattern. They are considerably more sensitive. When calibration runs with higher fixed values of non-linearity parameters are undertaken, the sensitivity gap between low and high sensitivity risks becomes even bigger. This is exactly what is required for calibration purposes when all runs are undertaken for non-linearity parameters that are equal in each run.

Running the SCRA model several times this way is aimed at pinning down particular non-linearity parameter's values when overall project durations or costs exactly correspond to the established thresholds (see Table 14.1).

Schedule and cost modules of the SCRA models are run semi-independently in calibration runs. Namely, when a non-linear cost module (N-CRA) is run for various values of cost-related non-linearity parameters, the schedule module is kept linear (L-SRA). This is a case when all schedule non-linearity parameters are equal to zero and the multipliers to one, correspondingly. This allows excluding schedule-driven costs from calibration of pure cost risk interactions.

It is important to keep in mind that the discussed calibration approach is clearly project specific. Identified values for schedule and cost non-linearity parameters and corresponding multipliers are directly associated with the project risk register developed for a linear case, the project base schedule and cost base estimate, and identified risk interactions.

In the case of project Zemblanity the cross-risk interaction mapping is defined by Figures 14.1 and 14.2. Apparently, if a different project with different schedule and cost base estimates, and a project-specific risk register including different risk interactions is considered, the calibration should be undertaken from scratch for that particular project. This should certainty produce different non-linearity parameters and non-linear multipliers that would satisfy to 10 per cent and 25 per cent thresholds.

This calibration approach is directly applicable to the risk interaction evaluation if the entire project system is characterized by a single non-linearity parameter. Such high-level assessment could be used as a-rule-of-thumb rough method. However, a more precise evaluation is required when risk interactions significantly vary in comparison with each other. This evaluation is developed in section 14.2.2.

14.2.2 Evaluation of Individual Risk Interactions

The sameness of non-linearity parameters used for calibration in each run should be removed as the next step when individual non-linearity parameters are being developed for particular instances of risk interactions. The same Table 14.1 can be used 'locally' for specific evaluations of individual risk interactions.

In a real project system, all non-linearity parameters are different as they reflect on specific instances of risk interactions (Chapter 13).

First of all, it can be easily figured out that the same summary impacts (for instance, ones corresponding exactly to 10 per cent or 25 per cent schedule slippages or overspendings), can be achieved not only in particular runs when all non-linearity parameters are equal. The same results can be yielded in multiple runs with various combinations of differing non-linearity parameters. Some parameters would correspond to weak interactions, some to medium, and some to strong.

Second, an individual set of non-linearity parameters can be derived that reflects on the specific chronic project system issues and the dynamic risk pattern. Individual non-linearity parameters should be selected keeping in mind guidelines of Table 14.1 used 'locally' for individual risk interactions as opposed to the project level (see section 14.2.1). For example, an engineering discipline might have a chronic project system issue related to quality or implementation of a particular procedure that corresponds to immature project system in this regard. This particular issue should correspond to a strong intra-risk non-linearity parameter according to Table 14.1, etc.

A practical calibration and individual evaluations of risk interactions is undertaken in Chapter 15. It should be reminded that non-linearity parameters belong to the continuum of values starting from zero (the linear case). Interactions belonging to the same category of Table 14.1 may be evaluated differently although in accordance with the allowed ranges derived from the calibration (see section 14.2.1).

Practically, evaluation of static and dynamic PSS–PDS mismatches and corresponding non-linearity parameters could be undertaken at three different levels as follows.

- An individual risk level: very detailed risk-specific and interaction-specific evaluation.
- A discipline level: risks pertaining to particular disciplines presumed having the same non-linearity parameters.
- A project level: an entire project is characterized by a generic non-linearity parameter.

As a matter of fact, the last project-level type evaluation is already used for calibration purposes in section 14.2.1. However this type of assessment might not be specific enough when individual risk interactions are evaluated and taken into account. It is believed that the discipline-level assessment should suffice in most cases.

The second project Zemblanity N-SCRA business case discussed in Chapter 15 demonstrates a practical application of the calibration and the non-linear modelling methodology discussed in this chapter.

14.3 Risk Interaction Addressing

As pointed out in Chapter 2, two risk assessments should be done: before addressing and after addressing. The second assessment factors in proposed addressing actions. Regarding the traditional PRM approach, risk addressing has two implications and it is aimed at reduction of either probability of risk occurrence (preventive actions) or risk impacts (recovery actions). As risk identification in a non-linear case also presumes identification of project system issues and risk interactions, this provides another dimension and possibility of risk addressing.

First of all, the traditional addressing of probabilities and impacts related to standalone risks should be kept in the non-linear case too. However, those causes that are qualified as project system issues require additional attention. Second, in addition, identified system issues and risk interactions should be addressed too. This could be done keeping in mind the four sorts of chronic and induced project system issues discussed in section 14.1. When intra-risk and cross-risk interaction become a part of the picture, corresponding bowtie diagrams (see Figures 9.2–9.4) should be used for risk interaction addressing. The traditionally considered dynamic risk-induced project system issues can be addressed using bowtie diagrams 2.2 and 9.1 as usual.

Apparently, in case of internal risk amplifications addressing actions should be aimed at increase of project system maturity by suppressing chronic system issues. Improvement of project development procedures and their implementation should be in the focus of this. As a rule, this is a long-term process that might not offer too many low hanging fruits. Moreover, it might be completed only in the next projects through implementation of valuable learned lessons.

In case of knock-on risk interactions, a key risk interaction addressing activity should be focused on those project procedures and their implementation that handle quality assurance and quality control (QA/QC). According to system dynamics rework cycle (see Chapter 7), a top QA/QC priority is to ensure that instances and implications of low work quality in particular phases of project development do not enter consecutive phases.

Concerning compounding, a key risk interaction addressing measure would be planning of project activities in a way that exclude simultaneous risks happening: where possible windows of risk occurrences should not overlap. High degree of flexibility of planning when simultaneous risks do happen should be exercised (see Figure 9.1).

Addressing of risk-induced PSS–PDS mismatches stemming from occurred standalone risks can be taken into account in traditional linear Monte Carlo modelling (see Chapter 5) although keeping in mind the amended bowtie diagram 9.1.

As discussed in Chapter 13, all assessments for risk interactions should be done for two cases: before and after addressing of risk interactions. Proposed addressing actions should tell apart corresponding non-linearity parameters and multipliers for these two cases. They should be included to the modified (non-linear) risk registers along with regular risk addressing actions (see Chapter 5).

In simple terms, addressing of risk interactions in a complex system zeroes in on a challenge to deprive a complex system of its traits at least as related to non-linear risk interactions. This has two recipes. First of all, it is aimed at destroying or disarraying the dynamic risk pattern it supports. This may downgrade a complex project system to complicated or even simple which still would be a non-linear system due to intra-risk interactions. Second, a more challenging task would be the eradication of chronic system issues. If this is also successfully done, the project system becomes not only simple but also the linear one. This is a case that is perfectly handled by the traditional Monte Carlo L-SCRA methodology.

14.4 Definition of a Generic L-SCRA Case

Linear and non-linear Monte Carlo models have been told apart on several occasions in this book. A legitimate question arises on how linear models should be defined in the context of the development of the non-linear models. What factors should be included in L-SCRA and N-SCRA models? The answer has two parts.

First, as it traditionally done, all risks should be identified as standalone in linear models. Dynamic patterns of risks should be ignored: all cross-risk interactions should be excluded except simple correlations.

Second, chronic systematic risk causes should not be distinguished being treated as regular risk causes that do not depend on project system maturity at all. This means that all risks should be treated as non-amplified. However, excluding internal risk amplifications consistently would be a much more difficult task than excluding cross-risk interactions. This actually points to a need to tell apart project system issues as systematic risk causes and all other regular causes of project risks that play roles in bowtie diagrams.

It looks like such delineation is easier to do before the end of FEL3 (prior to FID) when quantifying risks that could occur in execute. EPC contractors have yet to be selected and their characteristics stay unknown. Some averaged decent or rather optimistic characteristics of expected EPC contractors are normally implied in risk quantification prior to FID. However, if, for instance, a company has a proclivity to select the lowest bidders, this factor is usually ignored in an overoptimistic hope that a lowest bidder should miraculously bring forward somewhat decent and acceptable capabilities and skills. Similarly, chronic project system issues in FEL phases are easily ignored too due to presence of some types of organizational bias. All relevant project development procedures and practices are considered decently developed and well-implemented on average. As ignoring or unnoticing of the project system issues as a special type of systematic risk causes actually characterizes the traditional inputs to the linear Monte Carlo models in most cases, this is the second part of the answer.

Factually, ignoring or unnoticing of chronic system issues is a way to pretend that a selected risk quantification method has a low level of systematic errors (Figure 3.4) being highly accurate or, in other words, that the project system is highly mature. That is exactly the case of the traditional Monte Carlo modelling represented by the model L-SCRA in Chapter 5: all intra-risk and cross-risk interactions are written off.

However, most advanced risk analysts do identify project system issues when running traditional linear Monte Carlo models. They do include them in traditional bowtie diagrams (see Figure 2.2). However, these system-related causes make contributions only to probabilities of risk occurrences with impacts being unaffected. Although, this contribution cannot go beyond increasing of the risk probability up to $P = 1$, which is the natural ceiling in the linear case. No influence on risk impact is considered as a bowtie diagram of Figure 9.2 for internal risk amplification advices. So, all non-linear intra-risk multipliers $1 + \beta_{jj}^{int} \times R_j$ (see expressions (13.6) and (13.7)) are equal to one simply because of all internal amplification non-linearity parameters $\beta_{jj}^{int} = 0$ in the linear case. That is exactly what is required to develop the linear L-SCRA model case that would be used as a starting point in developing the non-linear N-SCRA model. Concerning cross-risk interactions, there is simply no place for them in the linear paradigm. This strongly resembles how parametric modelling relates to ANN modelling when expressions (4.1) and (4.2) are compared.

Chapter 15 is devoted to development of a Monte Carlo model that does take into account chronic project system issues ($\beta_{jj}^{\text{int}} \neq 0$) and cross-risk interactions ($\beta_{ij}^{knock} \neq 0, \beta_{ij}^{comp} \neq 0$) in the way discussed in Chapter 13. The linear L-SCRA model of Chapter 5 should be still utilized as it intakes modified risk consequences developed to reflect on the risk interactions in the complex project system.

14.5 Chapter's Markup and Takeaway

- Individual instances of intra-risk and inter-risk interactions can be introduced through corresponding non-linearity parameters and non-linear multipliers. The next steps are to visualize dynamic patterns of interacting risks. This is done by using risk interaction mapping.
- Schedule risk interaction mapping is based on regular project schedule logic with risks mapped to impacted tasks. Although additional routes for propagation of knock-on interactions and time windows for risk compounding should be identified.
- Cost risk interaction mapping is a simplified version of the schedule risk interaction mapping that does not contain schedule logic although it still shows knock-on and compounding interactions.
- Intra-risk interactions are not shown by risk interaction mapping although all risks are considered internally amplified where applicable.
- All individual instances of risk interactions should be represented in a modified (non-linear) project risk register that contains all required non-linearity parameters and non-linear multipliers.
- The concept of risk interaction mapping points to a need to expand traditional risk identification and addressing approaches to identification and addressing of risk interactions that should also be shown in the modified (non-linear) project risk register. Three bowtie diagrams for corresponding risk interactions should be kept in mind for these activities.
- To identify internal risk amplifications chronic project system issues that define project system maturity should be told apart from regular causes of project risks.
- To identify knock-on interactions all possible propagation paths of poor-quality impacts on down-the-line risks should be investigated.
- To identify risk compounding, all possible overlapping windows of risk occurrences should be pinned down.
- Apparently, addressing of the intra-risk interactions requires suppressing identified chronic project system issues to ensure the higher maturity of a project system.

- Addressing of knock-on interactions requires building of QA/QC barriers that not only suppress instances of poor quality of works but also prevent propagation of poor-quality impacts to down-the-line risks.
- Addressing of risk compounding calls for flexible project planning that prevents possible overlapping of risk window's occurrences where possible.
- When the modified project risk register is developed that contains all required non-linearity parameters, non-linear multipliers, and addressing actions it can be used for evaluations of risk consequences before and after addressing. The modified risk consequences that take into account all relevant risk interactions can be calculated and put to the existing linear L-SCRA model which becomes a non-linear N-SCRA model.
- Several runs of the newly built N-SCRA model could be carried out so that all non-linearity parameters are set up the same in each particular run, although they should be different in undertaken runs. This corresponds to rough evaluation of all risk interactions combined at the project level. This calibration helps establish thresholds that divide situations when risk interactions are weak, medium, and strong.
- Two calibration thresholds are established to tell apart 10 per cent and 25 per cent schedule slippage and overspending above a linear L-SCRA case.
- As a next step, more detailed assessments of non-linearity parameters are undertaken at either risk or discipline levels. Each individual risk-level or discipline-level assessment could fall in one of the categories (weak, medium, or strong). But each non-linearity parameter belongs to the continuum starting from zero (a linear case).
- Evaluation of non-linearity parameters is a subject to risk workshop discussions that should take into account proposed risk interaction's addressing measures and keep in mind the threshold criteria.
- Agreed non-linearity parameters are used to calculate modified risk consequences that are utilized as inputs to initially develope linear L-SCRA models resulting in non-linear N-SCRA models.
- Apparently, the linear L-SCRA case that is used as a foundation for development of the non-linear N-SCRA case does not tell apart chronic project system issues as systematic causes of project risks and regular project risk casus. Also all cross-risk interactions are missed out.

PART THREE

RISK QUANTIFICATION IN COMPLEX PROJECTS

15
Project Zemblanity Business Case (II): Non-linear Monte Carlo Schedule and Cost Risk Analysis (N-SCRA)

Chapter 5 introduces the project Zemblanity business case for the traditional integrated schedule and cost risk analysis referred to as the L-SCRA model. The L-SCRA results can be somewhat adequate if investigated is a project system that does not have much static and dynamic PSS–PDS mismatches, underlaying static or dynamic project system issues, and corresponding risk interactions. However, if a level of maturity of a project system is low (Chapter 12) and there are significant interactions among risks (Chapter 13), it would be preposterous to expect a high level of accuracy from the L-SCRA modelling. Simply, a very major part of risk exposure related to intra-risk and cross-risk interactions would be overlooked. This corresponds to a high systematic error (see Figure 3.4).

This chapter puts forward a second project Zemblanity business case devoted to the new Monte Carlo methodology that does take into account intra-risk and cross-risk interactions in the project system. The second business case describes two imaginary risk quantification workshops. The first one results in a N-SCRA-A model devoted to initial assessment of intra-risk and cross-risk interactions without their addressing. The second one gives rise to a N-SCRA-B model that takes into account proposed addressing of intra-risk and cross-risk interactions. Both models are based on after-risk-addressing assessments of standalone risks developed in Chapter 5.

15.1 N-SCRA Method Statement

As results of the initial schedule and cost risk analysis undertaken by the project team (see Chapter 5) have been challenged by representatives of Serendipity United PLC, a new risk quantification approach has been developed and applied by the consortium's PMC contractor—Risk Services & Solutions Inc.

On the one hand, several inputs used to build a linear L-SCRA model of Chapter 5 can be used to develop non-linear N-SCRA models. On the other hand, a different treatment of project system issues and cross-risk interactions is required.

The N-SCRA inputs previously used in L-SCRA modelling are as follows:

Modern Risk Quantification in Complex Projects: Non-linear Monte Carlo and System Dynamics Methodologies. Yuri G. Raydugin, Oxford University Press (2020).
DOI: 10.1093/oso/9780198844334.001.0001

- the project schedule (Figure 5.1 and Table 5.2),
- the project base estimate (Table 5.1),
- the identified standalone risks, their addressing, and their after-addressing assessments (Tables 5.3 and A.4),
- the general schedule and cost uncertainties (Tables A.2 and A3), and
- the agreed burn rates (Table A.3).

Additional or amended inputs to the N-SCRA modelling have been related to treatment of intra-risk and cross-risk interactions—internal risk amplifications, knock-on interactions, and compounding. Corresponding preparations have been undertaken by the project risk manager prior to holding the N-SCRA workshops. The following additions and amendments have been made (see Chapters 9, 13, and 14).

- Modified bowtie diagrams for three types of risk interactions have been used to identify relevant risk interactions in the project Zemblanity system (see Figures 9.2–9.4).
- Risk interaction mapping visualization has been carried out to describe dynamic risk patterns in the project system (see Figures 14.1 and 14.2).
- Identified risk interactions have been incorporated into the initial linear risk register template (Table A.4) to produce modified schedule and cost risk register templates for the non-linear case (Tables D.2 and D.3). The modified risk register templates show the associated non-linearity parameters and non-linear multipliers that quantify internal risk amplifications and cross-risk interactions (Table D.1).
- The calibration of non-linearity parameters has been carried out based on criteria established by Table 14.1 to identify those values of non-linearity parameters that establish dividers between low, medium, and high schedule slippages and overspendings. The calibration results are represented by Table D.4.

When the preparatory steps listed above have been completed, a first N-SCRA-A modelling could be undertaken.

15.2 A First N-SCRA-A Workshop

A two-day N-SCRA workshop to identify and assess risk interactions has been held in an office of a British consortium partner in Henley-on-Thames, Oxfordshire, UK. The same participants have attended this workshop as those who attended the initial workshop to develop the L-SCRA case (see Chapter 5). Key members of the integrated project team, delegates from all consortium partners (project owners), and representatives of the PMC consultancy Risk Services & Solutions Inc. have been in attendance.

Previously collected risk data as well as information about project schedule and cost baselines have been reviewed and validated by participants. After a briefing on non-linear Monte Carlo modelling, risk interactions, and project system dynamics participants have been involved in identification of project system issues and cross-risk interactions using bowtie diagrams for the three types of risk interactions (Figures 9.2–9.4) and assessments of interactions (Table D.1 of Appendix D) included to the modified non-linear risk register (Tables D.2 and D.3).

Several major project system issues have been identified as follows:

- stakeholder management and change management procedures are not implemented consistently,
- quality assurance and quality control (QA/QC) procedures in engineering and construction are not consistently implemented to reflect on features of the project location (Arctic area),
- presence of logistics difficulties related to the project location (Arctic area) are pointed out,
- commissioning and startup (CSU) procedures are not developed consistently to accommodate difficulties associated with the project location (Arctic area).

These project system issues are identified as chronic ones keeping in mind the bowtie diagram of Figure 9.2. They have been added to the initially developed risk register of the L-SCRA case (Table 5.3) resulting in Table 15.1.

Several cross-risk interactions have been discussed keeping in mind the knock-on and compounding bowtie diagrams (see Figures 9.3 and 9.4) and the preliminary developed risk interactions maps (14.1 and 14.2) as follows:

- risk related to stakeholder and change management may result in knock-on effects on engineering risks,
- risk related to weak QA/QC engineering procedures may result in knock-on effects on construction risks,
- risk related to weak QA/QC construction procedures may result in knock-on effects on commissioning and startup risks, and
- project planning and control procedures are developed and implemented in a way that allows compounding of several project risks due to overlapping of risk occurrence windows.

The pointed out cross-risk interactions have been included to the modified project risk register templates (Tables D.2 and D.3) in accordance with Table D.1. The added project system issues have been also reflected in the modified project register templates (Tables D.2 and D.3) to count internal risk amplifications (see Appendix D for details).

Tables D.5 and D.6 incorporate all above-discussed risk interaction identification results and contain a snapshot of assessed non-linearity parameters. These represent current project system issues and cross-risk interactions before any attempts to address them are undertaken. The results collected in Table D.5 and D.6 have been reviewed, verified, and endorsed by the workshop participants.

Table 15.1 Project Zemblanity risks including project system issues

Risk ID	Risk Title (Tag)	Cause(s)	Risk event	Impact(s)
R01	Design Basis Change (Proponent-Driven)	A. On-going reservoir simulations B. Changing stakeholders' requirements C. Overlooked components of scope D. Stakeholder management and change management procedures are not implemented consistently	Design basis may be changing by request of the proponent (after FEL2)	A. Rework and extra cost (engineering) B. Rework and schedule delay (engineering) C. Quality (engineering)
R02	Overlooked External Stakeholder's Requirements	A. Lack of communication with external stakeholders B. Recommendations of environmental impact assessment (EIA) and hazard and operability (HAZOP) studies C. Stakeholder management and change management procedures are not implemented consistently	Overlooking of major requirements of external stakeholders (government, media, etc.) and opposition from their side could take place	A. Rework and extra cost (engineering) B. Rework and schedule delay (engineering) C. Quality (engineering)
R03	Errors and Omissions (Engineering)	A. Interface management failure (among engineering disciplines) B. Inconsistent implementation of QA/QC procedures in engineering and failure to reflect on features of the project location (Arctic area).	Errors and omissions in engineering could take place	A. Rework and extra cost (engineering) B. Rework and schedule delay (engineering) C. Quality (engineering)
R04	Poor Work Quality (Construction)	A. Interface management failure B. Contractor's coordination difficulties C. Inconsistent implementation of QA/QC procedures in construction and failure to reflect on features of the project location (Arctic area).	Poor quality of construction works could take place	A. Rework and extra cost (construction) B. Rework and schedule delay (construction) C. Quality (construction)
R05	Delays in Material Delivery	A. Several delivery items on project's critical path B. Limited capacity of qualified vendors and competing projects C. Logistics difficulties related to the project location (Arctic area)	Some critical items might not be delivered to the construction site on time	A. Schedule delay (construction)
R06	Commissioning and Startup (CSU) Failure	A. Poor planning and unavailability of required CSU skilled labour B. Poor planning and unavailability of required spare parts C. Failed CSU equipment tests D. Commissioning and startup (CSU) procedures are not developed and implemented consistently to accommodate difficulties associated with the project location (Artic area)	Timely CSU might not be possible	A. Rework and extra cost (CSU) B. Rework and schedule delay (CSU) C. Quality (operations)

15.3 N-SCRA-A Modelling results

The endorsed information collected in Tables D.5 and D.6 has been used for development of modified inputs to the existing L-SCRA model after the workshop. As discussed in Appendix D, the developed modified impacts for the model N-SCRA-A are collected in Table D.9. They are to replace schedule and cost impacts of the linear

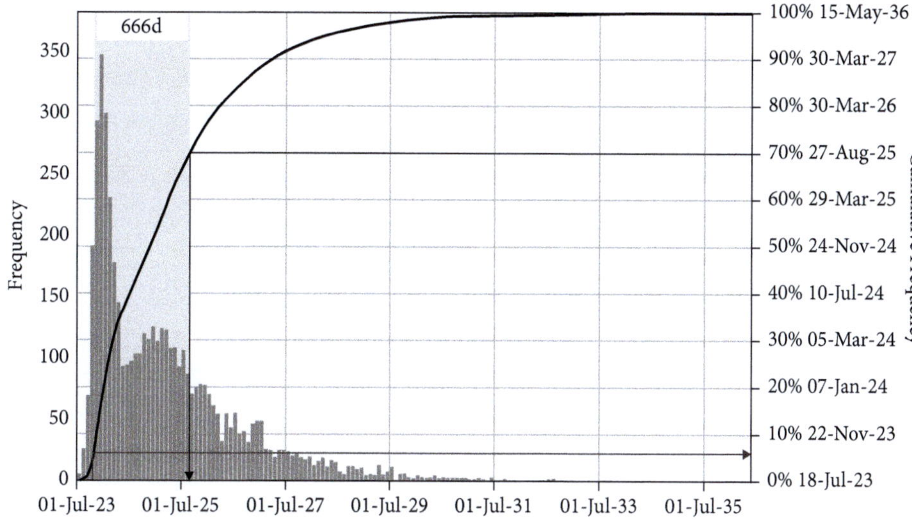

Figure 15.1 A distribution of project Zemblanity project completion dates (a non-linear N-SCRA-A case)

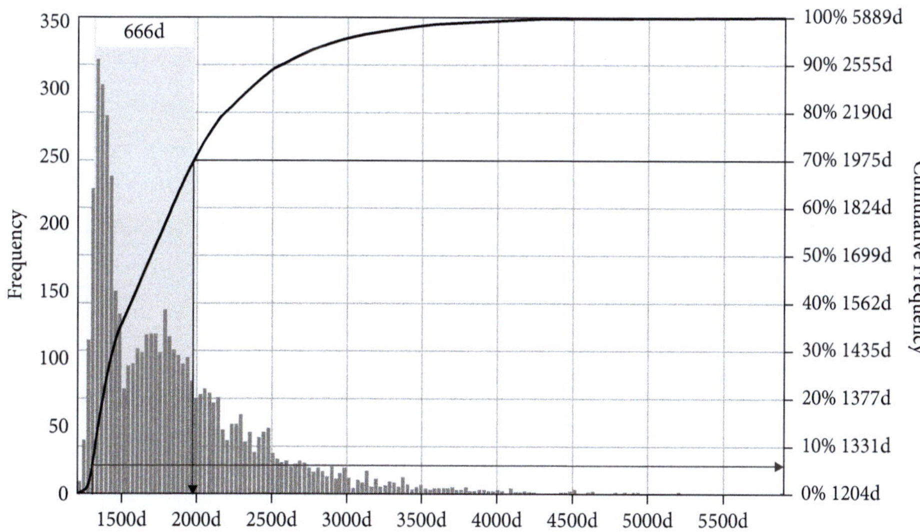

Figure 15.2 A distribution of project Zemblanity project durations (a non-linear N-SCRA-A case)

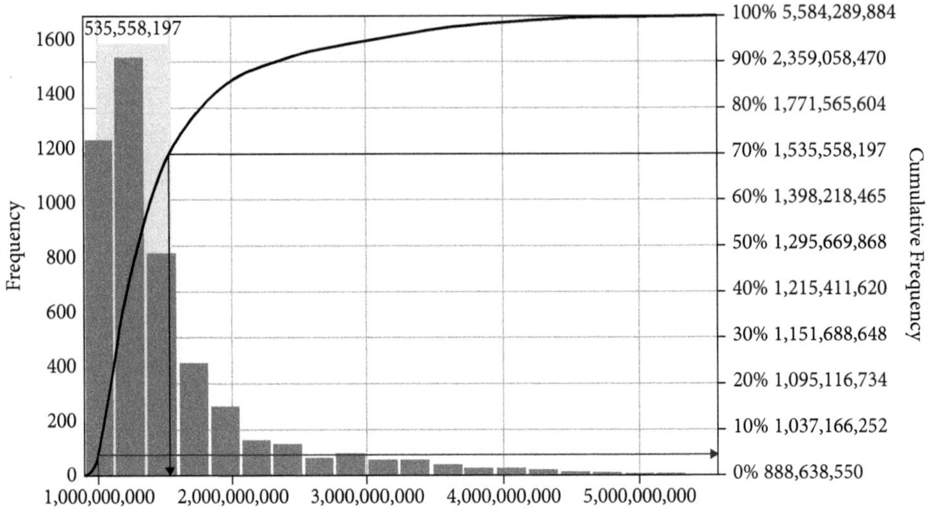

Figure 15.3 Project Zemblanity cost distribution including schedule-driven costs and cost general uncertainties (a non-linear N-SCRA-A case)

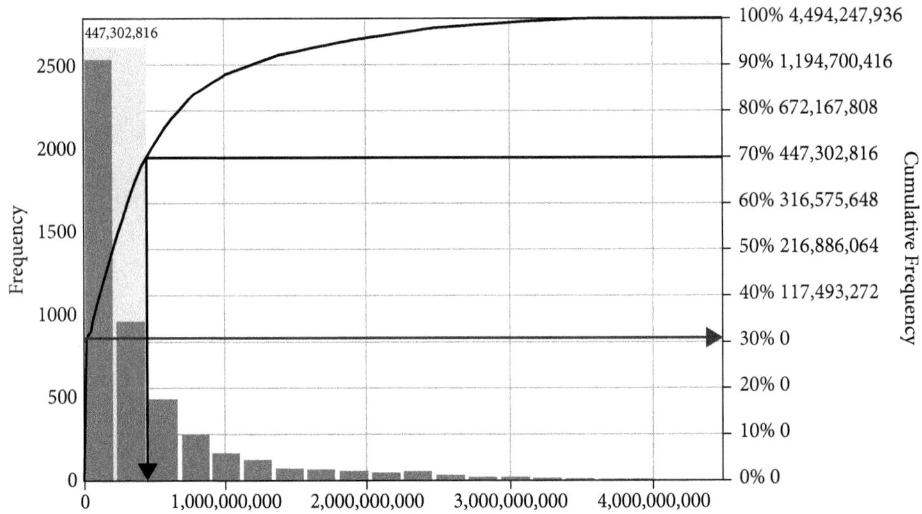

Figure 15.4 Project Zemblanity cost distribution for schedule-driven costs (a non-linear N-SCRA-A case)

L-SCRA risk register (Table A.5) in the L-SCRA model. This yields a non-linear N-SCRA-A model.[1]

Upon building, debugging, and running of the N-SCRA-A model results have been produced that are demonstrated by Figures 15.1–15.7.

[1] Non-linearity parameters should modify risk consequences $P \times I$, although Table D.9 still contains unmodified probabilities. As discussed in Appendix D, when both the unmodified probabilities and the modified impacts are put in Monte Carlo models, the models do handle products $P \times I$ that are properly prepared.

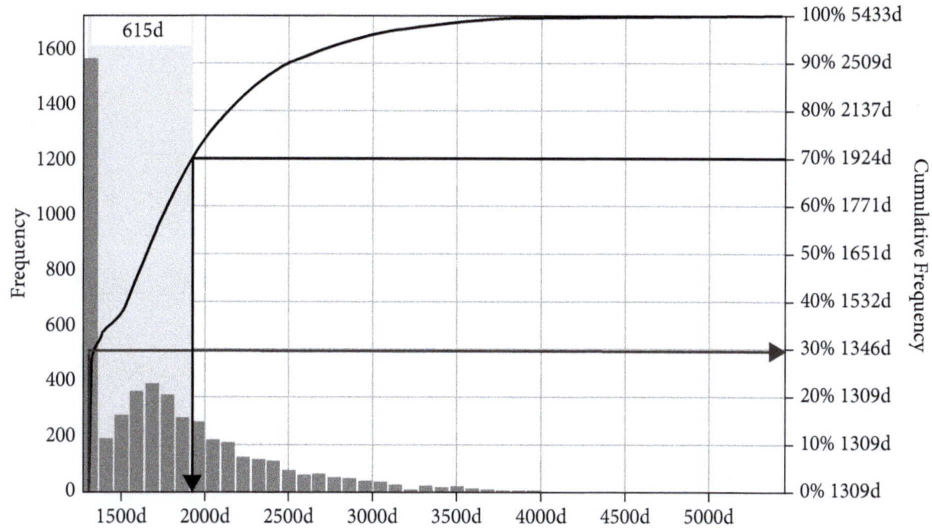

Figure 15.5 A project duration distribution when only risks are included (a non-linear N-SCRA-A case)

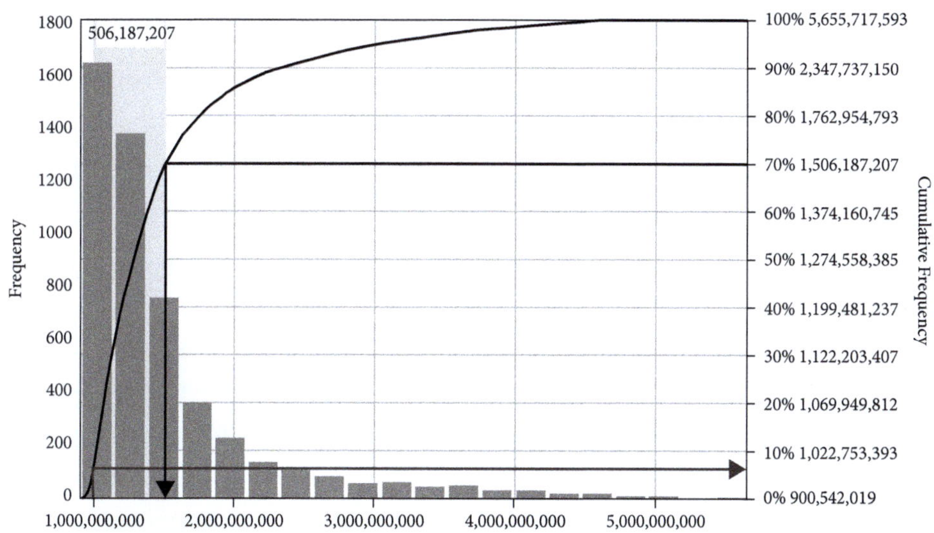

Figure 15.6 A project cost distribution when only risks are included (with schedule-driven costs) (a non-linear N-SCRA-A case)

In case of the model N-SCRA-A, the P70 duration (1975 days or 65.8 months, see Figure 15.2) and cost (1.54B, see Figure 15.3) correspond to about JCL55 (see Figure 15.7). To reach JCL70, the project completion should occur around 10 February 2026 being a durationof about 2100 days or 70 months. The JCL70 cost should be about 1.77B. These values correspond to about P80 duration and cost confidence level (see Figure 15.2 and 15.3). The JCL correction—the difference between JCL70 and P70 values—should be about 9.6 per cent of base schedule duration (1,309 days) and

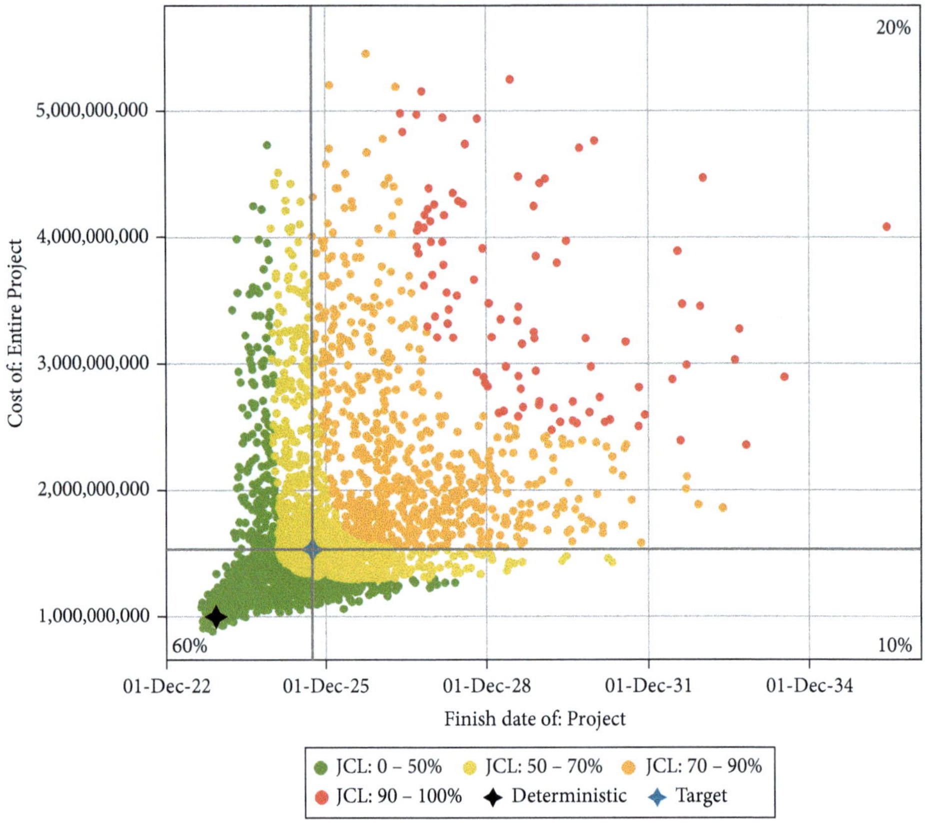

Figure 15.7 Joint confidence level (JCL) for project Zemblanity (a non-linear N-SCRA-A case): the crosshair shows P70 completion date (27 August 2025) and P70 cost ($1.54B)

23.4 per cent of base estimate ($1B). The P70 contribution of schedule-driven costs in case of the model N-SCRA-A is $447M or 44.7 per cent of base estimate (see Figure 15.4) which is almost twice as high as the cost JCL correction.

It should be pointed out that N-SCRA-A distributions for duration and cost general uncertainties are identical to ones for the model L-SCRA (see Figures 5.6 and 5.7) and are not shown in this chapter. Any realizations of intra-risk and cross-risk interactions are taken into account only though modification of risk events.

15.4 Discussion of N-SCRA-A Results

An overview of N-SCRA-A modelling results has been held by the executive team of Serendipity United PLC in a regional office of a Russian consortium partner in Ekaterinburg, Russia, two weeks after the first N-SCRA workshop held in Henley-on-Thames.

First, the participants have reviewed and validated the proposed non-linear Monte Carlo methodology and pointed out that all concerns related to the shortcomings of the traditional Monte Carlo methodology used previously (see Chapter 5) have been addressed through factoring in internal risk amplifications, knock-on interactions, and risk compounding.

Second, a serious concern has been expressed by all participants about anticipated P70 project Zemblanity duration and cost. The discussed P70 duration of 1975 days or 65.8 months (Figure 15.2) and cost of about $1.54B (Figure 15.3) have been considered by some consortium members a complete disaster and a show stopper for their participation in the project. As one of participants has pointed out, besides capital overspending of 54 per cent above the project case estimate the expected 51 per cent project delay could cause irreparable damage for the operational cash flow. It has been also pointed out that newly identified project system intra-risk and cross-risk interactions have resulted in a drastic increase of project duration contingency by 33 per cent and cost contingency by 34 per cent above the L-SCRA case.

If the JCL concept is applied to review the obtained N-SCRA-A results the expected schedule and cost outcomes could be even worse and completely appalling—project duration 70 months and project cost $1.77B at JCL70.

Third, to resolve the crisis, the participants have come up with measures to address identified static and dynamic PSS–PDS mismatches. All risk interactions qualified as high (see Tables D.5 and D.6) have been in the primary focus of the discussion. These measures have included the prompt and the consistent implementation of world-class:

- scope development and change management procedures that are accepted and followed by all consortium members without any exceptions,
- stakeholder management procedures to identify and manage all types of external stakeholders that is accepted and followed by all consortium members without any exceptions,
- commissioning and startup procedures to have adequate time and resources to become prepared for all aspects of commissioning and startup in the Arctic area,
- QA/QC procedures that take into account the project location and among other things prevents any impact of poor engineering quality on construction,
- project control and risk management procedures that allow flexible planning and adequate risk addressing including prevention of risk compounding.

It has been pointed out that implementation of the above-mentioned procedures should require a full commitment of all consortium members and an adequate manning of all main project disciplines.

The recommendation on implementation of the quality assurance and control procedure has a special provision that a final engineering model review prior to issuing of drawings should be undertaken by an independent third-party organization experienced in projects of this kind in the Arctic area. It has been recommended to avoid any time pressure on reviewers and allow an unrestrained amount of time for this. This should be a major action to avoid knock-on interactions of quality risks.

Concerning the last point on the recommendations list, the risk facilitator representing Risk Services & Solutions Inc. has pointed out that 'if a megaproject is considered a complex system it should be an adaptive system to be successful'.[2] This means a higher level of flexibility at work package level and below through adoptive short-term planning in both engineering and especially in construction. A close control (aka a dictate) of corporate planning and scheduling over execution of work should be refrained from. Corporate planning should be limited to package level (schedule level III), whereas a detailed rolling three-week plan of works should be developed by a local scheduler, site manager, and foremen. The purpose of the short-term planning is to avoid any possible work constraints including ones stemming from occurred project risks or their interactions. An advanced work packaging (AWP) methodology developed by Teknobuild of Calgary, Alberta, Canada (Teknobuild (2019)) has been recommended to manage work availability tactically.

The short-term planning should allow the avoidance of the majority of tactical work constraints unless they are driven by major risk events (see Figure 9.1). This should reduce inefficiencies and internal risk amplifications. Any possibility of risk compounding should be prevented through flexible planning too. Any potential possibilities of risk occurrence's overlapping should be monitored and avoided, where possible.

If work constraints are driven by major risk events a discussed level of flexibility should reduce time required to absorb impacts on productivity through redeployment of staff and changing work sequencing.

The proposed project control de-centralization has resulted in keen debates as corporate project development procedures of some of the consortium participants have insisted on high-level centralization of planning supported by development of very detailed schedules in execution. According to these procedures project teams should follow extremely detailed schedules developed a few months if not years beforehand. The proposed short-term flexible planning has not been heard of by the majority of participants, although a 40-year project management veteran attending the meeting has mentioned that the proposed short-term planning was quite usual in the beginning of his career.

Finally, all consortium participants agreed with a need to adopt the flexible AWP methodology in project Zemblanity.

A legally binding memorandum has been endorsed by the consortium members to implement agreed measures.

Upon formalizations of the agreed consortium partner's obligations, the PMC contractor has been authorized to rerun the SCRA workshop in order to develop updated schedule and cost contingencies.

15.4.1 A Second N-SCRA-B Workshop

A second one-day N-SCRA-B workshop has been held in an office of the PMC contractor Risk Services & Solutions Inc. in Calgary, Alberta, Canada, to update risk

[2] This could be a connotation of a phrase written by Niccolò Machiavelli in 1513 (Machiavelli (1996)): 'For a man who is accustomed to proceed in one mode never changes, as was said; and it must be of necessity that when the times change not in conformity with his mode, he is ruined.'

information regarding addressed intra-risk and cross-risk interactions. The purpose of the workshop has been established as a conversion of the agreed consortium member's obligations into modified inputs to the N-SCRA model. This post-addressing model has been referred to as the N-SCRA-B Monte Carlo model of project Zemblanity.

The same project schedule, base estimate, and the risk register of standalone project risks that have been used in the models L-SCRA and N-SCRA-A have been kept as key inputs to the N-SCRA-B model. The only amendments have been done to non-linearity parameters and non-linear multipliers.

The participants have reviewed the agreed measures, re-evaluated intra-risk and cross-risk interactions qualitatively, and assigned new non-linearity parameters to reflect on the agreed measures (Tables D.7 and D.8 of Appendix D). In a similar way as in the case of the model N-SCRA-A (sections 15.1 and 15.2), modified impacts for the model N-SCRA-B are collected in Table D.10. They are replacing schedule and cost impacts of the linear L-SCRA risk register (Table A.5) or the first non-linear N-SCRA-A risk register (Table D.9). After replacement of the inputs a non-linear N-SCRA-B model has been derived.

15.5 N-SCRA-B Modelling Results

The N-SCRA-B model that comprises the agreed addressing actions and actions to alleviate risk interactions produces results discussed in this section (Figures 15.8–15.14).

For the model N-SCRA-B, the P70 duration (1650 days or 55 months, see Figure 15.9) and cost ($1.31B, see Figure 15.10) correspond to about JCL58 (see Figure 15.14). To

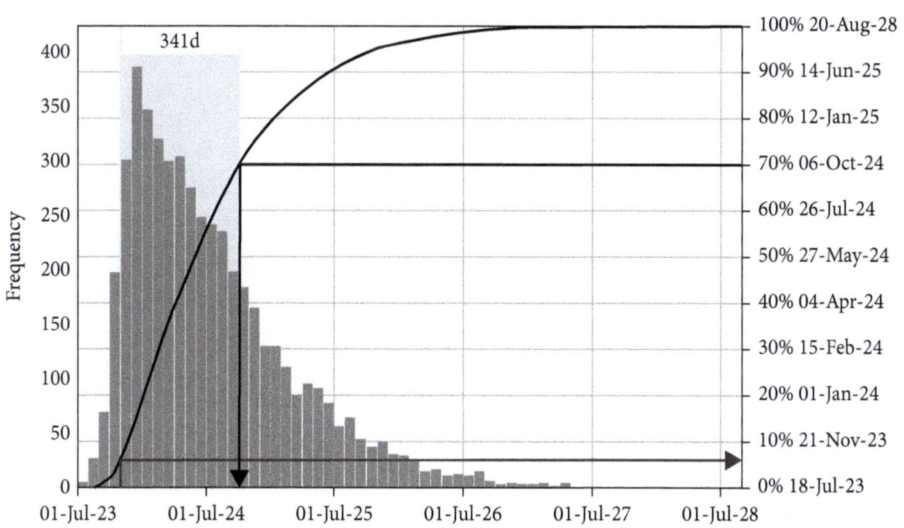

Figure 15.8 A distribution of project Zemblanity project completion dates (a non-linear N-SCRA-B case)

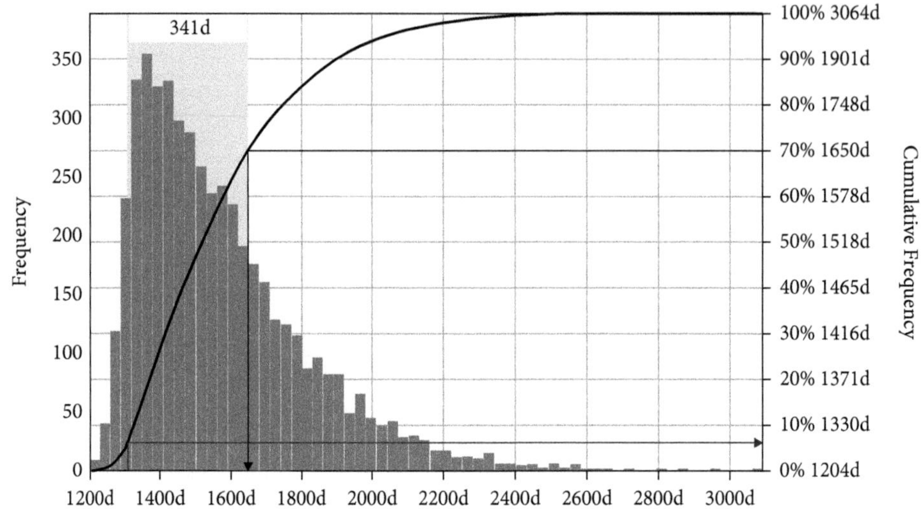

Figure 15.9 A distribution of project Zemblanity project durations (a non-linear N-SCRA-B case)

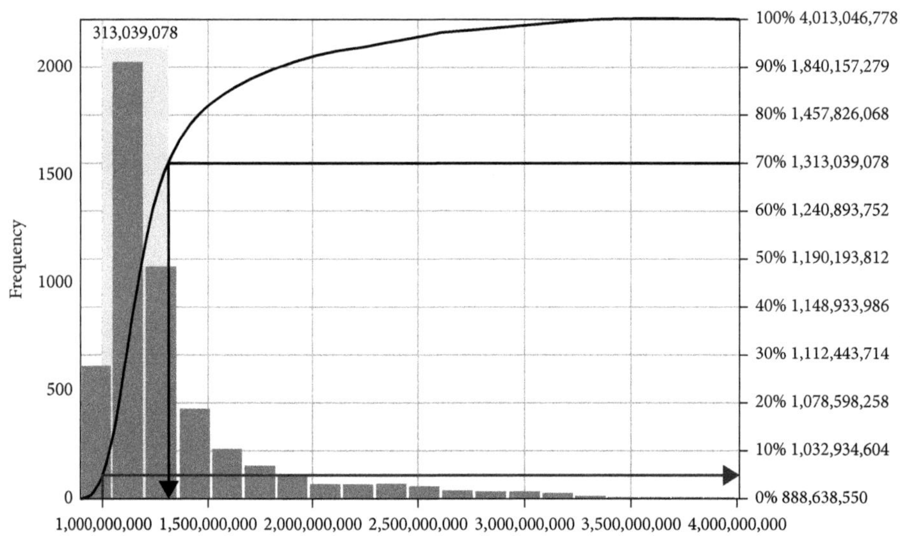

Figure 15.10 Project Zemblanity cost distribution including schedule-driven costs and cost general uncertainties (a non-linear N-SCRA-B case)

reach JCL70, the project completion should occur around 15 February 2025 having a duration of about 1760 days or 58.7 months. The JCL70 cost should be about 1.48B (see Figure 15.14). These values correspond to about P80 duration and cost confidence level (see Figures 15.9 and 15.10). The JCL correction factor—the difference between JCL70 and P70 values—should be 8.4 per cent of the base schedule duration (1309 days) and 16.7 per cent of the base estimate ($1B). The P70 contribution of

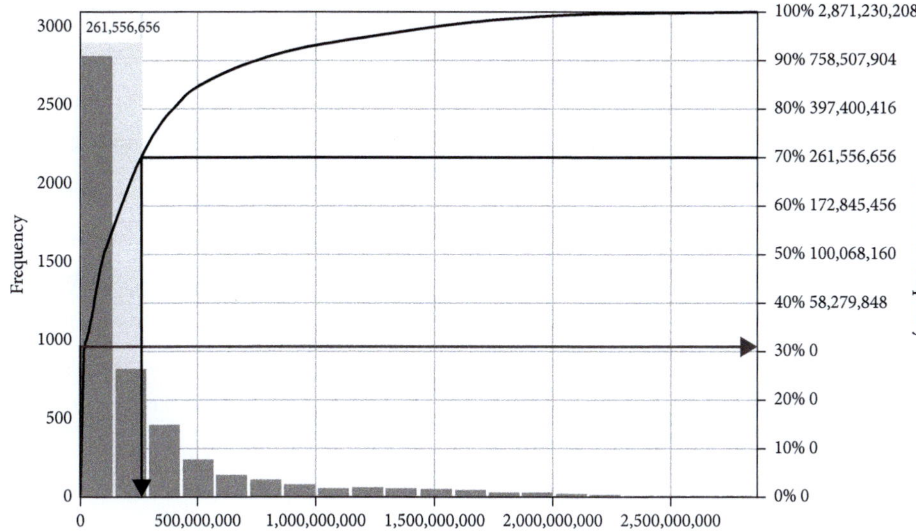

Figure 15.11 Project Zemblanity cost distribution for schedule-driven costs (a non-linear N-SCRA-B case)

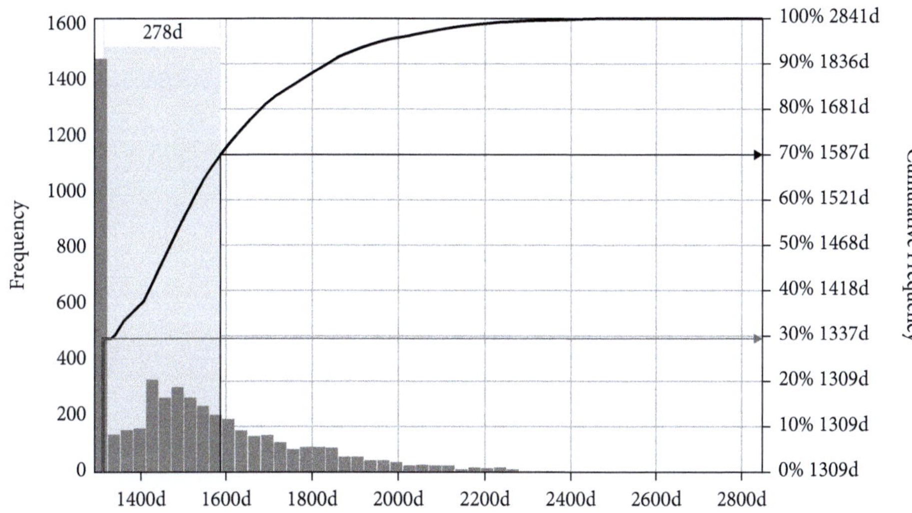

Figure 15.12 A project duration distribution when only risks are included (a non-linear N-SCRA-B case)

schedule-driven costs in case of the model N-SCRA-B is $262M or 26.2 per cent of base estimate (see Figure 15.4) which is 9.5 per cent higher than the JCL cost correction factor of 16.7 per cent.

The same as in the case of the model N-SCRA-A, the N-SCRA-B distributions for duration and cost general uncertainties are identical to ones for the model L-SCRA (see Figures 5.6 and 5.7) and are not shown in this chapter.

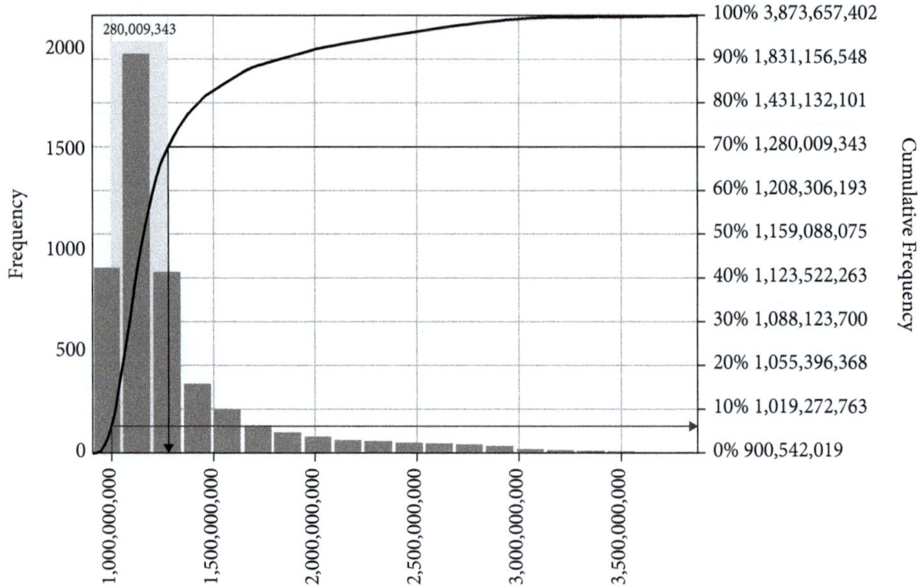

Figure 15.13 A project cost distribution when only risks are included (with schedule-driven costs) (a non-linear N-SCRA-B case)

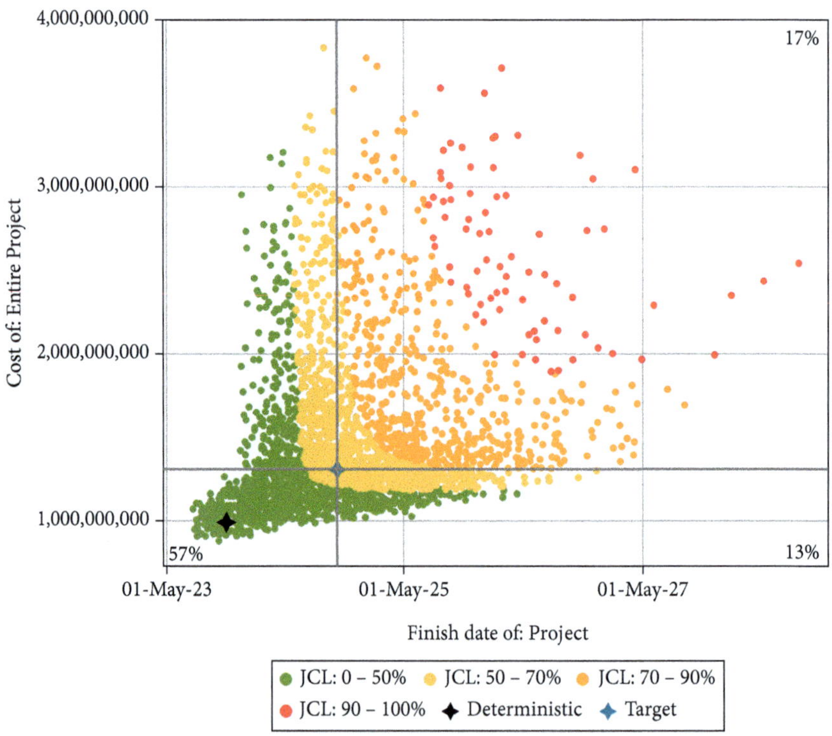

Figure 15.14 Joint confidence level (JCL) for project Zemblanity (a non-linear N-SCRA-B case): the crosshair shows P70 completion date (6 October 2024) and P70 cost ($1.31B)

15.6 Discussion of N-SCRA-B Results

As expected, the results for the N-SCRA-B model belong in-between the results for the L-SCRA and N-SCRA-A models. This is directly explained by substantial although not complete addressing of static and dynamic PSS–PDS mismatches pertaining to the model N-SCRA-A.

According to Figures 15.8–15.14 for the N-SCRA-B model, it would required 1650 days or 55.0 months to complete the project with the confidence level P70 on 6 October 2024 with expected total cost of $1.31B. This is only 11.1 per cent above the duration and 14.4 per cent above the cost of the L-SCRA model at the confidence level P70 (see Table 15.2), whereas the overall P70 schedule contingency above the base schedule is 26.1 per cent and cost contingency above the base estimate is 31.3 per cent. The differences in P70 schedule and cost contingencies with the L-SCRA models (11.1 per cent and 14.4 per cent, correspondingly) can be attributed to remaining intra-risk and cross-risk interactions even after the undertaken addressing.

15.7 Comparison of Results for the Models L-SCRA, N-SCRA-A, and N-SCRA-B

15.7.1 Comparison of General Results

Table 15.2 summarizes general results of Monte Carlo modelling for the three models L-SCRA, N-SCRA-A, and N-SCRA-B developed and discussed in sections 5.4, 5.5, 15.3, 15.4, 15.6, and 15.7. Table 15.2 clearly positions the model N-SCRA-B in-between the other two models in terms of required schedule and cost contingencies.

15.7.2 Comparison of Schedule-driven Costs

Figures 5.5, 15.4, and 15.11 compare schedule-driven costs for the three models. In the linear L-SCRA case schedule-driven costs may be evaluated as $37M or 3.7 per

Table 15.2 Comparison of P70 results for the models L-SCRA, N-SCRA-A, and N-SCRA-B

Model	Completion date	Duration, d	Duration above baseline, %	Duration above L-SCRA, %	Cost, 'B$	Cost above baseline, %	Cost above L-SCRA, %
Deterministic baseline	31-Oct-23	1309	0.0%	−11.9%	1.000	0.0%	−12.9%
L-SCRA, P70	25-Apr-24	1486	13.5%	0.0%	1.148	14.8%	0.0%
N-SCRA-A, P70	27-Aug-25	1975	50.9%	32.9%	1.536	53.6%	33.7%
N-SCRA-B, P70	6-Oct-24	1650	26.1%	11.0%	1.313	31.3%	14.4%

Table 15.3 Comparison of schedule-driven costs for the models L-SCRA, N-SCRA-A, and N-SCRA-B

Model	Schedule-driven cost, '$M	% of base estimate
Deterministic baseline	0	0
L-SCRA, P70	37	3.7%
N-SCRA-A, P70	447	44.7%
N-SCRA-B, P70	262	26.2%

cent of deterministic base estimate at P70 confidence level (Figure 5.5). The non-linear case N-SCRA-A yields astonishing $447M schedule-driven costs (44.7 per cent of the base estimate). A N-SCRA-B case still results in substantial $262M schedule-driven costs (26.2 per cent of the base estimate). The comparison of schedule-driven costs with overall project costs at P70 confidence level demonstrates that they are the very top contributors to overall project costs when substantial intra-risk and cross-risk interactions are taken into account (Table 15.3). Indeed, in the case of the model N-SCRA-A the overall project contingency is 53.6 per cent, whereas schedule-driven cost's contribution is assessed as 44.7 per cent. For the model N-SCRA-B with the overall cost contingency of 31.3 per cent, the schedule-driven cost's contribution is assessed as 26.2 per cent.[3] Needless to say, disproportionally high contributions of the schedule-driven costs include multiple non-linear factors such as internal risk amplifications of all risks, their interactions, and delivery of the resulting schedule risk impacts to project costs through burn rates.

This observation should be another reason to discourage any intent to conduct standalone cost risk analyses (CRA): due to missed schedule-driven costs any CRA model would have very high systematic errors (see Figure 3.4).

15.7.3 Comparison of JCL Correction Factors

A remarkable JCL trend is observed for models L-SCRA, N-SCRA-A, and N-SCRA-B (Figures 5.10, 15.7, and 15.14). The JCL charts for the models N-SCRA-A and N-SCRA-B do not resemble American footballs any longer. They rather take a shape of a funnel or a plume. This could be explained by a role of non-linear risk interactions in general and their impact on schedule-driven costs in particular. Namely, smaller schedule delays lead to smaller schedule-driven costs. These iterations correspond to modest realizations of non-linearity in the models and they are collected in the lower left corners of the charts around the base schedule and base estimate values (1309 days and $1B, correspondingly). It could be imagined that these values still represent the rudimentary American-football distribution of Figure 5.10. However, when iterations result in large schedule delays, they should inevitably inject high schedule-driven costs along

[3] Contributions of various risk factors are additive only if mean-value confidence-level criteria are adopted. If the criteria significantly differ from the mean-value ones (as P70), the contributions cannot be simply added up.

Table 15.4 Comparison of JCL correction factors for the models L-SCRA, N-SCRA-A, and N-SCRA-B.

Model	Completion date and duration, d	Cost, 'B$	Actual JCL	Date and duration, d, required to reach JCL70	Duration JCL correction factor	Cost required to reach JCL70, 'B$	Cost JCL correction factor
Deterministic baseline	31-Oct-23 (1309)	1.00	n/a	n/a	n/a	n/a	n/a
L-SCRA, P70	25-Apr-24 (1486)	1.15	JCL57	1-Jun-2024 (1523)	2.8%	1.18	3.4%
N-SCRA-A, P70	27-Aug-25 (1975)	1.54	JCL55	10-Feb-2026 (2100)	9.6%	1.77	23.4%
N-SCRA-B, P70	6-Oct-24 (1650)	1.31	JCL58	15-Feb-2025 (1760)	8.4%	1.48	16.7%

with high random errors. This phenomenon is represented by astonishing scattering (explosion) of results in the right upper corners of the charts and corresponds to much higher random errors of the model N-SCRA-A and N-SCRA-B than random errors of the model L-SCRA.

Table 15.4 contains a summary of JCL results for the three discussed models. It could be noticed that the duration JCL correction factors change relatively slowly when project non-linearity is growing. At the same time cost JCL correction factors change quite dramatically. This can be explained by the fact that a project duration is an independent parameter in terms of schedule-driven costs, whereas a project cost is not. As a result, both schedule and cost risk exposures including all risk interactions are accumulated in project costs.

Table 15.4 summarizes the information about the JCL for the three models discussed in sections 5.6.3, 15.3–15.6.

15.7.4 Main Factors of Project Risk Exposure: A Hierarchy of Importance

Quite interesting conclusions can be derived if JCL assessments for the three considered models (Table 15.4) are compared with corresponding schedule-driven costs (Table 15.3).

The more intensive are risk interactions, the bigger the difference between contributions of schedule-driven costs and JCL correction factors, although both grow at a different pace and cannot be ignored. In the linear L-SCRA case their contributions are about equal and very small. However, this conclusion is based on P70 confidence level that differs significantly from mean-value criteria. If corporate confidence level criteria are selected close to mean values, for example P50 (see expressions 5.2 and 5.3 of Chapter 5), the JCL corrections would be negligibly low in all cases.

At the same time, the logic of the Monte Carlo model's development and running dictates that all non-linear phenomena in a project system should be taken into account first.

The results should be amended afterwards using JCL correction factors that can be easily obtained as a by-product of the non-linear Monte Carlo modelling. The JCL correction factors can be treated as dependent parameters: they depend on schedule-driven costs.

In the case of the L-SCRA model general uncertainties play a dominant role as the other risk factors seem rather feeble. Notably, standalone risks including schedule-driven costs contribute just a few percentage points to project contingencies.

The real power of project risks can be discovered only when intensive enough risk interactions are taken into account: the top contributing factor to risk exposure in complex projects comes from risk interactions. However, these factors give rise to schedule-driven costs and eventually to JCL correction factors as by-products.

15.8 Consortium Participants' Conclusions

Based on insights coming from the discussion and comparison of results for the three models—L-SCRA, N-SCRA-A, and N-SCRA-B—participants of Serendipity United PLC have agreed to proceed with project Zemblanity based on legally binding obligations to improve the project delivery subsystem (PDS).

It has been argued that in accordance with the N-SCRA-B results, a revised approved project duration should be 55 months, completion date 6 October 2024, and cost $1.31B based on P70 confidence level (Table 15.2). These correspond to 26 per cent of schedule and 31 per cent of cost contingencies above deterministic baselines.

This preliminary conclusion has been challenged by several participants arguing that the proposed duration (55 months) and cost ($1.31B) are based on individual schedule and cost confidence levels P70. It has been insisted that joint confidence level JCL70 should be used instead. Hence, instead of base duration of 1309 days or 43.6 months and base estimate of $1B following project completion date and cost have been put forward for discussion purposes:

- Project completion date: 15 February 2025 (overall project duration 1760 days or 58.7 months),
- Project contingency above base schedule duration: 35 per cent,
- Project cost: $1.48B,
- Project cost contingency above base estimate: 48 per cent.

After a comprehensive discussion the following decision has been made that has taken into account both positions.

The stretch targets have been established to be communicated to the project team based on P70 results of the model N-SCRA-B as following:

- Duration: 55 months (contingency 26 per cent),
- Completion date: 6 October 2024,
- Total cost: $1.31B (contingency 31 per cent).

It has also been decided that duration and cost management reserves should be established that reflect on the differences between P70 and JCL70 results (duration and cost JCL70 correction factors). This decision should stay confidential; the JCL70 targets can be used in corporate planning.

The JCL70 targets have been established as follows:

- Duration: 58.7 months (contingency 35 per cent),
- Completion date: 15 February 2025,
- Total cost: $1.48B (contingency 48 per cent).

Despite both targets being much higher than initial base duration and estimate, it is recognized by the consortium partners that the proposed duration and cost targets reflect well on real complexity of project Zemblanity.

It has been clearly accentuated by the representative of Risk Services & Solutions Inc. that both approved targets should be a subject to full implementation of agreed risk-addressing actions including ones for standalone risks and risk interactions. If some of the addressing actions fail, project schedule and cost contingencies should be increased correspondingly. If all risk-addressing actions fail, a most gruesome outcome based on before-risk-addressing case should be expected that would mean a complete project failure.

In addition, recommendations have been put forward by representatives of Risk Services & Solutions Inc. on making the project Zemblanity system more adaptive and flexible. The recommended short-term work availability planning should be adopted to better prevent and react on occurred risks and their interactions as the approved duration and cost are subject to successful implementation of risk and risk-interaction addressing. These recommendations have been approved by the consortium partners.

Before the review meeting was concluded, one of the consortium members proposed to change the name of the consortium from Serendipity United PLC to Serendipity LTD.

15.9 Chapter's Markup and Takeaway

- A full implementation of the non-linear Monte Carlo concept is undertaken to conduct a second risk evaluation for project Zemblanity.
- The initially developed linear model L-SCRA is used as a foundation to build the non-linear N-SCRA model. In essence, this is the same model although with modified inputs that take into account all relevant risk interactions. Re-calculations of the inputs are done using the modified risk register that contains all relevant instances of risk interactions as well as all required non-linearity parameters and non-linear multipliers.
- Risk interaction mapping is used to capture all relevant risk interactions forming dynamic risk patterns and represent them in the risk register.

- After-addressing risk assessments used in the linear model L-SCRA are kept in the model N-SCRA. Risk interaction addressing was considered twice—before and after addressing—resulting in the two models N-SCRA-A and N-SCRA-B.
- Legally bounding obligations to suppress considered chronic project Zemblanity system issues are taken into account in the model N-SCRA-B.
- As expected, results of the model N-SCRA-A (before risk interaction addressing) are unacceptable: the P70 project duration and cost are too high.
- The after-addressing case N-SCRA-B produces somewhat acceptable results that belong in-between the results for the models L-SCRA and N-SCRA-A.
- It is decided that P70 results for the model N-SCRA-B should be established as stretched targets communicated to the project team.
- It is decided that duration and cost management reserved should be established based on JCL70 results for the model N-SCRA-B. This decision should stay confidential; the JCL70 target may be used in corporate planning.
- It is recognized by the consortium partners that the proposed duration and cost targets reflect well on the real complexity of project Zemblanity.
- Short-term work availability planning is approved to make project Zemblanity truly flexible and adaptive.

16

Project System Dynamics: A Non-linear Case

Chapter 7 introduces system dynamics models for project Zemblanity when none of the non-linear interactions among project risks are considered. The model SD1 describes a trivial risk-free case that is fully aligned with the deterministic schedule baseline of project Zemblanity. The model SD2 introduces two standalone engineering and construction rework cycles. This allows the factoring of two quality risks, R03 and R04 related to poor work quality in engineering and construction in these two rework cycles, correspondingly, although the other four risks R01, R02, R05, and R06 are ignored in the model SD2. The model SD3 mirrors the linear L-CSRA case as all six project risks R01–R06 are taken into account although no interactions among them are considered. The models SD1–SD3 are conditionally labelled linear as they ignore any risk interactions. As such they represent quite feeble system dynamics yielding minuscule project duration contingencies.

Now it is time to pay heed to the risk interactions in the SD modelling.

This chapter brings forward two additional system dynamics models which could be conditionally labelled as non-linear as these mimic two non-linear Monte Carlo models of Chapter 15.

A model SD4 mirrors the non-linear Monte Carlo model N-SCRA-A. A model SD5 imitates features of the model N-SCRA-B when risk interactions are addressed. These two SD models accommodate all schedule risk interactions of the models N-SCRA-A and N-SCRA-B. One additional model is brought forward to demonstrate a role of risk compounding. This is based on the model SD4 although with different risk timing.

16.1 A System Dynamics Model SD4: Mirroring the Model N-SCRA-A

This section introduces the model SD4 that mirrors the Monte Carlo model N-SCRA-A.

The model SD3 that mirrors the Monte Carlo model L-SCRA takes into account all six risks R01–R06, although their internal amplification and interactions among them are ignored. This section brings forward a model that counts risk interactions including possibilities (see Figure 14.1) for:

- risks R01 and R02 to get compounded,
- risks R01 and R02 to impact risk R03 (knock-on interactions of risks R01 and R02 with risk R03),

Modern Risk Quantification in Complex Projects: Non-linear Monte Carlo and System Dynamics Methodologies. Yuri G. Raydugin, Oxford University Press (2020). © Yuri G. Raydugin.
DOI: 10.1093/oso/9780198844334.001.0001

- engineering errors and omissions to enter construction (a knock-on interaction of risks R03 and R04),
- construction risk R04 to influence commissioning and startup risk R06 (a knock-on interaction of risks R04 and R06),
- risks R04 and R05 to become compounded,
- risks R01–R06 to be internally amplified before interacting with the other risks.

The corresponding model SD4 is shown by figure 16.1.[1]

Risk information of the model N-SCRA-A can be used to calculate required inputs to the model SD4. Table D.9 for the models N-SCRA-A is used to recalculate impacts for the model SD4. The results of the recalculation are shown in Appendix E (Table E.2).

The recalculation is done the same way as in the case of models SD2 and SD3 (see Appendix C). All risk impact values take into account all relevant risk interactions and corresponding multipliers M_{ij} captured in Table D.1 of Appendix D. One major exception relates to recalculation for the 'poor EC-work quality' parameter. This parameter is not part of the SD3 model: the SD3 model describes two independent engineering and construction rework cycles. By this reason the information for risk R04 in table E.2 is split into two parts.

The first part takes into account two multipliers for risk R04 related to internal amplification M_{44}^{int} and compounding M_{45}^{comp} with risk R05 (see expressions (E.1) and (E.2)). A share of construction units requiring rework (the 'poor C-work quality' parameter) is introduced through a product of two multipliers $M_{44}^{int} \times M_{45}^{comp}$. This parameter is estimated as 3.4 per cent (see Table E.2).

The second part takes into account the knock-on impact of risk R03 (poor engineering quality) on risk R04 (poor construction quality) though introduction of the knock-on multiplier M_{43}^{knock} (see expression (E.3)).

Hence, the 'poor EC-work quality' parameter is introduced by the multiplier M_{43}^{knock} as a share of drawings with errors and omissions entering construction undetected. The 'poor EC-quality' parameter is equal to 2.1 per cent (see Table E.2).

Table E.2 also shows that number of drawings with errors and omissions is 9.2 per cent which serves as the 'poor E-work quality' parameter in the model SD4.

It is important to reiterate the role of the 'poor EC-work quality' parameter which is interpreted as a representative of the knock-on impact of engineering quality risk R03 on construction quality risks R04. Namely, when engineering and construction modules (the two rework cycles) of a project SD model become linked through the 'poor EC-work quality' parameter, it brings forward true system dynamics through establishing a new major feedback loop and EC-rework cycle (Figure 16.1). An additional level 'Undiscovered Rework (EC)' and an associated 'rework discovery rate (EC)' are introduced too.

[1] This is a simplified demonstration version of the model SD4. All workable versions of the models discussed in this chapter are available on the book's companion website www.oup.co.uk/companion/Raydugin.

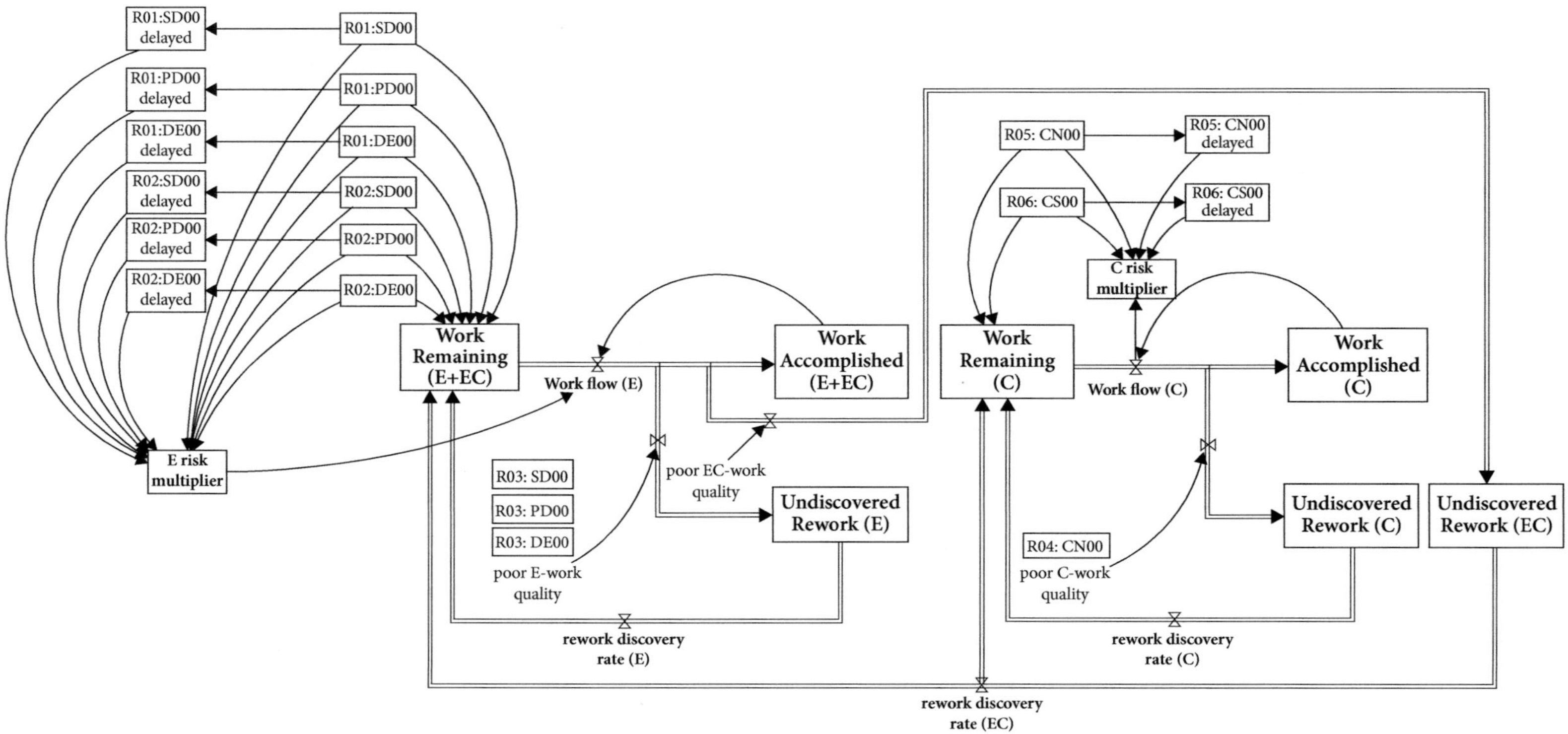

Figure 16.1 Models SD4 and SD5 that combine engineering and construction rework cycles

Upon discovery of drawings with errors and omissions in construction they move to the 'Work Remaining (E+EC)' stock. A delay of the discovery is defined by the 'rework discovery rate (EC)'. The 'Work Remaining (E+EC)' stock includes:

- drawings returned without exiting engineering (marked by letter (E)),
- drawings with errors and omissions that have entered construction and later returned (marked by a combination of letters (EC)), and
- additional quantities of drawings injected by risks R01, R02, R05, and R06.

Apparently, this newly introduced EC-rework cycle that combines engineering and construction rework cycles is not taken into account in the L-SCRA model (Chapter 5) and in the model SD3 (Chapter 7). Risks R03 and R04 that do not interact in the model SD3 start interacting in the model SD4: risk R03 finally finds a way to knock on risk R04.

Any project risks could be internally amplified due to various internal engineering and construction inefficiencies (aka chronic project system issues) before they take part in knock-on and compounding interactions. In the case of risks R03 and R04, the higher the internal amplification, the higher the 'poor E-work' and the 'poor C-work' quality parameters.

The role of risk compounding is considered specifically and separately in a modified model SD4 in section 16.3.

It is important to reflect on the level of complexity of the models discussed in this chapter. All risk interactions—internal amplifications, knock-on interactions, and compounding—could be modelled as dependent parameters using a detailed SD modelling approach, although such models are very complex and could easily reach a level of complexity of SD models developed by Cooper (1980). They require to take into account multiple stocks and variables organized in prolific feedback loops that describe a project system (procedures, skills, implementation, workers' morale, burnout, workers' hiring and firing, workers' skills, stakeholders, schedule pressure, etc.). For simplicity purposes, all risk interactions are introduced in the models SD4 and SD5 as independent parameters through recalculation of the Monte Carlo model's parameters (see Appendices D and E).

The model SD4 does take into account temporary impacts of

- risks R01 and R02 on engineering 'workflow (E)', and
- risks R05 and R06 on construction 'workflow (C)'.

Impacts of risks on productivity are missed out in the models SD2 and SD3. This is done in the model SD4 in accordance with the understanding that occurred risks should result in induced work constraints, in out-of-sequence work, and temporary loss of productivity until a project team regroups and rearranges its activities (see Chapter 9).[2]

[2] In more advanced SD models quality of engineering (risk R03) and construction (risk R04) could be impacted by other risks too. Impact of other risks on quality risks was taken into account in initially developed SD models. Although for simplicity purposes, the impact on quality is left out in the models discussed in this book.

The temporary risk impacts are introduced in connection with spans of time when occurred risks may impact productivity (see Table E.1). It is believed that productivity of works should be smoothly restored to the pre-impact levels after the defined spans of time.

The impacts on productivity are considered abrupt initially when risks occur. However, those impacts are smoothly vanishing. This is based on an assumption that the project team should be able to handle productivity drops resulting from risk-induced work constraints and out-sequence work eventually. So-called risk multipliers for risks R01 and R02 in engineering ('E risk multiplier'), and R05 and R06 in construction ('C risk multiplier') that impact on workflows are introduced (see section E.4). They are represented by Figure 16.1 too.

The risk multipliers are clearly time-dependent as shown by Figures 16.2 and 16.3. These multipliers influence engineering 'Undiscovered Rework (E)' (see Figure 16.4),

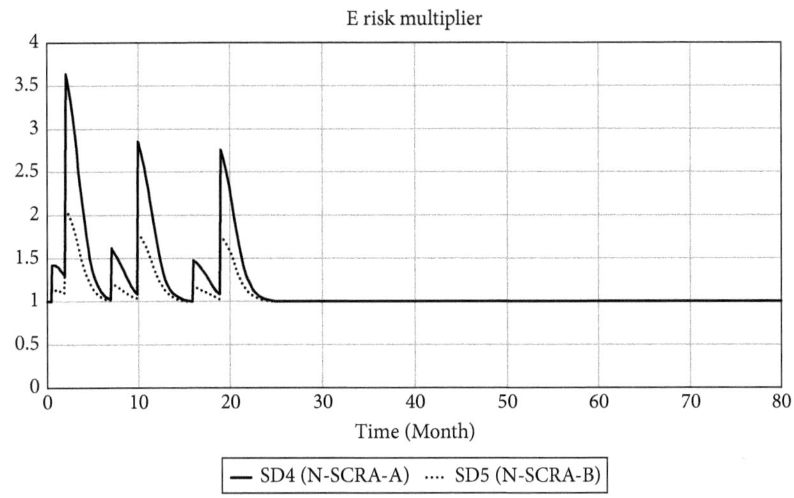

Figure 16.2 Engineering 'E risk multiplier' for the models SD4 and SD5

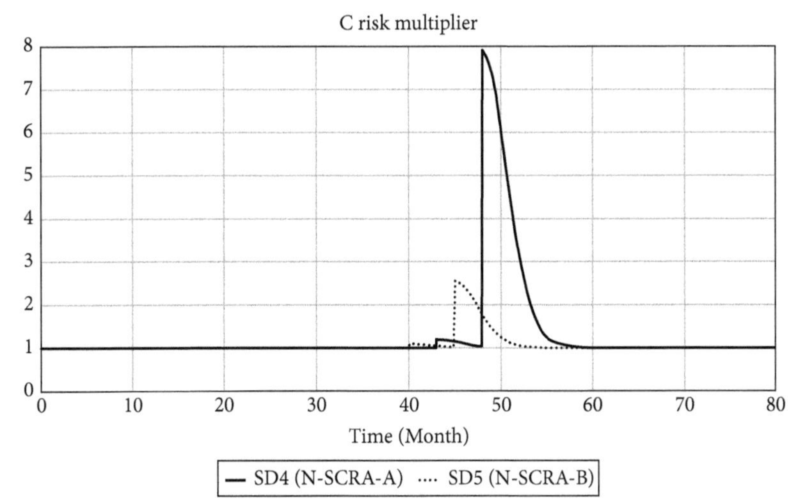

Figure 16.3 Construction 'C risk multiplier' for the models SD4 and SD5

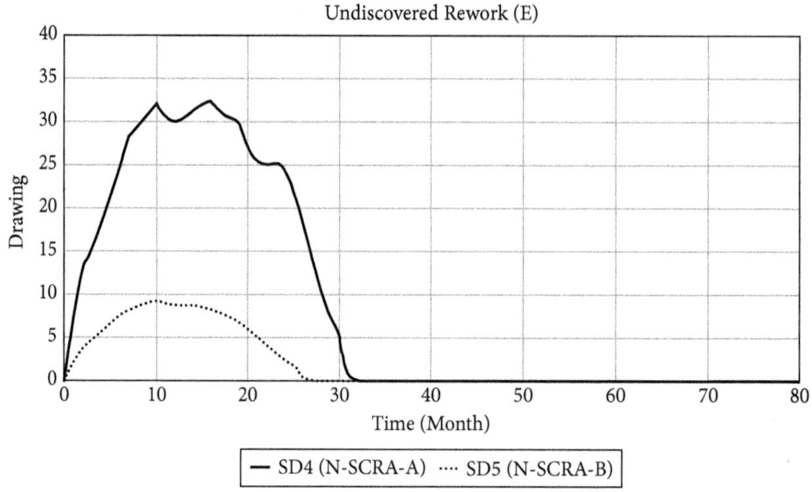

Figure 16.4 Engineering 'Undiscovered Rework (E)' for the models SD4 and SD5

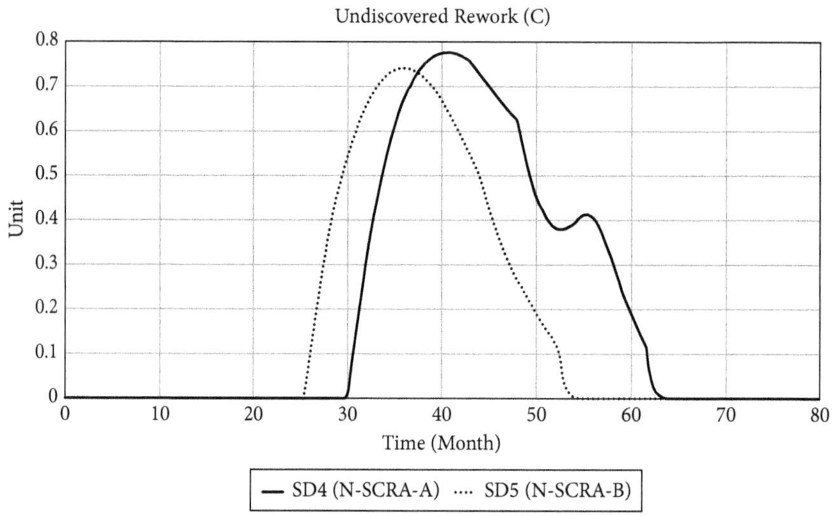

Figure 16.5 Construction 'Undiscovered Rework (C)' for the models SD4 and SD5

construction 'Undiscovered Rework (C)' (see Figure 16.5), and engineering-construction 'Undiscovered Rework (EC)' (see Figure 16.6). Their direct impacts on engineering and construction workflows are clearly shown by Figures 16.7 and 16.8, correspondingly.

Figure 16.6 introduces a new stock unheard of previously—'Undiscovered Rework (EC)'. It shows a stable buildup until regular engineering including engineering rework cycle is completed. This buildup is impacted by engineering risks. When construction begins upon completion of engineering, the EC-rework discovery begins. The 'Undiscovered Rework (EC)' stock becomes fully depleted closer to the end of the project.

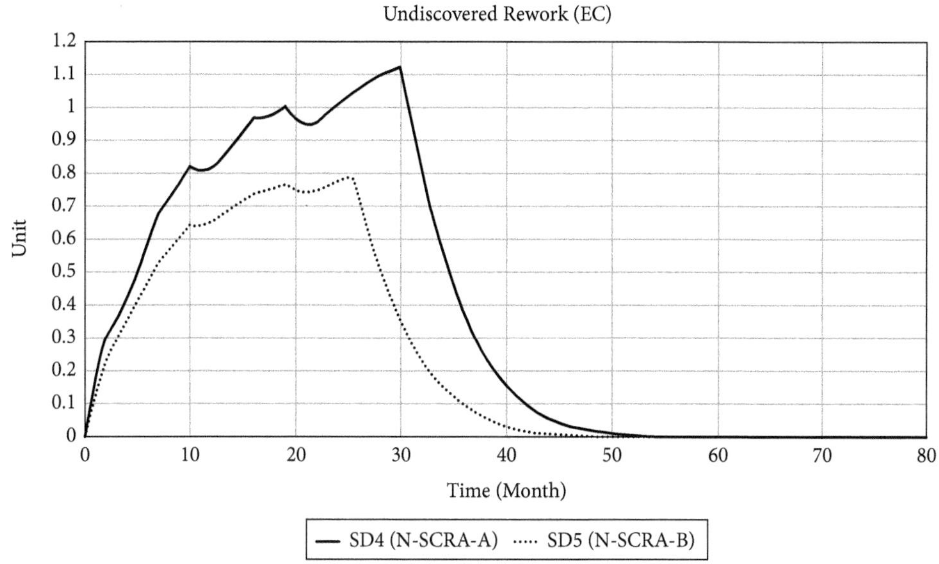

Figure 16.6 Undiscovered engineering–construction rework ('undiscover rework (EC)') for models SD4 and SD5

Figures 16.7 and 16.8 bring forward more dramatism showing engineering and construction workflows. Engineering workflow of the model SD4 shows temporary reductions due to impacts of risks. It also has a spike before engineering completion. These are quite typical for project system dynamics modelling as spikes like this are coming from drawings with errors and omissions returning from construction upon discovery in final batches. The number of spikes and their locations directly depend on characteristic times to detect errors and omissions (see Table C.1).

Figures 16.7 and 16.8 point to the fact that introduction of risk interactions result in significant increase in the project duration. In case of the model SD4 the expected duration of engineering is 30.3 months, which is 43 per cent longer than the engineering duration for the base model SD1 (21.2 months). Overall duration of the project for the model SD4 is 61.4 months (see Figure 16.8) which is 41 per cent above the duration of 43.6 months for the base model SD1.

The comparison of results for the models SD4 and N-SCRA-A shows their good alignment. Mean value of the distribution containing only risks should be compared with the SD4 model results as SD modelling does not include general uncertainties. A distribution based only on the risk register of the Monte Carlo model N-SCRA-A (see Figure 15.5) has a mean-value duration of 1784 days or 59.5 months.[3] Hence, the SD4 and N-SCRA-A results are in a relatively good alignment: 61.4 months vs. 59.5 months.

[3] Full model's statistics can be obtained when running workable models available on the book's companion web-site at www.oup.co.uk/companion/Raydugin.

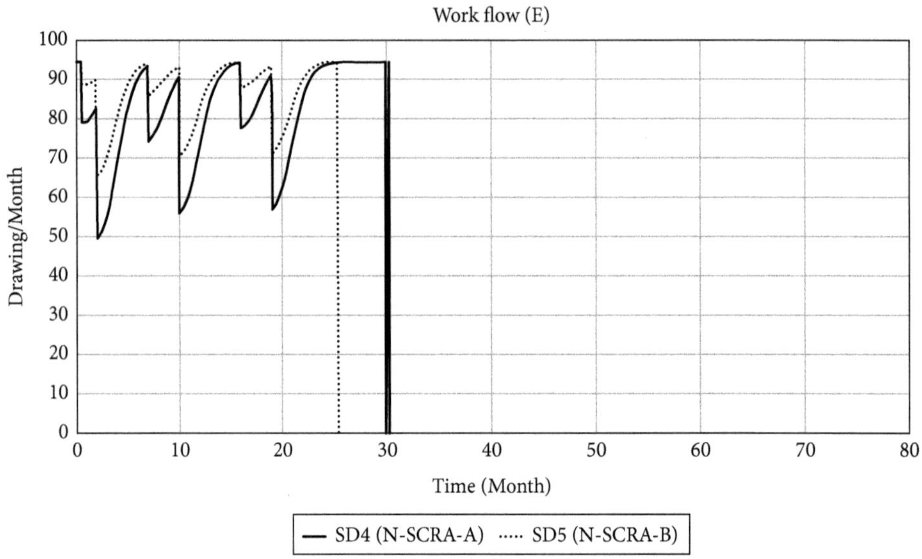

Figure 16.7 Engineering 'work flow (E)' for the models SD4 and SD5

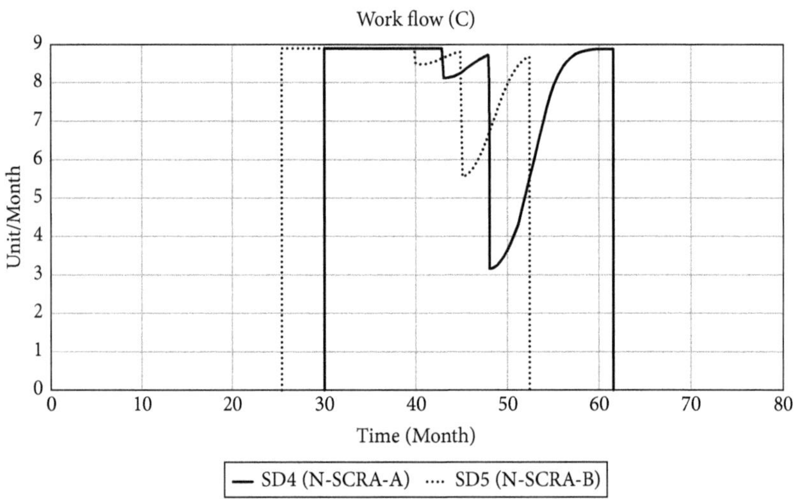

Figure 16.8 Construction 'work flow (C)' for the models SD4 and SD5

16.2 A System Dynamics Model SD5: Mirroring the Model N-SCRA-B

The business case of project Zemblanity discussed in Chapter 15 has a second model N-SCRA-B devoted to the scenario when additional risk addressing measures to reduce risk interactions have been proposed. This scenario is based on the same non-linear model introduced by Figure 16.1 although risk impacts have been recalculated to accommodate less severe project risk exposure: Table E.3 should be used in place of Table E.2.

The same way as in case of Table E.2, Table E.3 derives values for the 'poor E-work quality', 'poor C-work quality', and 'poor EC-work quality' parameters that are equal to 2.5 per cent, 2.0 per cent, and 1.5 per cent, accordingly.

As in the case of the model SD4, timing of risks R05 and R06 has been adjusted so that these risks would actually belong to construction. It is assumed that these risks would occur after 40 and 45 months since the beginning of the project, correspondingly (see Table E.1).

Results for the model SD5 are represented by Figures 16.2–16.8 along with results for the model SD4. The direct comparison of the results for the two models reveals that dynamics of the model SD5 becomes feebler in comparison with the model SD4. According to Figures 16.7 and 16.8 engineering should be fully completed in 25.5 months, whereas the entire project would require 52.4 months. These numbers correspond to 20 per cent longer engineering duration and 20 per cent longer overall project duration than in the case of the model SD1. The numbers should correspond to the mean-value confidence level in the Monte Carlo N-SCRA-B case when only risks are taken into account (Figure 15.12). Actually, the distribution of Figure 15.12 has a mean value of 1520 days or 50.7 months.

The SD5 modelling result (the project duration of 52.4 months) is in a relatively good alignment with the result produced by the Monte Carlo model N-SCRA-B (the project duration of 50.7 months).

16.3 A System Dynamics Model SD4 with Compounding

Timing of risk occurrences in the models SD4 and SD5 is defined in a way that avoids overlapping of risk impact windows (see Table E.1). In other words, the models SD4 and SD5 actually do not comprise compounding at all.

An additional model based on the model SD4 is considered in this section that possesses some risk compounding (see Table E.1). This model referred to as 'the model SD4 with compounding' has the same risk impacts after interactions as applicable to the model SD4 (see Table E.2).

Figure 16.9 demonstrates explicitly the effect of compounding. As a result, the entire project should take 62.3 months in the case of the model SD4 with compounding as opposed to 61.4 months in the case of the model SD4. This adds one month to overall project duration due to the additional impact on work productivity when the risks occur simultaneously.

Timing of risk occurrences in Monte Carlo models is fully defined by mapping of risks to the ends of the impacted normal tasks. In the case of high-level schedules this might not reflect on timing of the actual risk occurrences. On the contrary, the SD models can establish precise timing of risk occurrences as demonstrated by Table E.1. This should be considered an additional loose end that cannot allow a full Monte Carlo–SD model's alignment.

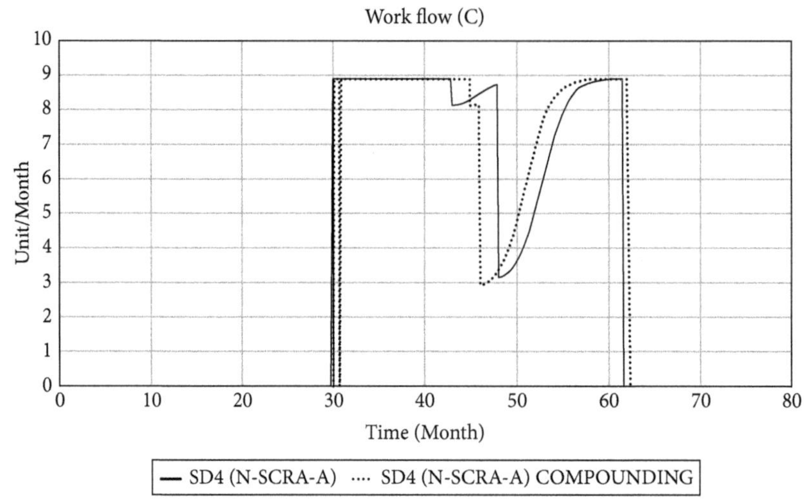

Figure 16.9 Construction 'work flow (C)' for the model SD4 with and without compounding

16.4 Project System Dynamics: Lessons Learned

When looking at the Table of Contents of this book prior to reading, a question could be asked on why a book on risk quantification contains three chapters on system dynamics. Hopefully, Chapter 16 answers this question. This answer has two parts.

First, because system dynamics can be used for risk quantification on its own although with some restrictions. Second, because it uncovers and points to non-linear features of project systems that are missed out but should be taken into account in Monte Carlo models.

Project system dynamics can become a standalone risk quantification method if limitations listed below (see section 16.4.1) are taken care of. One of the general technical constraints is exceedingly high levels of complexity of system dynamics models when these limitations are addressed. Even if system dynamics modelling is not used on its own for risk quantification this may play a supportive although still very important role to the Monte Carlo methodology. Sections below explore both aspects.

16.4.1 Limitations of Project System Dynamics Modelling

Limiting factors of system dynamics modelling in projects are mentioned on several occasions in this chapter and in Chapter 7. These natural limitations are collected and listed below.

- It does not model general project uncertainties: it is limited to quantification of uncertain events.

- It has to take into account uncertain event's probabilities P as factors in one-point expected values $P \times I$: this difficulty does not exist when dispute resolution and claim modelling is undertaken where all risks already occurred and became issues with one-point known impact values and absolutely certain probabilities $P = 1$.
- It models uncertain events as one-point mean values: no distributions can be used as inputs and no other confidence level parameters (e.g. P70) can be used as outputs except mean values.[4]
- It has difficulties to model internal risk amplifications: corresponding models would easily become too complex and impractical.
- It has difficulties to model cost risks and especially schedule-driven costs: corresponding models also could easily become too complex and impractical.
- It has difficulties with risk occurrence timing as the exact timing should be imposed (see Table E.1): this represents a special difficulty with modelling of risk compounding.
- To be practical project system dynamics models keep just a few rework cycles: being high-level models they might miss some important aspects of project system modelling.
- It does not accommodate the concept of joint confidence level (JCL).
- It is insensitive to feeble risk interactions: as Figures 7.3–7.6 show, the difference between results for near-linear models SD2 and SD3 is minuscule.
- It has always a number of variables that do not have one-to-one relations with parameters used in other project models, and first and foremost, in Monte Carlo models.

These limiting factors should be kept in mind when undertaking SD modelling and comparing results of SD and Monte Carlo models.

16.4.2 A Quest for Calibration

The last point on the list of system dynamics limitations requires a special clarification and an honest confession. It may seem that despite the above-mentioned SD modelling limitations, Table 16.1 demonstrates a satisfactory alignment of Monte Carlo models and their system dynamics analogues. It should be confessed that this alignment is reached through assumptions imposed on several loose-end parameters of the SD models. Among these parameters there are as follows (appendix E).

- Selected timing of risk impacts (see section E.1).
- Characteristic times to detect errors and omissions in engineering and poor quality in construction (see section E.2).

[4] Some system dynamics tools do have limited Monte Carlo functionality based on ranging of input parameters, although it is not that helpful to model uncertain events.

- Characteristic times to detect drawings with errors and omissions that enter construction undetected (see section E.2).
- Period of time when occurred risks impact on workflows (see sections E.3 and E.4).
- Mathematical expressions used for modelling of 'E risk multipliers' and 'C risk multipliers' (see section E.4).
- Share of drawings with errors and omissions and construction units with poor quality that require full rework (see sections E.5–E.8).
- A share of drawings with errors and omissions detected in construction that could be resolved right in construction (without rework in engineering) (see section E.9).

In other words, the presence of these loose-end SD parameters brings forward benefits despite them being associated with SD modelling limitations: a possibility of calibration of both system dynamics and Monte Carlo models. Even better alignment of results produced by these two methodologies could be achieved through additional fine-tuning of the loose-end parameters which is not undertaken.

The mutual calibration of the two types of models undertaken here has several features. In both models SD4 and SD5, the shares of engineering drawings required full rework[5] out of all drawings with errors and omissions established as 10 per cent. After recalculation (see Tables E.2 and E.3) this results in 9.2 per cent of all drawings with errors and omissions (the 'poor E-work quality' parameter) in the case of the model

Table 16.1 Comparison of results for the models SD1–SD5 with their Monte Carlo analogues

System dynamics models		Monte Carlo models		SD modelling: project duration		Monte Carlo modelling: project duration (risks only, mean value)	
Model	**Feature**	**Model**	**Feature**	**Duration, month**	**% above SD1**	**Duration, month**	**% above deterministic**
SD1	Risk free	Deterministic	Base schedule	43.6	0.0%	43.6	0.0%
SD2	Two rework cycles	n/a	n/a	45.5	4.4%	n/a	n/a
SD3	L-SCRA analogue	L-SCRA	Linear model	46.4	6.4%	46.3	6.2%
SD4	N-SCRA-A analogue	N-SCRA-A	Non-linear model A	61.4	40.8%	59.5	36.5%
SD5	N-SCRA-B analogue	N-SCRA-B	Non-linear model B	52.4	20.2%	50.7	16.3%

[5] A full-rework equivalent is meant here. There could be no drawings that require full rework at all. Although time spent on partial rework of all of them should be recounted as if several new drawings have to be produced.

SD4, and 2.5 per cent in the case of the model SD5. The shares of construction units with deficiencies requiring full rework are established as 16 per cent for both models. This resulted in 3.4 per cent of units requiring full rework (the 'poor C-work quality' parameter) in case of the model SD4 and 2.0 per cent in case of the model SD5 (see also Tables E.2 and E.3). If the shares of drawings or units that require full rework is much higher, there could be a situation that either engineering or construction would never get finished. This QA/QC topic is further discussed in Chapter 17.

Apparently, calibration discussed here is about alignment of two types of models: system dynamics and Monte Carlo. No calibration of the models based on available empirical data is undertaken as this task is outside of this conceptual study undertaken in this book based on a case study of the imaginary project Zemblanity. The empirically-based calibration would be possible when two conditions are met. First, the corresponding models are built for a real project. Second, as soon as a required database of relevant completed projects that also contains information about maturity of project systems and occurred risks is available, the empirical calibration of both models for a given upcoming project would be straightforward and easy.

16.4.3 A Possibility of Monte Carlo Model Input's Modelling

The logic of this book so far is that Monte Carlo models may be recalculated as inputs to corresponding system dynamics models. The linear L-SCRA for project Zemblanity is developed in Chapter 5. Its inputs and results are used to build the corresponding model SD3 in Chapter 7. In the second round, the non-linear Monte Carlo models N-SCRA-A and N-SCRA-B are developed in Chapter 15. They are recalculated to the input to models SD4 and SD5 in this chapter.

This section explores a possibility to use the discussed Monte Carlo–SD link the other way around: to model inputs to Monte Carlo models using SD modelling. However, as there are no one-to-one relations among system dynamics and Monte Carlo model's parameters, recalculation between the models would still require some assumptions that should be supported by empirical data or unbiased expert judgements.

A critical part of the Monte Carlo modelling does relate to collection of required input data. Usually three methods are used for this as follows.

- Utilization of relevant empirical data on previously occurred risks in comparable projects, if available.
- Expect judgements as an amalgam of experts' previous experiences on similar relevant risk, if unbiased.
- Risk modelling, if possible.

Empirical data to be used in risk assessments require well-established company or industry databases of occurred risks in previously completed projects. These databases should be developed in specific risk management format to be useful. Proponents of regression analysis methodology are proud of the availability of some relatively relevant databases which development and normalization were originated sometime in the middle of the last century and took dozens of years in many cases. Some of the available databases may be used to extract required risk data. This is not possible in many cases as format of these databases does not allow the collection of useful risk information.

The main issue with many available empirical project databases is that they were developed for two particular purposes—first, to undertake regression analyses (descriptive statistics), and, second, to try parametric modelling (inferential statistics). They normally do not contain required information about project context differentiators, occurred risks, maturity of project systems, etc.

That is why it would be better if databases for occurred risks were developed from scratch with required fields—to contain traditional detailed fields pertaining to project risk registers. These fields on risks as future or occurred events should be maintained on the individual risk level. Any attempts at data averaging should be avoided. In addition, in the light of the discussion about the static and dynamic PSS–PDS mismatches undertaken in this book, corresponding databases should contain records about observed project system issues and risk interactions in completed projects. Actually, the modified non-linear risk registers for upcoming projects should use this information too. Creating and maintaining of such databases is a viable although a long-term challenge. Should all Monte Carlo risk quantification activities be stopped until required high-quality risk databases are developed? Certainly not. Expert judgement and risk modelling are still two viable and practical options.

Expert judgements can be considered an inevitable substitute to missing empirical data in a right format being based on practical experience of specialists, although this method may easily succumb to various realizations of bias. Despite this, expert judgement is a most-often used method in both Monte Carlo and parametric modelling (see Chapters 1–4).

Modelling of risks especially technical risks is possible in some instances. The modelling could be aimed at assessments of both probabilities and impacts of standalone risks. In practice this is quite a rare occurrence. Instead some simple arithmetic's assessments are used to evaluate risk probabilities and impacts.

Using project system dynamics may open new possibilities to model some risks and their interactions. This can reduce the importance of expert judgements or at least should help to rationalize and debias them. This also may partially compensate the existing gaps in availability of required empirical risk data.

Precision of system dynamics models could be amazingly high. In the case of disputes and claims, their results could be accepted by arbitrations and courts (see Chapter 7). Being based on modelling of rework cycles, system dynamics modelling is a direct fit for purpose to model impacts of risks related to poor quality and lower productivity. This is also an efficient tool to model knock-on and compounding risk interactions as demonstrated in this chapter. Besides this SD models can put forward valuable insights on the mechanisms of risk impacting and interacting and provide guidelines for their assessments and addressing even in the case of standalone risks (see Chapter 9).

Figure 9.1 explains a detailed mechanism of impacts of schedule and cost risks. Two primary factors influencing the impacts are reworks and work constraints. These factors usually work hand-in-hand as in many cases they are both activated by the same disturbances (aka occurred risks). However, in the light of the PSS–PDS mismatch discussion, these two factors represent two sides of a coin. They contribute to two different sides of dynamic PSS–PDS mismatches.

Namely, any rework (new work, adjustment of a finished work, scrapping of a finished work, or scrapping of an unfinished work) can be interpreted as a dynamic part of the PSS–PDS mismatch emerged in the project structure subsystem (PSS). On the contrary, induced work constraints should be interpreted as a dynamic part of the PSS–PDS mismatch coming from the project delivery subsystem (PDS). These two dynamic mismatches are exacerbated by any other previously existing static or dynamic PSS–PDS mismatches.

The knowledge about the nature of rework cycles and work constraints can be used when assessing standalone risks and their interactions. This provides additional insights when assessing risks as inputs to Monte Carlo models. That is why along with the basic bowtie diagram for a standalone risk (see Figure 2.2), a more sophisticated diagram of Figure 9.1 should be used: it should be considered a substitute to the right part of the traditional bowtie diagram. In turn, this knowledge should allow integration of factors related to rework and work constraints into risk assessments used as input to Monte Carlo models. For example, when assessing cost risks, scrapping of finished and unfinished works should be directly considered a contribution to fixed sunk costs from a risk impact in addition to variable schedule-driven costs. Similarly, information about induced work constraints—lack of information, labour, materials, equipment, and permits (see Chapter 9)—is a valuable guideline to assess resulting delays (Figure 9.1).

The discussion in this section accentuates a value of SD methodology for modelling some inputs to Monte Carlo models including cross-risk interactions. What about modelling of internal risk amplifications of standalone risks? Chapter 17 provides some insights on the origin of internal amplifications and ways to model them.

16.5 Chapter's Markup and Takeaway

- The system dynamics model SD3 that mimics the Monte Carlo model L-SCRA is based on the two non-interacting rework cycles and it ignores risk interactions. It yields inadequately low project contingency being feeble in terms of realizations of non-linearity in system dynamics.
- Two additional models are developed to combine the two rework cycles and take into account all relevant risk interactions. The model SD4 mimics the Monte

Carlo model N-SCRA-A (before risk interaction addressing). The model SD5 follows the Monte Carlo model N-SCRA-B (after risk interaction addressing).

- Non-linear multipliers developed for the non-linear Monte Carlo models are used for recalculations of the Monte Carlo model's inputs to the SD models.
- A cardinal addition to the model SD4 and SD5 in comparison with the models SD2 and SD3 relates to introduction of the link between the engineering and construction rework cycles due to the knock-on impact of risk R03 (poor engineering quality) on risk R04 (poor construction quality).
- A key feature of models SD4 and SD5 relates to the impact of occurred risks on productivity of works. Engineering and construction risk multipliers are introduced to aggregate impacts of occurred risks on productivity of engineering and construction works.
- Due to risk interactions and aggregated impact of occurred risks on productivity of works, the models SD4 and SD5 demonstrate substantial non-linear system dynamics and yield results that are in a relatively good alignment with the results produced by the models N-SCRA-A and N-SCRA-B, correspondingly.
- If risks occur simultaneously, the resulting impact on project duration would be higher than impacts of standalone risks due to risk compounding.
- Monte Carlo and system dynamics models do not have one-to-one relations among their parameters. Some assumptions are required to align these two types of the models. A possibility of mutual calibration of the two types of models can be used due to this.
- Project system dynamic modelling can be used as a standalone risk quantification method. But it has several restrictions and constraints. These should be kept in mind when comparing Monte Carlo and SD models.
- Project system dynamics can be used for modelling of some Monte Carlo inputs.
- Even if SD modelling is not used as a standalone risk quantification tool or a tool for modelling Monte Carlo inputs, it provides valuable insights on non-linear risk interactions in project systems. Some of the valuable insights related to the rework cycles and non-linear risk interactions should be incorporated in the Monte Carlo models which is done in Chapter 15.

17

Spin-off Discussions on Project Complexity

A primary goal of this book—risk quantification in complex projects—is reached in Chapters 15 and 16. Two workable and closely related methods—non-linear Monte Carlo SCRA and project system dynamics—have demonstrated their power and alignment. However, this journey brings forward several valuable spin-off topics that might not be directly used in practical project risk quantification although are worth being shared. Few observations presented in a form of essays can be treated as a contribution to the academic complexity-of-project and complexity-in-project discussion initiated in Chapter 8.

Chapter 6 on project system dynamics basics that considers the bathtub model is designed for 'mathematical gourmets'. This chapter should be enjoyable to 'physics gourmets' as well as advanced project specialists that are interested in the broader view on risk management in complex projects.

17.1 Projects as Complex Adaptive Systems and Over-planning

As discussed in Chapter 8, complex systems should be adaptive. This well-known mantra that is almost written in stone finally deserves converting to a practical recipe which is a primary purpose of this section.[1]

First of all, in practice, a reverse trend can be observed: the more complex a project is, the more rigid and tight controls are imposed on it which is detrimental to project success. How could this trend be thoughtfully reversed based on the general complexity principles?

It could sound like a project management heresy to many, but one key project system issue in complex projects is over-planning. Project owners preach for tighter project controls over project teams and project deliverables in an attempt to increase chances for project successes. Corporate schedulers develop detailed project schedules (levels IV, V, etc.) with thousands of normal tasks several months or even years before a real job would be done by completely different organizations! One needs to get exposed to centralized commanding planning system for a while to fully fathom implications of such proclivity.

[1] The author of this book is grateful to Michael Mac Guinness, a brilliant co-worker and a great friend, for multiple discussions on the topic. Michael has tons of practical experience in planning, scheduling, and dispute resolutions that began way before modern scheduling tools came of age. Due to his highly-valuable insights it feels sometimes like Michael has got at least 70 years of diversified project planning experience, which is actually true if a (−40%; +40%) confidence interval is used for the assessment.

Modern Risk Quantification in Complex Projects: Non-linear Monte Carlo and System Dynamics Methodologies. Yuri G. Raydugin, Oxford University Press (2020). © Yuri G. Raydugin.
DOI: 10.1093/oso/9780198844334.001.0001

This is a classical situation of putting the cart before the horse, where a cart alludes to modern project scheduling software packages, and the horse refers to the project development context and the healthy planning process. (The similar situation with PRM tools is discussed in Chapter 2 as related to PRM systems.) Playing with sophisticated scheduling tools that do not have any technical limitations to break project activities down to 15-minute tasks results in a situation when planners simply do not see the forest for the trees.[2]

Developing detailed schedules with several thousand activities that actually would be carried out in several months or years by different organizations in execution phases does bring up a false illusion of control over a project outcome.[3] On the contrary, this is a way to completely lose control: due to inevitable multiple interacting risks leading to deviations from project objectives, the detailed rigid schedules might quickly lose credibility in the eyes of project team members and any connection with reality.

This is an everyday project reality that any project encounters various changes non-stop. This statement is not even about major scope changes initiated by stakeholders every now and then. This is simply about ongoing perturbations and disruptions (aka occurring risks of various origin and severity) organized in dynamic patterns. Any project is constantly on a bumpy road. In this context, change management should be understand as a methodology to manage occurred risks including out-of-sequence works. If a project as a complex system has a rigid detailed schedule that should be followed no matter what, this would mean that:

- the suppressed flexibility prevents the project system becoming adaptive, and
- the failed change management is replaced by crisis management.

This just means no schedule at all and a complete chaos emerging due to over-planning and rigid control. Long story short, this means the project is ruined. The only hope is that a new more adaptive project delivery system emerges from the chaos through self-organization and owing to strong leadership.

From the angle of project system maturity discussed in Chapter 12, the over-planning is a top chronic project system issue responsible for failure of complex projects.[4] Following

[2] Actually, a standard minimal time planning unit in the modern scheduling tools is an hour. However, minutes with two decimals can be also used as time planning sub-units. This means that the actual minimal time planning unit is 0.01 minutes or 0.6 seconds! This provides a lot of room for the advanced schedule granulation.

[3] Recently the author of this book came across a remarkable mega-project in FEL3 that had about 20,000 normal tasks in its schedule. A PMC contractor that certainly would not take part in the project execution developed that grandeur and granular schedule arguing that a better control over the project outcome should be achieved. However, after a prompt investigation a real 'innocent' root cause of this 'control passion' and abnormality was uncovered: the PMC contractor simply burned man-hours to keep their schedulers busy.

[4] One example that does justify very detailed rigid scheduling relates to managing short shutdown windows (usually 40–80 days) required for installation or connecting of equipment at refineries. However, the best industry practice also prescribes having three major decision milestones beforehand to ensure the work availability during the windows. A notice about planned works should be submitted to a facility owner, say, eighteen months before the installation work starts; engineering activities should be completed, say, twelve months before the start; and all required equipment and spare parts are delivered to the site, say, six months before the start. Failure to obey any of these milestones is considered unacceptable and results in termination of the entire planning process. In this case any detailed schedule will not be helpful.

this logic, over-planning can be acceptable in simple projects that do not have much risk interactions and should shrug harmful over-planning off easily.

Decentralization and flexibility in planning at the tactical level should become a practical norm to make complex project systems truly adaptive and successful. In practical terms corporate scheduling should be stopped at a package level. Schedules of the kind are usually developed in FEL3 and referred to as level III schedules. Normally they contain just a few hundred normal tasks. A few-week-ahead rolling planning should be undertaken in execution—both in engineering and especially in construction. A site manager, a local scheduler, and foremen who actually work at a project site should be doing the short-term rolling planning. This is how things worked successfully in the 1970s and 1980s!

Returning to Chapter 9 (see Figure 9.1), the purpose of the short-term rolling planning is to ensure that a project is flexible enough and capable of properly managing:

- emerging rework,
- emerging work constraints,
- out-of-sequence work,
- occurred risks,
- possible risk interactions including knock-on and compounding interactions, and
- schedule-driven costs.

In simple terms, the short-term planning should ensure proper work availability and maintain productivity: required information, labour, materials, equipment, and permits should be timely available whenever possible. If it is not possible, out-of-sequence work could be undertaken temporarily instead as a band-aid solution.

Figure 9.1 is developed as the right-side part of bowtie diagrams shown by Figures 2.2, 9.2–9.4. In this context, managing the above-mentioned items simply means recovery risk-addressing actions when interacting risks do occur. In addition, the short-term planning should result in being proactively prepared for and reduce probability of work constraint's emergence.

The logic of the short-term rolling planning leads to a conclusion that level III schedules can be converted to more detailed level IV schedules but for next few (maybe three or four) weeks as a part of the rolling planning. This type of short-term planning may be accompanied by a mid-term planning to have a broader planning horizon of one to three months. Although once again this type of planning should be undertaken by the on-site workers, not by corporate planners sitting in corporate towers.

This rolling planning provides much more real control than the long-term rigid and illusive scheduling by a simple reason: a complex project system must be adaptive and flexible to be successful.

The discussion above does not mean that modern scheduling software packages should not be used at all. They should be used responsibly and should not take over the entire project planning process. The planning context and the planning process of complex projects should be served by scheduling tools and not the other way around.

In this regard, it would be worthwhile to remember the difference between a good planner and a good scheduler. A good planner should be capable of carrying out successful project planning even without any scheduling tool. Conversely though, a good scheduler might not be ready to do a good project planning. Project planning is too important to be left to the schedulers.

17.2 Single and Multiple Project Failure Modes

As mentioned by Merrow (2011), projects tend to be either very good or very bad in their performance, almost none of them belong in-between. Leo Tolstoy (Tolstoy (2016)) also writes on project failures if his words 'happy' and 'family' are replaced by 'successful' and 'project' in his following citation: 'Happy families are all alike; every unhappy family is unhappy in its own way.'

In mathematical terms there should be so-called strange attractors of good and bad performance that define where completed projects end up, as argued in Chapter 8. A possibility for projects to end up around one of two strange attractors can be tagged as 'bi-polar project disorder' if corresponding probability distributions have two distinctive probability peaks.

This section is devoted to further investigation of this topic. A notion of a project single failure mode should be introduced for this. This points to a single risk or a small distinctive group of risks that are responsible for poor project schedule or cost outcome. This mode should be associated with the second probability distribution's peak that may be tagged as a bad project performance strange attractor: when these risks occur, the result is a project of chaos and failure.

As opposed to a single failure mode, there could be multiple risks of comparable severity that, if occurred, lead to a project failure altogether. This situation constitutes a multiple failure mode and is represented by a single probability distribution peak. (In the case of the two-peak distributions the first peak could be interpreted as a place that harbour the multiple failure modes.) This peak is a solo strange attractor as the second one does not exist. Occurred risks of comparable severities are accumulated one by one in the course of project development and execution giving rise to the final actuals covered by the single peak.

The statements in the previous paragraph can be treated as hypotheses that are investigated and proven below in this section. For simplicity purposes, this discussion is continued by considering schedule risks only.

Risks in a complex project system act as a part of a dynamic risk pattern due to multiple risk interactions. Some standalone risks could be significantly enhanced by these interactions; some could keep low severity even after taking part in several interactions. It would be logical to presume that top risk players should be most sensitive to a project. Based on this assumption, sensitivity analysis charts produced as a standard outcome of any Monte Caro modelling—so-called tornado diagrams on the risk jargon—should help identify the top risk players.

An additional model N-SCRA-C is introduced below to study the failure modes. For demonstration purposes this model is built in a simple way without close considerations of particular risk interactions. Table D.11 introduces the inputs whatever risk interactions are behind them. These inputs are used in the basic Monte Carlo model that stands behind the previously run models L-SCRA, N-SCRA-A, and N-SCRA-B. The same general uncertainties are applied to the model N-SCRA-C as in the other models (see Table A.2). The purpose of this demonstration is just to get two peaks for the schedule distribution whichever risk interactions are behind them (see Figure 17.1).

The sensitivity chart that corresponds to the two-peak distribution has two dominating risks R01 and R02 (see Figure 17.2).

It is easy to prove that risks R01 and R02 are responsible for the second peak. If risks R01 and R02 are excluded from the model N-SCRA-C, the second peak vanishes

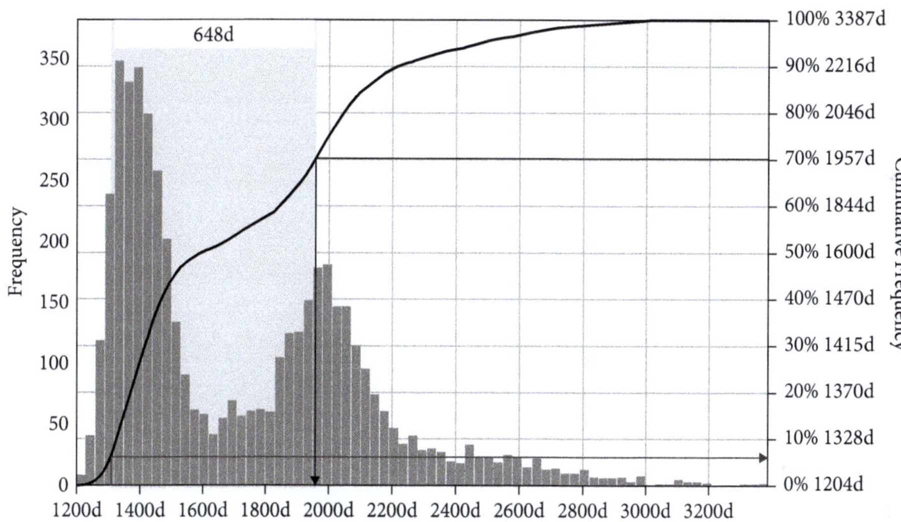

Figure 17.1 A two-peak distribution of the model N-SCRA-C

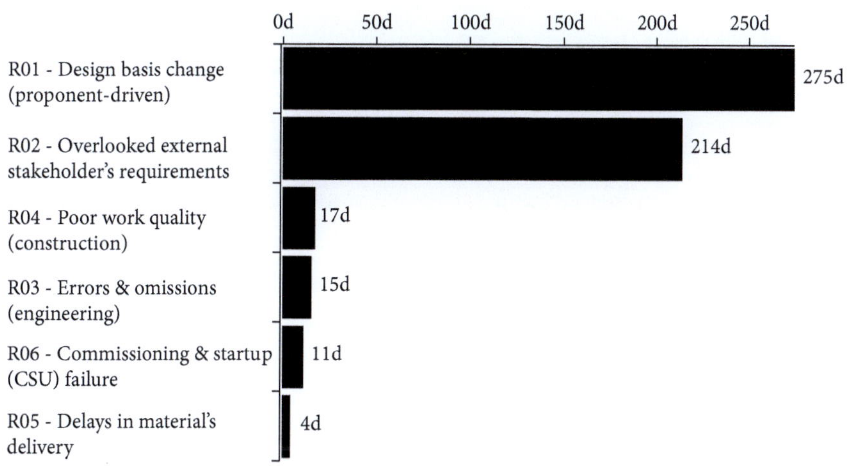

Figure 17.2 The sensitivity chart for the two-peak distribution of the model N-SCRA-C

(see Figure 17.3). In this case the tornado diagram contains three comparable risks R03, R04, and R06 in terms of sensitivity (see Figure 17.4).[5]

It should be pointed out that general uncertainties that are a part of the model N-SCRA-C do play a certain role in forming duration distributions and cannot be ignored. They smooth down the distributions. For example, Figure 17.5 shows a rough

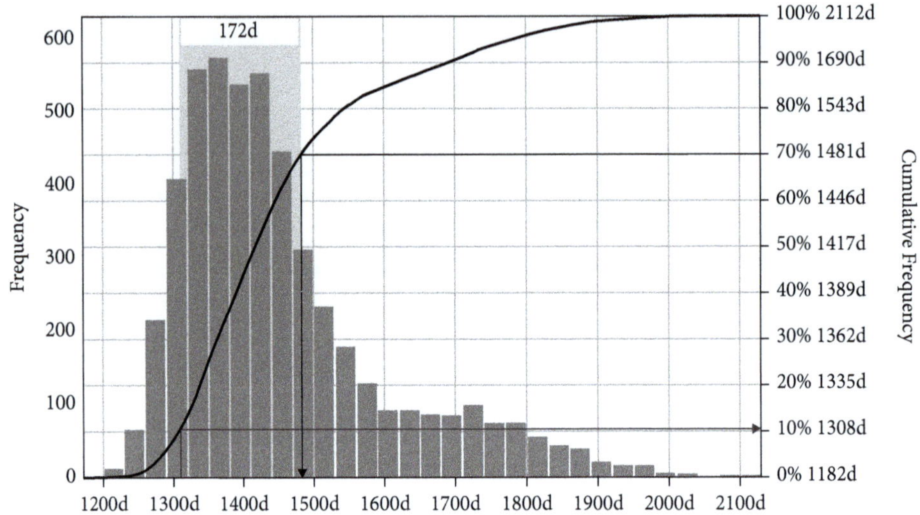

Figure 17.3 A schedule distribution of the model N-SCRA-C if risks R01 and R02 are excluded

Figure 17.4 The sensitivity chart for the model N-SCRA-C if risks R01 and R02 are excluded

[5] Risk R05 is kept having a low sensitivity in all models considered in this book on purpose to show the role of low-sensitivity risks.

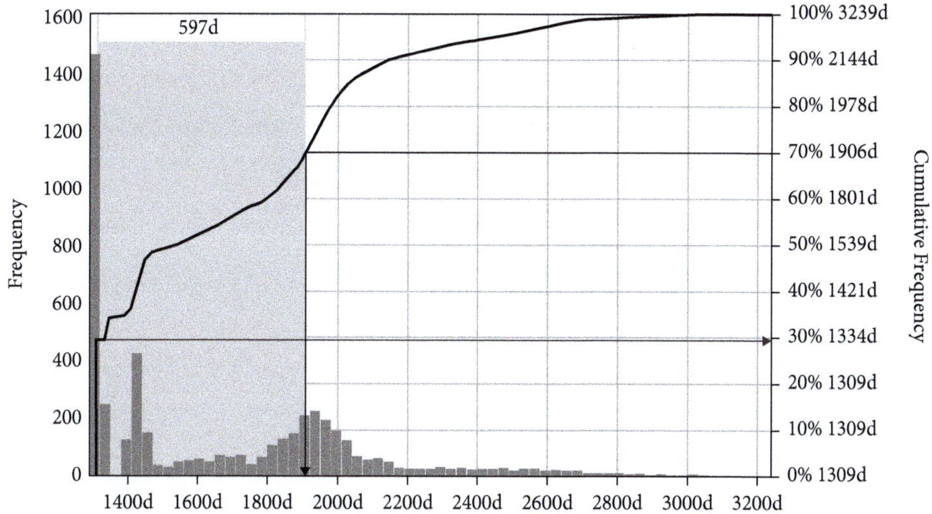

Figure 17.5 A two-peak distribution of the model N-SCRA-C if all general uncertainties are excluded

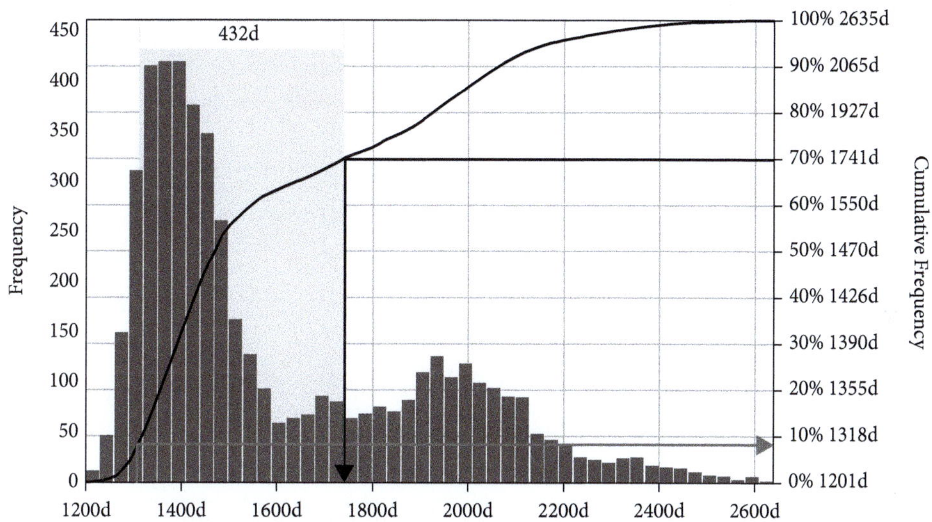

Figure 17.6 A two-peak distribution of the model N-SCRA-C if risk R02 is excluded

two-peak distribution if all general uncertainties are excluded. If just one of the two most sensitive risks R01 or R02 is excluded when the other is kept, the results are comparable with ones represented by Figures 17.1 and 17.2: the second peak still exists although it is less prominent. This example when risk R02 is excluded is shown by Figure 17.6.

The direct comparison of Figures 17.1 and 17.2 with figures 17.3 and 17.4 allows a conclusion to be made that the second peak is associated with two highly sensitive risks R01 and R02 that form a single failure mode.

There is a fundamental difference between the two two-peak distributions represented by Figures 17.1 and 17.6. In the case of Figure 17.1 the P70 duration clearly

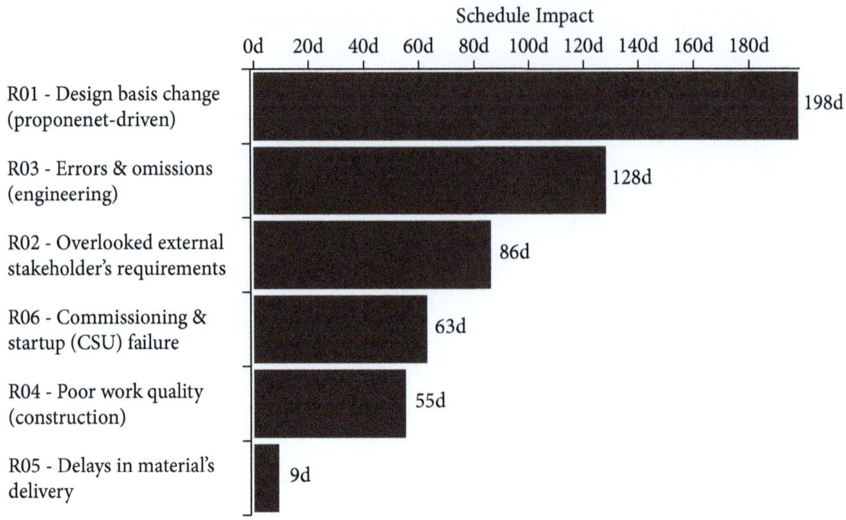

Figure 17.7 A sensitivity chart for the model N-SCRA-A

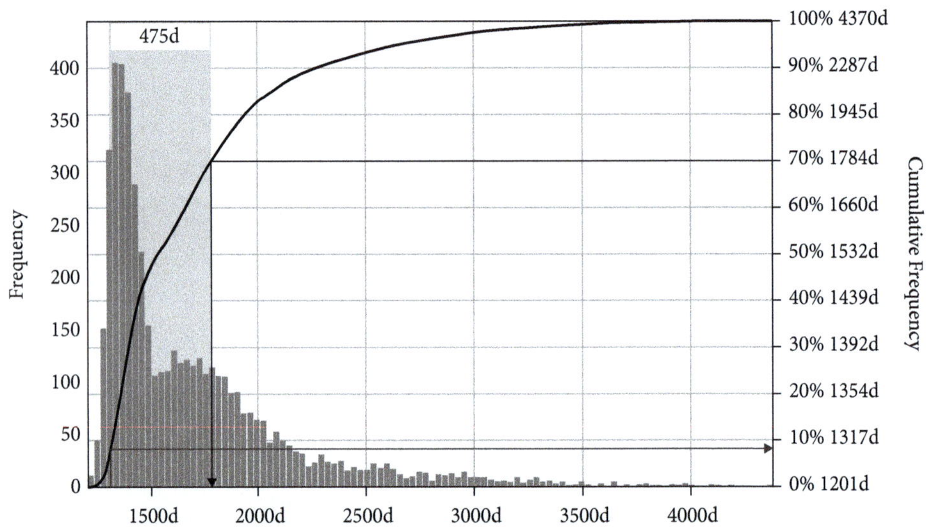

Figure 17.8 A distribution of the model N-SCRA-A if risk R01 is excluded

belongs to the second peak. However if risk R02 excluded, P70 duration moved to the area in-between the peaks. In the first case with a confidence level P70 the project should fail due to a single failure mode formed by two combined risks R01 and R02. In the other case the confidence of this is much lower.

Figure 17.3 shows the one-peak case when both risks R01 and R02 are excluded which corresponds to the multiple failure mode formed by risks R03–R06 and by general uncertainties.

At this point a reader should raise a question about the model N-SCRA-A that also has two peaks (see Figures 15.1 and 15.2). Does this mean that project Zemblanity should succumb to a single failure mode?

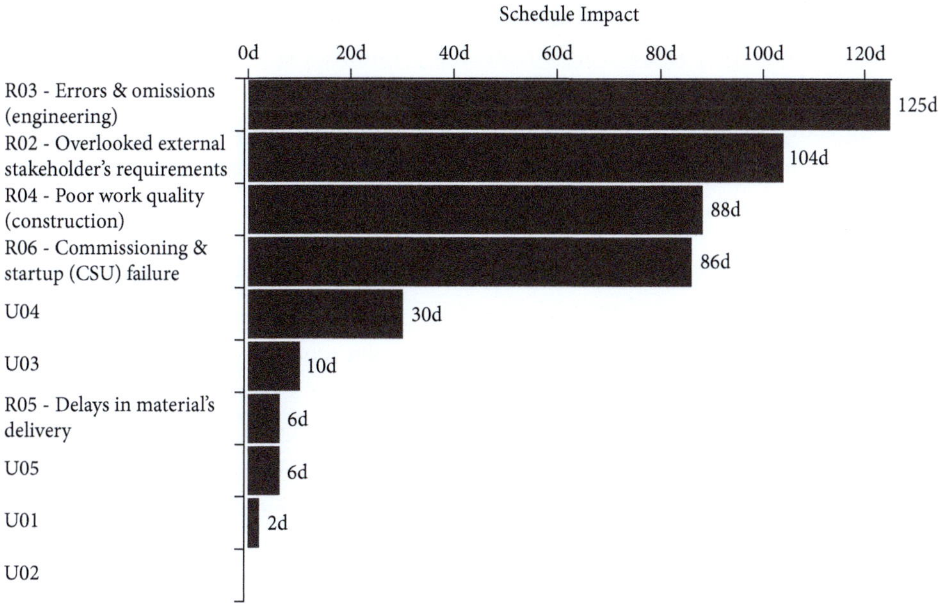

Figure 17.9 A sensitivity chart for the model N-SCRA-A if risk R01 is excluded

A sensitivity chart of the model N-SCRA-A is shown in Figure 17.7. It does not show any risk that has a distinctive sensitivity. A suspicion that risk R01 can generate the second peak and, hence, form a single failure mode can be easily played down if the model N-SCRA-A is run without risk R01. Figure 17.8 still demonstrates two peaks despite risk R01 being gone. The corresponding sensitivity chart is shown by Figure 17.9. It does not show any risk that has distinctively high sensitivity despite the distribution of Figure 17.8 still containing the two peaks.

How can this discrepancy be explained? Or maybe the hypothesis that a single failure mode should be justified by a single risk (or a small group of risks) of extremely high sensitivity is wrong? Figure 17.9 helps to solve this conundrum.

As mentioned above, general uncertainties play a certain role in forming the considered distributions. Figure 17.9 shows general uncertainties U01–U05 and their positions in the hierarchy along with the risks R01–R06. Apparently, all general uncertainties have low ranks in the sensitivity chart and play a modest role. Whereas, all risks except R05 have comparable top ranks. It is clear that all general uncertainties along with risk R05 form a group that is responsible for the first peak. It becomes clear that the other risks R01–R04 and R06 form the second group that are responsible for the second peak. However, as the second peak is formed by the blend of almost all the risks except risk R05, it cannot correspond to a single failure mode. The two-peak distribution of the model N-SCRA-A is just a result of segregation of general uncertainties and almost all uncertain events. In a risk-free case there will be a single first peak representing general uncertainties. The second peak plays a role of a single peak in the general-uncertainty-free case, if risk R05 is ignored (see Figure 17.5). No jump from the first peak to the second peak is expected as most likely the system belongs to the second peak by default. A true single failure mode based on one or very few sensitive risks that presumes the transition to the second peak is represented only by the model N-SCRA-C.

In the course of project development and execution a single failure mode can be formed if the situation with a particular risk or a small group of risks deteriorates. For example, new static or dynamic project system issues can emerge. As a result, one risk might become extremely sensitive and become solely responsible for throwing a project towards chaos and beyond. In mathematical terms such risk should form the second peak.

The situation could be different regarding cost failure modes. As CRA models receive outputs from SRA models to take into account schedule-driven costs, it can be expected that the resulting cost distribution should be a blend of multiple failure modes formed by schedule and cost risks and schedule and cost general uncertainties. These may become tangled in the resulting intricate cost distribution. As a result, possible multiple single failure modes get intertwined resulting in the cost multiple failure mode. It is difficult, therefore, to expect that a cost distribution has two peaks, although it is still thinkable to construct some exotic cost models with two-peak distributions. The fact that Figure 3.3 that represents empirical data on actual costs has one peak does not contradict this discussion. Notably, Figure 3.2 that demonstrates empirical data on actual project durations does possess two peaks.

A Monte Carlo model might result in a distribution that contains even more than two peaks in some cases. As a matter of fact, initially the model N-SCRA-C has been built purposely having three peaks. However, it has been decided to keep the more straightforward and practical two-peak distribution example.

17.3 A Single Failure Mode: An Analogy from Physics

Section 8.3 is devoted to review of complexity in projects where several analogies from physics are mentioned to fathom project complexity. This section may be considered a contribution to the complexity-in-project domain. This is based on reviewing a phase transition model when a physical system travels from one state to another. A full mathematical description of the phase transition model is provided by Strogatz (2015). A double-well potential is considered to define two equilibrium states of a physical system (see Figure 17.10): bottoms of the two wells correspond to the equilibrium states A and B.[6]

[6] States of a quantum system are defined by so-called wave function ψ that is a solution of the Schrödinger equation. It may be expressed in its most simple time-independent form as $\left\{ {}^{d^2\psi(x)}\!/\!_{dx^2} \right\} = \varphi \times \left[V(x) - E \right] \times \psi(x)$, where E is the system's energy, $V(x)$ is a potential energy, φ is a physical constant. Sometimes $V(x)$ is called a potential pit or well of a depth V_0 where a system is trapped. It is remarkable that squared module of the wave-function $|\psi(x)|^2$ represents a probability for the system to be in a particular state. A Monte Carlo distribution also represents probabilities of a project system to obtain a particular project duration or cost outcome. This is the foundation for the discussed analogy and that is what Figure 17.10 tries to envisage.

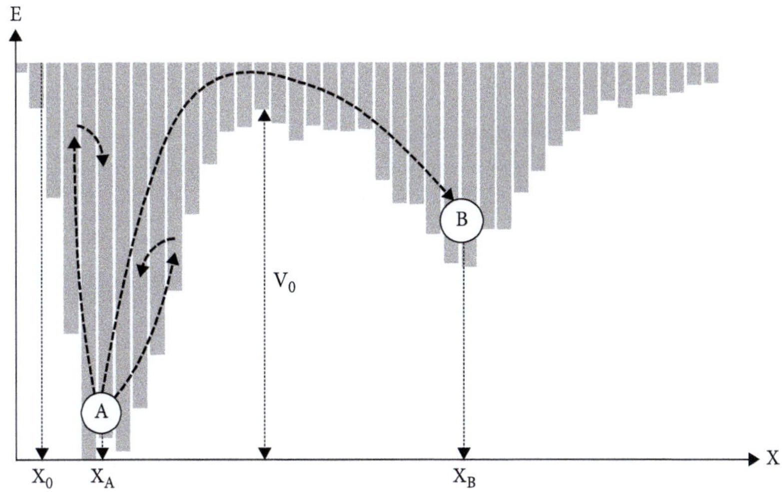

Figure 17.10 A system in a two-well energy potential

A transition $A \rightarrow B$ between the two states is possible if an external periodic force is applied that shakes up and pushes a system from the initial equilibrium state A (Strogatz (2015)). As shown by Figure 17.10, if fluctuations around the initial equilibrium state A are low (aka linear) as the external force is weak, a system may return to the initial equilibrium state A or it may stay nearby in the same well. In the case the force is strong enough the system could overcome the barrier between the two wells and end up in the second well. This apparently corresponds to high (aka non-linear) fluctuations.

If a project system has some similarities with a physical system in terms of complexity and implications of non-linearity, then these features could be further extended as follows.

- Occurred risk events that happen in a project system every now and then can be treated as the external periodic force that brings a project out of equilibrium.
- Any one-peak or two-peak duration or cost distribution could be understood as a representation of a one-well or two-well potential where a project system resides. A Monte Carlo probability distribution chart could be simply rotated by 180°: actually, an attentive reader may have already noticed that Figure 17.10 is obtained from Figure 17.1 this way.
- The potential $V(x)$ could have just one well (see, for example, 180° rotation of Figure 17.3) or two wells (see Figure 17.10).
- A two-well situation points to a dominant risk or a small group of dominant risks that are obvious from the sensitivity chart (see Figure 17.2). This supposedly happens when some risks could be critically enhanced due to internal amplifications and cross-interactions which is quite possible in complex

projects.[7] They should have a capacity to throw the project system to the second well.

- The value x_0 of Figure 17.10 can be interpreted as a deterministic base duration. Values x_A and x_B may be understood as project durations when a project system ends up either in the equilibrium state A or B, correspondingly. Hence, $x_A - x_0$ and $x_B - x_0$ can be interpreted as corresponding duration contingencies.
- Actually, a project system might not end up in these two equilibrium states: it may end up nearby although this would have lower probability than equilibrium states A or B have.
- A transition from one well to the other could be treated as a situation when a project becomes chaotic due to occurrences of the corresponding most sensitive risks. It jumps out of the first equilibrium state A and ends up in the second state B which is triggered by a single failure mode. This explains the analogy with two strange attractors discussed in Chapter 8. In physics abrupt transitions of the kind are called the first-order phase transitions.
- A one-well situation corresponds to a case when there is no dominant risk or a small group of risks characterized by extremely high sensitivities. The project system just accumulates consequences of occurred risks forming a final actual project outcome. Continuous phase transitions of the kind are called the second-order phase transitions in physics.
- Does this mean that all complex projects must jump from one well to the other to a chaotic state inevitably? Hopefully not.[8] A likelihood of this event could be easily assessed using an adopted confidence level criterion. For example, if P70 confidence level is used and its value does belong to the second peak, the project will end up in the second chaotic state B with probability 70 per cent (see the example of Figure 17.1).

17.4 Quality in Complex Project Systems: Practical Aspects

The importance of quality controls in projects are discussed in this book on several occasions. Quality risks create the backbone of the project rework cycles that serve as foundations of project system dynamic models. This topic inevitably highlights two implications. One relates to modelling and risk quantification in complex projects, the other to practical QA/QC implementation.

[7] A two-well potential can be formed only in clearly non-linear systems: to get two wells a potential should take a form $V(x) \sim x^4 - x^2$ (Strogatz (2015)). A trivial linear situation when $V(x) \sim x^2$ corresponds to the one-well potential. The system can afford small linear deviations from the equilibrium but stays trapped inside the well: there is nowhere to escape to as the second well does not exist. This always corresponds to a one-peak probability distribution. It could be imagined that a dynamic risk pattern supported by a project system can be interpreted as the potential $V(x)$ formed by the affinity of interacting risks where a project system resides.

[8] If the iron law of megaproject is recalled at this point again (Flyvbjerg (2017)), this desperately optimistic expression 'hopefully not' should be replaced by 'almost certainly yes'.

These are natural constraints to poor work quality and poor-quality detection time parameters which should be kept in mind when developing and calibrating system dynamic (SD) models. These two types of parameters should be properly evaluated as part of the assessment of existing project system issues using available relevant empirical data.

If these parameters are selected out of a realistic range, corresponding SD models will produce very pessimistic results: simply, a project might be never completed. Any project work would require more rework. Unfortunately, this could be a realistic result in some notorious cases. This is a reminder about the importance of proper SD model's calibrations.

Second, QA/QC implementation has three main challenges. The QA/QC function should:

- exclude or minimize any possibility for propagation of poor quality of works to down-the-line project phases. This has a goal to decouple rework cycles and avoid the knock-on impacts,
- detect as many instances of poor quality engineering and construction works as possible to decrease rework inside rework cycles, and
- detect instances of poor quality of works as soon as possible.

The points above inevitably call for a healthy trade-off between QA/QC improvements aimed at capturing as many quality defects as possible and the improvements that zero in on capturing the quality defects as soon as possible. A case when 100 per cent of poor-quality instances are always detected although with a substantial delay could be less preferable than a case when the detection is done promptly although not with exactly 100 per cent of them.

Major QA/QC road blocks on the way to poor-quality propagation should be built between project phases. Their purpose is to stop delayed knock-on impacts of up-the-line quality risks on down-the-line quality risks. The main burden to stop the knock-on interactions should be carried by corresponding thorough and persistent technical reviews. For example, the final engineering model reviews should be allowed an excessive amount of time and carried out by highly-qualified checkers. In essence, these road blocks are an attempt to reduce complexity of a given project through cutting the possibilities of cross-risk interactions at large and knock-on interactions in particular. If this is successfully done a complex project will obtain characteristics of complicated or even simple projects as most risks are reduced to non-interacting although amplified risks (see Chapter 8).

17.5 Nature and Origin of General Uncertainties and Internal Risk Amplifications

Standalone non-interacting risks are discussed in the first part of the book. According to Chapter 2, four types of risks can be told apart depending on uncertainties of occurrences and impacts associated with them—uncertain events, discrete uncertain events, general uncertainties, and issues. The situation becomes complicated in the second part of the book when risks of all four types are not considered standalone any longer: they start interacting being exposed to internal risk amplifications, knock-on interactions, and compounding.

Along with the considered interactions, a role of project system issues is accentuated in internal risk amplifications. Chronic project system issues give rise to the static PSS–PDS mismatches and serve as systematic risk causes (see Figure 9.2).

Besides this an origin of the dynamic PSS–PDS mismatches is discovered. First, internally amplified standalone risks when occurred induce rework (belongs to PSS) and work constraints (belong to PDS) as Figure 9.1 shows. Second, previously occurred risks may provide new dynamic project system issues to down-the-line risks, which defines knock-on risk interactions (see Figure 9.3). Third, if some risks occur simultaneously, this additionally impacts productivity of works to carry out both regular work and required new rework which constitutes compounding risk interactions (see Figure (9.4)),

One missing piece of the puzzle is better understanding of the origin of internal risk amplifications, although it would not hurt to better understand the origin of general uncertainties to start with.

17.5.1 The Nature and Origin of General Uncertainties

A limiting case when an entire project risk exposure is reduced to general uncertainties corresponds to old and unglamorous activity ranging and line-item ranging Monte Carlo methodologies discussed in Chapter 3. These methodologies are quite remarkable as they do not presume existence of uncertain events whatsoever. However, another limiting case may be thought out when the entire project risk exposure is represented by uncertain events only: no general uncertainties are contemplated at all.

Hypothetically it is possible although not quite practical to identify hundreds if not thousands of project uncertain events that include both very major risks of high severity and all other possible lower-severity risks. A schedule risk analysis example can be considered to illustrate this.

If a project schedule consists of 500 normal tasks and each normal task is impacted by ten project risk events of various severities, it would be required to identify, assess, and manage 5,000 project risks. This number could be somewhat reduced if each risk is mapped to, say, five normal tasks on average. One thousand is still a very high number of risks to manage, which does not make any practical sense.

What should be done to handle this hypothetical situation? What factually happens in real projects instead?

Only a few top project risks (tier-one risks) are actually identified, assessed, and addressed. The usual number of tier-one risks belongs, say, to the two-to-three-dozen range. They are on the project team's radar screen being kept in a project risk register.

What about the remainder of the less important multiple tier-two risks? Examples of engineering tier-two risks may include as follows.

- A key engineering specialist might go down with a flu and call in sick for a couple of days.
- A computer system used for development of engineering drawings could be temporarily down.
- There could be communication difficulties among engineering offices located on different continents due to the language barrier and the difference in time.
- etc.

Despite it being a complete waste of effort and time to identify, assess, and address a myriad of lower-severity tier-two risks, they cannot be completely ignored. This justifies the origin of general uncertainties as follows.

It could be imagined for a moment that all tier-one risks are ignored and only tier-two risks are considered in modelling. In the SRA example, they are all mapped to corresponding normal tasks. It is possible to consider each normal task separately and build an individual Monte Carlo model for each task. Following this logic there should be 500 micro Monte Carlo models. Each normal-task-based Monte Carlo model with several tier-two risks attached can be run independently. As always, an outcome for each Monte Carlo model will be a duration distribution that provides an outcome for a given normal task. Needless to say, each such distribution can be treated as a general uncertainty around its normal task's duration.

Due to the complete impracticality to build and run several hundred micro Monte Carlo models of this kind, this step is passed over. Instead, assessments of general uncertainties for each normal task are undertaken. This is what happens in Monte Carlo risk quantification workshops along with discussions on important tier-one risks. These activity ranges (line-item ranges in the case of cost risk analyses) along with tier-one risks mapped to corresponding normal tasks play the role of inputs to traditional Monte Carlo models. In other words, general uncertainties play a role of a dumping ground for tier-two uncertain events.

If all tier-two risks can be treated and handled the same way as the tier-one risks, there would be no need for general uncertainties as inputs whatsoever. Practical usage of general uncertainties is justified by a possibility to assess them specifically one by one for each normal task without modelling. Their assessments are done 'as-is': this is a subject of their snapshot audit in a given project phase which does not require predicting their future impacts.

All tier-two risks discussed above are understood as future uncertain events that produce general uncertainties. In reality some tier-two risks might have occurred by the moment when general uncertainty assessments are undertaken. The same as in the case of occurred tier-one risks they should be taken into account in a project base schedule or cost base estimate.

As argued in Chapters 3 and 13, some tier-two risks that are hidden in general uncertainties might become important enough in the course of project development. This may be recognized first as a need to inflate corresponding general uncertainty ranges. Instead a new risk should be recovered from general uncertainties (tier-two risks) and transferred to a project risk register to become a new tier-one risk.

17.5.2 General Uncertainties and the Hybrid Model

A similar thinking experiment can be undertaken for another limiting case: modelling of general uncertainties at the project summary level when the entire project schedule is replaced by one summary task. It could also be imagined that all tier-two risks counted in hundreds are mapped to that single summary task. Such single-summary-task Monte Carlo model can be run hypothetically and might produce a duration distribution for that summary task. The difficulty with this approach is the transition from that hypothetical single-summary-task Monte Carlo model to actual justification and assigning of ranges around that summary task.

Whereas assigning of ranges at the activity level is well-structured and can be specific enough for each normal task, assigning ranges to the single summary task representing an entire project schedule is not structured at all and cannot be specific. Project logic will be completely ignored in this case. The only way to assign the ranges to the summary task that looks somewhat reasonable is using empirical data and expert judgements based on information for somewhat similar projects. As a next step that mammoth summary task with ranges is added to a regular Monte Carlo model for tier-one risks.

Actually, this is exactly the hybrid Monte Carlo model reviewed in Chapter 4 where its deficiencies are discussed in detail. Besides the other shortcomings, it suffers from double-dipping. Empirical information on completed projects used to assess the summary task ranges already contains information about all occurred risks including tier-one risks that are counted again in the Monte Carlo module of the hybrid model.

The same method is applicable to cost modelling by proponents of the hybrid model. An additional difficulty of the hybrid cost risk modelling is that it impedes counting schedule-driven costs. Schedule-driven costs are general uncertainties by their nature. A structured method to count them is to derive duration distributions for normal tasks from an SRA model and multiply them by applicable burn rates. This means they should be counted at the line-item level as variable costs though using specific relevant burn rates (see Chapter 3). This is not possible in case of the hybrid model. If an input from a hybrid schedule risk model is used to quantify schedule-driven costs, what burn rate should be applicable for this at the project summary level? As discussed in Chapter 3, undertaking a CRA without consistent counting of schedule-driven costs does not make much sense. Hence, this statement is even more applicable to the hybrid risk modelling.

17.5.3 The Nature and Origin of Internal Risk Amplifications

The intent of this section is to better understand the origin of internal risk amplifications in the non-linear case and compare it with the nature of general uncertainties in the linear case.

What is the exact nature of internal risk amplifications? These should exist even in a simple project in absence of knock-on and compounding interactions among identified risks. This is not exactly right. Internal risk amplifications of tier-one risks are a summary knock-on impact of multiple tier-two risks.

This discussion has a time-related aspect. Any risk events—both tier-one and tier-two—if they have already occurred in the past, should be taken into account either as a part of a project base schedule or cost base estimate or as causes of new risks. The first option means accrual of schedule delays and extra costs when schedule and cost objectives are impacted. The second option is relevant to impacts on other objectives, for example, on quality objective.

The difficulty with the second option in traditional Monte Carlo models is that risk interactions are not somewhat consistently considered: all risks are identified as standalone (see Figure 2.2). Realizing the importance of identifying risk interactions smooths the path towards non-linear Monte Carlo modelling. In this case occurred risks especially quality risks can become dynamic project system issues themselves. If the occurrence is a tier-one risk, this is a situation when a dynamic project system issue emerges resulting in the dynamic PSS–PDS mismatch. What if what has occurred are tier-two risks? They are not managed and not explicitly considered by project teams. Actually, non-interacting tier-two risks are already taken into account as general uncertainties in the linear case (see 17.5.1). They are part of traditional linear Monte Carlo models. However, the tier-two non-linear risk interactions are still missing out.

Tier-two risks may interact with tier-one risks through cross-interactions. This situation is fully described by a corresponding bowtie diagram for knock-on (Figure 9.3) and compounding (Figure 9.4) risk interactions. They yield myriads of dynamic risk-interaction-induced project system issues that serve as additional dynamic causes of tier-one risks. However, as the number of tier-two occurred risks and the induced dynamic causes can be counted in hundreds, their impacts can be combined resulting in a few chronic project system issues that serve as systematic causes of tier-one risks. These causes are reasons for internal amplifications of the tier-one risks (see Figure 9.2). The conversion of multiple dynamic project system issues induced by tier-two risk interactions looks somewhat comparable with the conversion of multiple standalone tier-two risks to general uncertainties. That is a mechanism as to how multiple occurred lower-severity risks knocking on tier-one risks can be aggregated as few chronic project system issues resulting in internal tier-one risk amplifications. Apparently, in contrast to the linear mechanism of the general uncertainty aggregation, the emergence of internal amplifications of tier-one risks have the non-linear nature.

Apparently, in the case of traditional linear Monte Carlo modelling regular causes of project risks that do not depend directly on features of project systems (e.g. geotechnical conditions, weather events, etc.) and chronic project system issues are not told apart. Any causes could become included in the traditional bowtie diagram (see Figure 2.2). Regular causes of risks and systematic causes that reflect on features of a project system should be told apart in the non-linear case. This leads to a technical definition of systematic causes of project risks: they represent summary impacts of those tier-two risks that directly dependent on characteristics of a project system.

Despite the linear and non-linear Monte Carlo methodologies being completely delineated as two separate methodologies in this section, the linear Monte Carlo modelling is fully included to the non-linear models as discussed in Chapters 14 and 15. The non-linear methodology is built on the linear Monte Carlo modelling just through adding risk interactions.

Figure 17.11 recapitulates the discussion above. Although it does not show one of the most important factors of the risk exposure—schedule-driven costs. These are presumed to be a part of both the linear and the non-linear Monte Carlo methodologies.

The idea to aggregate factors of secondary importance in bulk is widely used in quantum mechanics and mathematical physics when main components of models or equations (tier-one factors) are treated meticulously similar to tier-one risks, whereas secondary factors (tier-two factors) are considered second-order perturbations. The

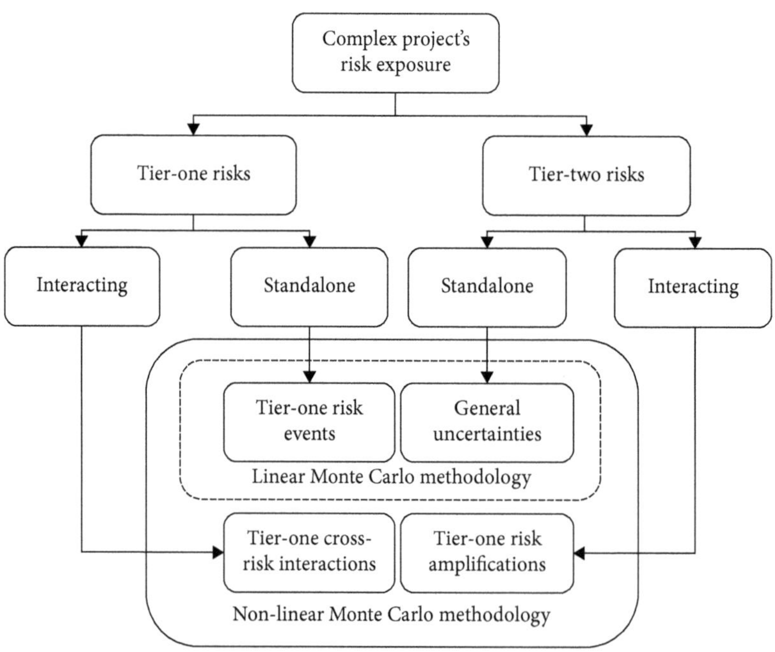

Figure 17.11 Risk exposure of a complex project and its quantification

corresponding approach is called 'the perturbation theory' (see Eisberg and Resnik (1985) or any other book on quantum mechanics or mathematical physics). Simply speaking, this is the way to keep most informative components in a model and to aggregate all substantially less important factors as secondary corrections although without getting rid of them. Contributions of tier-two risks to general uncertainties in the traditional linear Monte Carlo models and to internal risk amplifications in the non-linear models are good examples of this approach—they are kept as perturbations. However, perturbations to these small perturbations can be ignored as negligible. By this reason non-linear corrections to general uncertainties are not considered in this book.

17.6 Expected Project Outcome: Reverse Engineering and Stretched Targets

This final section of Chapter 17 zeroes in on two risk handling strategies that are beyond traditional linear risk addressing discussed in Chapter 2—avoid, mitigate—prevent, mitigate—recover, transfer, and accept. Moreover, they are even beyond newly established strategies to address three types of risk interactions—internal amplifications, knock-on interactions, and compounding—discussed in Chapters 14 and 15. These two additional strategies are based on reverse engineering and establishing attainable stretch targets.

17.6.1 Expected Project Outcome: Reverse Engineering

The traditional risk addressing thinking is based on 'if X, then Y' logic as follows. If identified project risks are addressed in a particular way using a set of addressing actions X_1, then a project schedule or cost outcome should be Y_1. If the risks are addressed in a particularly different way using a set of actions X_2, then an outcome should be Y_2, etc. This approach fully depends on two factors. First, chances for success to implement developed addressing actions. Second, adequacy of the selected risk quantification methods (see Chapters 3 and 4). This traditional logic could be treated as one of usual what-if planning scenarios when two main risk quantification cases—before addressing and after addressing—are run.

This logic should be reversed to become 'if Y, then X'. Namely, if desirable is a project outcome Y, then a set of addressing actions X should be reverse-engineered: if a project success is really expected, a certain set of risk addressing actions including addressing of chronic and dynamic project system issues is mandatory, unless a miracle is expected.

The second business case of project Zemblanity (see Chapter 15) actually demonstrates this approach partially. For simplicity purposes only chronic and dynamic project system issues responsible for intra-risk (internal amplification) and cross-risk (knock-on and compounding) interactions have been addressed to tell apart the models N-SCRA-A and N-SCRA-B. The same set of traditional risk addressing actions is used in the models L-SCRA, N-SCRA-A, and L-SCRA-B (see Chapters 5 and 15).

In a general case, it could be reversely modelled what sets of addressing actions should be applied to ensure that project schedule or cost contingency does not exceed 25 per cent, 20 per cent,15 per cent, 10 per cent, etc. of base duration and estimate. What chronic and dynamic project system issues should be eradicated at least partially for this? Technically this means defining corresponding non-linearity parameters (see Chapters 13–15) that stay behind the risk interactions that should ensure yielding the required modelling outcomes. Needless to say, actual risk-interaction addressing actions should stand behind these parameters as discussed in Chapter 15.

Going back to the discussion on the Hiding and Revealing Hands, this is an additional measure to replace the Hiding Hand by the Revealing Hand. However, if the reverse engineering is not proactively pursued, the Revealing Hand could be successfully replaced by the Hiding Head. This is an analogue of or a connotation to playing ostrich in project risk management.

17.6.2 Expected Project Outcome: Stretched Targets

The happy example when the project team beat a P50 project duration stretched target is discussed in Chapter 3. It should be reminded that the project team was quite strong, project developed procedures were well developed and consistently implemented. Major stakeholders including federal and provisional governments fully supported the project, etc. In terms of chronic project system issues and static PSS–PDS mismatches the project system was highly mature (see Chapter 12). Apparently, few project risks did occur and they probably interacted weakly with each other. However, this was a good example of the near-linear Monte Carlo modelling case. This was one of two key factors why the traditional Monte Carlo modelling turned out to produce extremely accurate results (Raydugin (2017a)).

The second factor was a type of bias that could be qualified as positive. Various realizations of psychological and organizational bias are discussed in Chapters 1–4 that all are treated as systematic errors in various steps of PRM process including risk assessments. They all are understood as realizations of negative bias. In terms of Figure 3.4 they all increase the methodology's systematic error Δ. On the contrary, positive bias could be distinguished. Positive bias is about the motivation and leadership to beat established goals.[9] This has a direct link to the topic on project shaping and project leadership discussed in 11.1.2, although the positive bias can be embroidered to any aspect of PDS. This is a sort of positive or upside project system issue or anti-issue as discussed in 13.4.6.

[9] Real-life stories of dozens of great leaders—from Winston Churchill and Steve Jobs, to Albert Einstein, to Peter the Great—could be relevant in this regard. In the context of this discussion they all could be characterized by a strong mostly positive bias that could be quite irrational at times. The positive bias discussed here is usually referred to as 'a vision'.

Positive bias can play an important role only if it can outweigh negative project system issues. Namely, if a project system is immature (see Chapter 12), any realizations of the positive bias would be suppressed or dispersed by overwhelming chronic and dynamic project system issues. If the project system is mature or becomes mature, any realizations of positive bias would be able to play a definitive fine-tuning role as project system issues become less dominant. In this case the positive bias has a capacity to reduce the systematic error Δ of the adopted risk quantification method: when a stretched target is established, it helps to mobilize and fine-tune PDS to beat the target. Any possible moderate systematic error could be screened out by the positive bias.

That is exactly what happened in the discussed example (Raydugin (2017a)). Once again, this is possible only in the case of quite mature project systems that are responsive to stretched targets. The positive bias in cases of immature project systems cannot find a required response to beat a stretched or any other target. It will be simply bogged down in project system issues.

17.7 Chapter's Markup and Takeaway

- A practical recipe is required to convert the well-known mantra that complex projects should be adaptive to practice.
- A reverse trend is observed that complex projects are being made less adaptive through imposing total control and over-planning. Excessively detailed schedules are developed for projects that will be executed in a few months or even years. This represents a situation when scheduling tools take over the entire project planning context and process.
- In reality any complex project suffers from on-going disruptions and perturbations non-stop, which makes a rigid and detailed schedule to lose any link with reality.
- Decentralization and flexibility based on short-term project planning to ensure work availability is the proposed recipe for success of complex projects.
- Some projects have two-peak schedule and cost probability distributions that are associated with two strange attractors. Coming from physics the strange attractors may be understood as two most likely duration or cost project outcomes. Transition from one state of a project system to the other is interpreted as a phase transition. The second state is associated with chaos and failure.
- It is demonstrated that a single most sensitive risk should be responsible for a jump from one state (or phase) to the other. The abrupt transition of the kind is called a first-order phase transition in physics.
- If a distribution has a single peak, multiple risks comparable in sensitivity are responsible for the final project outcome due to multiple mixed failure modes. A slow transition to the final state (or phase) is called a second-order phase transition in physics.
- Sensitivity charts can immediately identify risks with distinctively high sensitivities that could be responsible for the first-order transitions to chaos.

- Single failure mode transition (two-peak distributions) is likely in SRA modelling. CRA models take in SRA results and have their own failure modes. It is less likely that a single failure mode survives in a CRA model.
- Quantum physics provides a direct analogy for the system's transition based on a behaviour of a system in a two-well potential described by Schrödinger's equation. It could be imagined that the dynamic risk pattern supported by a project system is interpreted as the potential formed by the affinity of interacting risks where a project system resides. Non-linearity stemming from risk interactions ensures that the potential has two wells.
- System dynamics experience points to two important aspects of QA/QC: numbers of identified defects and times required to identify them. A healthy trade-off should be established between the two. If the number of identified defects is critically low and it takes too much time to identify them, a project might never be completed as rework will be snowballing.
- These two parameters—numbers of identified defects and the identification times—could be used along with the other system dynamic parameters for mutual calibrations of system dynamic and Monte Carlo models as well as calibration based on empirical data.
- General uncertainties are an aggregate product of all tier-two non-interacting risks that did not make it to the project risk register containing tier-one risks. They are considered in a linear Monte Carlo case when risks are non-interacting.
- Internal risk amplifications (intra-risk interactions) are an aggregate product of all the tier-two interacting risks in the non-linear case when risks are interacting.
- A basic logic used when developing project contingencies should be reversed: from defining risk addressing and then wondering what would be a resulting contingency, to establishing the desired contingency and then defining required addressing. This reverse-engineering contingency development approach should help avoid some Hiding Head implications.
- Stretched targets are aimed at triggering positive bias and motivation to beat them. This strategy is viable only when a project system is mature enough. Stretched targeting and the positive bias help to reduce systematic errors of the Monte Carlo methodology.

18
Conclusion

This book is devoted to risk quantification in complex projects as a branch of political mathematics.

The political aspects of risk quantification are defined by various realizations of psychological and organizational bias linked with:

- features of risk quantification method selection,
- data developed and used as inputs to risk quantification models,
- experience and training of risk data collection facilitators,
- experience and education of risk quantification modellers,
- robustness of the risk quantification models and risk procedures, and
- traits and proclivities in practical usage of obtained risk quantification results in decision-making.

The mathematical aspects of risk quantification are defined by the magnitude of a systematic error pertaining to a selected risk quantification method stemming from:

- features of risk quantification method selection,
- data developed and used as inputs to risk quantification models,
- experience and training of risk data collection facilitators,
- experience and education of risk quantification modellers,
- robustness of the risk quantification models and risk procedures, and
- traits and proclivities in practical usage of obtained risk quantification results in decision-making.

What a surprise! These two lists related to the bias factors and to the systematic errors are identical. The bias (both psychological and organizational) as a driver behind the political aspects has a twin brother which is a systematic error of a selected risk quantification methodology also known as the statistical bias. The Bias brothers are two sides of the same coin called project risk quantification.

The political aspects of project risk quantification comprise a reach variety of notions and terms from the Hiding Hand and strategic misrepresentation to *bahramdipity*, to zealotry.

The mathematical aspect cannot be understood without comprising of various terms from convenience and judgement sampling, to 10 per cent syndrome, to the Revealing Hand.

Modern Risk Quantification in Complex Projects: Non-linear Monte Carlo and System Dynamics Methodologies. Yuri G. Raydugin, Oxford University Press (2020). © Yuri G. Raydugin.
DOI: 10.1093/oso/9780198844334.001.0001

The political aspects of bias should be addressed by political means. The mathematical aspects—by mathematical.

Namely,

1. all political aspects of bias pertaining to risk quantification should be handled by either government (in the public sector) or corporate (in the private sector) policies and procedures that ensure that decision-makers handle all relevant risk quantification aspects—from selection of risk quantification methodologies to proper use of risk quantification results—in a consistent, structured, and transparent way;
2. all mathematical aspects of risk quantification should be delegated to vetted risk managers, facilitators, and modellers that possess required experience, training, and education.

In both cases when either political or mathematical aspects of project risk quantification are not settled down properly the entire risk quantification efforts would be perfectly futile, guaranteed. In some particular practical cases this is an expected and a desired outcome in accordance with the Hiding Hand principle. To cut a long story short, the cases when political aspects of risk quantification are handled by 'mathematicians', or mathematical aspects are handles by 'politicians' should be avoided, if possible.

A reasonable match between political and mathematical aspects of project risk quantification should be found. In physics language, both aspects should enter a resonance. Or, simply, they should get on the same wavelength.

The cornerstone mathematical aspect of the proper risk quantification is about a selection of a methodology that has acceptably low level of the systematic error. In most cases this is all about risk exposure factors that are included and the way they are included in risk quantification modelling.

The traditional linear Monte Carlo risk quantification methodology takes into account all major risk exposure factors in its L-SCRA version but risk interactions pertaining to complex projects. Apparently, the magnitude of a systematic error is proportional (apparently in a non-linear manner) to the project complexity including intensity of risk interactions. The traditional methodology should be acceptable in cases when risk interactions in a project system can be ignored. These are mostly the cases of simple projects. Alarmingly, in the other cases—complicated, complex, and chaotic projects—the magnitude of the systematic error of the traditional Monte Carlo modelling becomes intolerably high. Hence, the development of the modified non-linear Monte Carlo methodology undertaken in this book is pretty logical and, in a way, overdue.

Following the traditional project risk management process, risk interactions as major contributions to project risk exposure of complex projects should be identified, assessed, and addressed. This book provides required concepts and tools for these main project risk management process steps.

Identification for the three main sorts of non-linear risk interactions—internal risk amplification, knock-on interactions, and compounding—is supported by the notion of the static and dynamic PSS–PDS mismatches and related types of chronic and dynamic project system issues. Risk interaction mapping that visualizes the dynamic risk pattern supported by a project system is developed to identify the aggregated impact of an affinity of interacting risks on project schedule and cost objectives. The three modified bowtie diagrams are developed for identification of particular instances of the risk interactions.

A method of the assessment of risk interactions forming the dynamic risk pattern is developed. This is the N-SCRA methodology. As it takes into account the last missing component of project risk exposure—risk interaction—it promises to have the quite low systematic error.

Addressing of risk interactions is based on the idea to disarray the dynamic risk pattern supported by a system of a complex project. On the one hand, the Monte Carlo methodology could be used as a tool for optimization and simplification of the PSS. On the other hand, if PSS and its complexity for a particular project are given, the only way to reduce project complexity is to address risk interactions.

In the case of intra-risk interactions chronic long-term systematic issues should be addressed. In the case of knock-on interactions, routes for propagation of knock-on risk interactions should be blocked. In the case of compounding, windows for simultaneous risk happening should be closed.

Simply speaking, these addressing measures are aimed at depriving a complex project system of its dynamic complexity traits. If most risk interactions are eliminated, the previously complex system could be downgraded to complicated or even simple.

The development of the non-linear Monte Carlo risk quantification methodology was supported by insight from system dynamics. Project system dynamics modelling undertaken in parallel allowed features of project systems related to non-linear risk interactions that enrich Monte Carlo modelling to be uncovered. Despite several shortcomings, the project SD modelling could be utilized as a standalone project risk quantification tool or a tool to model some Monte Carlo input.

Although the developed non-linear Monte Carlo methodology works well, for the time being it is not supported by existing Monte Carlo software packages that zero in on traditional Monte Carlo methodology by obvious reasons. This makes building of non-linear Monte Carlo models slightly onerous. As soon as the modern Monte Carlo tools are modified and allow factoring in project complexity and corresponding non-linear risk interactions, the non-linear Monte Carlo modelling would become a pleasant although boring and repetitive routine.

The paramount goal of this book is to convince project and project risk practitioners of the compelling need to finally take project complexity into account in practical terms when managing complex projects and when developing their schedule and cost contingencies. The pertinacious risk quantification method that meets both challenges turns out to be the non-linear Monte Carlo SCRA methodology. The author of this book cannot exclude a hypothetical situation when an alternative modelling approach is put forward that also:

- takes into account project complexity including risk interactions, and
- has an acceptably low systematic error.

If these two conditions are obeyed, such method should be fully supported.

As this work is conceptual and all models—both Monte Carlo and SD—are based on the imaginary project Zemblanity business case, no practical calibration of the models based on empirical data is undertaken, although certainly this topic deserves attention.

The discussion about the risk quantification methods is led in this book as a collision of the contesting Hiding and Revealing Hands. The author is not completely sure that everything is wrong with the Hiding Hand in all situations. There is also no complete confidence that everything is right with the Revealing Hand by default. Despite the situation not being exactly black and white, the author's preferences are clearly on the side of the Revealing Hand.

The newly-developed non-linear Monte Carlo methodology representing the Revealing Hand should help avoid 'unpleasant unsurprises' and 'unhappy, unlucky and unexpected discoveries occurring by design' in complex projects. Would this not also help to decline the pushy invitation to attend the 'banquet of consequences' in a proper way?

APPENDIX A

The Linear Monte Carlo Schedule and Cost Risk Analysis Model (L-SCRA)

This appendix provides details on inputs to integrated schedule and cost risk analysis model L-SCRA developed for the Zemblanity project business case of Chapter 5.

Tables presented in this appendix have multiple rows and columns. Due to space constraints all tables are reduced to minimum size through table 'normalization'. Namely, all repetitive columns are left out. For example, in all risk registers Risk ID plays a role of a unique risk identifier, so that risk titles and addressing actions are represented only once in Table A.1. Risk causes, events, and impacts are shown by Table 5.3.

Only after-addressing risk assessments are discussed in this appendix as all Monte Carlo models are built for the after-addressing cases only.

All models represented in this book are developed for demonstration purposes. They are largely oversimplified versions of models that would be required for risk quantification in real projects.

Table A.1 Deterministic risk register of project Zemblanity (after risk addressing)

Risk definition		**Addressing Action**	**After-addressing assessment scores**			
Risk ID	**Risk Title**		**Severity**	**Probability**	**Schedule**	**Cost**
R01	Design Basis Change (Proponent-Driven)	Internal stakeholder's communications Operations and design alignment Change management protocol	12	3	3	4
R02	Overlooked External Stakeholder's Requirements	External stakeholder's identification External stakeholder's communication External stakeholder's follow-up	9	3	3	3
R03	Errors and Omissions (Engineering)	Timely hiring required engineering specialists Implementation of the engineering review procedure Engineering interface management planning	6	2	3	3
R04	Poor Work Quality (Construction)	Timely hiring required construction specialists Construction interface management planning Implementation of QA/QC procedure	8	2	4	3

Continued

Table A.1 Continued

Risk definition		Addressing Action	After-addressing assessment scores			
Risk ID	**Risk Title**		**Severity**	**Probability**	**Schedule**	**Cost**
R05	Delays in Material Delivery	Long lead item's delivery planning	4	2	2	2
R06	Commissioning and Startup (CSU) Failure	Commissioning and startup planning Timely planning for tie-ins Interface management and communications with contractors	9	3	3	3

For example, only six project risks are included in the risk register of project Zemblanity (see Table A.1), whereas risk registers of real projects may contain a few dozens of risks, etc.

A.1 Outcome of Deterministic Risk Analysis

Risk register of project Zemblanity developed using deterministic (scoring) method is shown by Table A.1 Only risk scores for two quantifiable project objectives—cost and schedule—are shown that are a subject of the risk quantification due to the space constraints. The other project objectives—quality, safety, environment, reputation, etc.—that are not quantifiable by Monte Carlo methodology are left out. A corresponding version of the project RAM is shown by Figure 3.1.

A.2 Inputs to the Linear Schedule Risk Analysis (L-SCRA) Model

A schedule of project Zemblanity used in the L-SCRA modelling is shown in Figure 5.1.

General uncertainties associated with the normal tasks are described by Table A.2. Base estimate as well as burn rates used for calculations of variable costs and cost ranges are shown by Table A.3.

Table A.2 General schedule uncertainties of project Zemblanity

ID	Description	Duration, d	Start	Finish	Min, %	Min, d	ML, d	Max, d	Max, %
MS01	Project Start	0	1-Apr-2020				0		
SD00	Scope Definition (FEL2)	140	1-Apr-2020	18-Aug-2020	95	133	140	154	110
PD00	Project Definition (FEL3)	226	19-Aug-2020	1-Apr-2021	90	203	226	260	115
MS02	FID	0	2-Apr-2021	2-Apr-2021			0		
DE00	Detail Engineering	270	2-Apr-2021	27-Dec-2021	85	230	270	351	130
CN00	Construction	550	28-Dec-2021	30-Jun-2023	85	468	550	715	130
CS00	Commissioning and Startup	173	12-May-2023	31-Oct-2023	85	147	173	225	130
MS03	Project Completion	0		31-Oct-2023			0		

Table A.3 A base estimate, burn rates, and general cost uncertainties of project Zemblanity

ID	Summary Activity	Duration, d	Fixed cost, $'000	Burn rate $'000/d	Variable cost, $'000	Total cost, $'000	Min, %	Max, %
	Project Zemblanity		472,380		527,620	1,000,000		
SD00	Scope Definition (FEL2)	140	2,000	30	4,200	6,200	95	110
PD00	Project Definition (FEL3)	226	5,580	70	15,820	21,400	90	115
DE00	Detail Engineering (incl. Procurement)	270	422,800	160	43,200	466,000	85	130
CN00	Construction	550	36,000	750	412,500	448,500	85	130
CS00	Commissioning and Startup	173	6,000	300	51,900	57,900	85	130

A risk register of project Zemblanity with inputs required for the linear schedule and cost risk analysis (L-SCRA) is represented by Table A.4. For simplicity purposes it contains only data for the after-risk-addressing case.

Table A.4 A L-SCRA risk register (a linear case)

Risk ID	Correlated with	Correlation coefficient, %	Mapped to	Probability, %			Schedule impact, d			Cost impact, $'000		
				Min	Max	P_{after}	Min	ML	Max	Min	ML	Max
R01	R02	90%	SD00	20	30	25	20	40	70	2,500	10,000	25,000
	R03	90%	PD00									
	R04	90%	DE00									
	R05	90%										
	R06	90%										
R02	R01	90%	SD00	15	25	20	10	20	40	3,500	7,500	12,500
	R03	90%	PD00									
	R04	90%	DE00									
	R05	90%										
	R06	90%										
R03	R01	90%	SD00	10	20	15	15	30	50	3,500	5,000	6,500
	R02	90%	PD00									
	R04	90%	DE00									
	R05	90%										
	R06	90%										
R04	R01	90%	CN00	15	25	20	30	50	70	20,000	30,000	50,000
	R02	90%										
	R03	90%										
	R05	90%										
	R06	90%										
R05	R01	90%	CN00	10	20	15	10	20	40	5,000	7,000	10,000
	R02	90%										
	R03	90%										
	R04	90%										
	R06	90%										

Continued

Table A.4 Continued

Risk ID	Correlated with	Correlation coefficient, %	Mapped to	Probability, %			Schedule impact, d			Cost impact, $'000		
				Min	Max	P_{after}	Min	ML	Max	Min	ML	Max
R06	R01	90%	CS00	10	20	15	20	30	50	10,000	15,000	30,000
	R02	90%										
	R03	90%										
	R04	90%										
	R05	90%										

Table A.5 Probabilities and risk impacts used as inputs to the model L-SCRA

L-SCRA	Probability, %	Schedule impact, d			Cost impact, $'000		
		Min	Most likely	Max	Min	Most likely	Max
R01	25	20	40	70	2,500	10,000	25,000
R02	20	10	20	40	3,500	7,500	12,500
R03	15	18	30	50	3,500	5,000	6,500
R04	20	30	50	70	20,000	30,000	50,000
R05	15	10	20	40	5,000	7,000	10,000
R06	15	20	30	50	10,000	15,000	30,000

Table A.5 rewrites Table A.4 through dropping information about mapping and correlations, which are the same in all Monte Carlo models of this book. The abridged format of Table A.5 is also used in Appendix D for non-linear Monte Carlo modelling inputs.

A.3 The Workable Model L-SCRA

The workable model L-SCRA is available on the book's companion website at www.oup.co.uk/companion/Raydugin.

APPENDIX B

Analytical Solutions for the Bathtub (BT) Model

B.1 A General Differential Equation for the BT Model

Equation (6.3) can be re-written in a differential equation form for a BT model. It takes a form of a first-order differential equation that describes water level changes over time as a difference between inflows and outflows as follows.

$$dL(t)/dt = I(t) - O(t), \tag{B1.1}$$

$$L(t=0) = L(0), \tag{B1.2}$$

where $L(t)$ is a level of water in a bathtub at any moment of time t, $I(t)$ is water inflow, $O(t)$ is water outflow, $L(0)$ is an initial water level. Apparently, water inflow $I(t)$ and outflow $O(t)$ are measured in [metre/minute] units being flow rates by their nature. Specific solutions to equation (B1.1) depend on characteristics of inflows and outflows as well as on imposed initial and boundary conditions.

Boundary conditions could be imposed on BT model solutions $L(t)$ to ensure that $L(t) \leq L_{max}$ and $L(t) \geq 0$. In some cases, they are required to introduce realism to modelling. In other cases, boundary conditions are obeyed without imposing but due to structures of corresponding models. For instance, in case of a generic model BT1 no boundary conditions are imposed which leads to unrealistic unconstrained growth or decay. Model BT2 is the same as model BT1 but with the imposed boundary conditions. Model BT3 does not have a boundary condition on exponential growth although its solution does allow a smooth decay to zero level. Model BT4 complies with the growth boundary condition through smooth growth approximating a maximum level. Model BT5 modifies model BT4 in a non-linear way to ensure that unconstrained exponential growth is replaced by smooth S-shape sigmoid curve.

Five versions of the BT model are introduced in 6.1. Their analytical solutions are discussed in sections B2–B6. Visualized results of system dynamics modelling for models BT1–BT5 are discussed in 6.2–6.4.

B.2 An Analytical Solution for the Model BT1

A most simple version of the BT model BT1 is realized when inflow and outflow rates $I(t) = r_{in}(t) = r_{in}$ and $O(t) = r_{out}(t) = r_{out}$ are constant:

$$dL(t)/dt = r_{in} - r_{out} \tag{B1.3}$$

Direct integration of the equation (B1.3) yields a solution for the model BT1 as follows.

$$L(t) = L(0) + (r_{in} - r_{out}) \times t \tag{B1.4}$$

If $r_{in} > r_{out}$, an unlimited linear growth of the water level is expected from the initial level $L(0)$. *If* $r_{in} < r_{out}$, an unlimited linear decline of the level is expected including negative values. If $r_{in} = r_{out}$, the level stays unchanged over time being equal to $L(0)$. A graphical representation of the solution (B1.4) for the model BT1 is shown by Figure 6.4.

B.3 An Analytical Solution for the Model BT2

As the model BT1 demonstrates either unlimited and unrealistic growth or decline, a model BT2 is introduced that imposes boundary conditions on equation (B1.3). As a result, it has the following modified solution (B1.5)–(B1.7).

$$L(t) = L(0) + (r_{in} - r_{out}) \times t, \text{when } 0 < L(t) < L_{\max} \quad \text{(B1.5)}$$

$$L(t) = 0, \text{ when } t > L(0)/(r_{in} - r_{out}) \text{ and } r_{in} < r_{out} \quad \text{(B1.6)}$$

$$L(t) = L_{max}, \text{ when } t > (L_{max} - L(0))/(r_{in} - r_{out}) \text{ and } r_{in} > r_{out} \quad \text{(B1.7)}$$

This solution also demonstrates a linear growth or decline but only in the interval $0 < L(t) < L_{max}$ defined by the boundary conditions. The level L(t) is time-independent outside of this interval. A graphical representation of the solution (B1.5)–(B1.7) for the model BT2 is shown in Figure 6.5.

B.4 An Analytical Solution for the Model BT3

In a more sophisticated case inflow and outflow rates at any given moment may depend on water level at that particular moment: $I(t) = L(t)/t_{in}$ and $O(t) = L(t)/t_{out}$. This yields the following equation.

$$dL(t)/dt = \left(1/t_{in} - 1/t_{out}\right) \times L(t). \quad \text{(B1.8)}$$

The solution of this equation is an exponential function as follows.

$$L(t) = L(0) \times exp[(1/t_{in} - 1/t_{out}) \times t]. \quad \text{(B1.9)}$$

If $\left(1/t_{in} - 1/t_{out}\right) < 0$, the water level will decline to the zero level. If $\left(1/t_{in} - 1/t_{out}\right) > 0$, the solution corresponds to the exponential growth of the water level that must be constrained at level L_{max} due to the boundary condition $L(t) < L_{max}$. A graphical representation of the solution (B1.9) for the model BT3 is shown in Figure 6.6.

B.5 An Analytical Solution for the Model BT4

The exponential growth abruptly stopped at the L_{max} level due to the imposed boundary conditions does not seem natural and calls for a sort of smoothing.[1] Hence, the fourth variant of the BT model that is based on the following equation:

$$dL(t)/dt = (1/t_{in} - 1/t_{out}) \times (L_{max} - L(t)), \quad \text{(B1.10)}$$

It has a solution for $\left(1/t_{in} - 1/t_{out}\right) > 0$ as follows.

$$L(t) = Lmax - \left(Lmax - L(0)\right) \times exp\left[-\left(1/t_{in} - 1/t_{out}\right) \times t\right]. \quad \text{(B1.11)}$$

It ensures that the water level *L(t)* will be approaching the maximum level L_{max} smoothly. To obey the boundary condition $L(t) < L_{max}$, feedback loops in the model BT4 play the smoothing role as described by equation (B1.10). A graphical representation of the solution (B1.11) for the model BT4 is shown by Figure 6.7.

B.6 An Analytical Solution for the Model BT5

The first-order differential equations (B1.3), (B1.8), and (B1.10) are linear as they contain linear expressions for the level *L(t)*. In some applications a non-linear character of first-order equations is assumed. A non-linear version of the BT model equation may look as follows.

$$\frac{dL(t)}{dt} = \left(\frac{1}{t_{in}} - \frac{1}{t_{out}}\right) \times L(t) \times \left(L_{max} - L(t)\right) \quad \text{(B1.12)}$$

[1] It would worthwhile to mention that the boundary condition $L(t) < L_{max}$ may be dropped in some economics, demographics, etc. applications depending on contexts of developed models.

Equation (B1.12) was developed in the nineteenth century as a part of a population growth model by Verhulst (1845). Its purpose was to challenge the theory of Malthus on geometric progression of the population growth. Integration of the equation (B1.12) provides the following solution[2] when $\left(1/t_{in} - 1/t_{out}\right) > 0$.

$$L(t) = L_{max} \times L(0) / \left\{ L(0) - \left(L_{max} - L(0)\right) \times exp\left[-\left(\frac{1}{t_{in}} - \frac{1}{t_{out}}\right) \times t \right] \right\} \qquad \text{(B1.13)}$$

The solution for the model BT5 is called a logistics distribution and shown by Figure 6.8. When $L(0) \ll L_{max}$ the solution takes a form of a S-curved sigmoid function (Strogatz (2015)). Otherwise, when $L(0) \preceq L_{max}$ (i.e. when $L_{max} - L(0) \approx L_{max}$) the resulting solution resembles a smooth growth curve of model BT4 shown by Figure 6.7.

B.7 The workable models BT1–BT5

The workable models BT1–BT5 are available on the book's companion website at www.oup.co.uk/companion/Raydugin.

[2] This is easy to check if the solution (B1.13) of equation (B1.12) is placed to equation (B1.12).

APPENDIX C

Inputs to the System Dynamics Models SD1–SD3

This appendix provides supporting information required for development of the system dynamics models SD2 and SD3 that are discussed in Chapter 7. As the model SD1 is risk-free it is not in the primary focus of this appendix. Some important details of the models SD2 and SD3 are left out and not shown in Figures 7.1, 7.2, and 7.7. Most of them are overviewed in this appendix. Corresponding information for the model SD4 and SD5 (see Chapter 16) that mirror Monte Carlo models N-SCRA-A and N-SCRA-B (see Chapter 15) is collected in Appendix E.

In all system dynamics models for project Zemblanity, total number of drawings produced is established as 2000 (the 'initial project definition (E)' parameter); and the total number of construction work packages is set as 200 (the 'initial project definition (C)' parameter).

C.1 Activation of the Construction Rework Cycle

One module that is not shown in Figures 7.1, 7.2, and 7.7 is standalone engineering rework cycle that is used to activate construction rework cycle. This is shown in the real models SD1–SD3 that are available on the book companion website at www.oup.co.uk/companion/Raydugin.

The models SD1–SD3 presume activation of their construction rework cycles as soon as engineering works are completed. In terms of work flows this means that when engineering work is done (i.e. the engineering 'work flow (E)' stops), the construction 'work flow (C)' that has been previously at a zero-rate level should be switched on.

Mathematically this trigger is represented as an 'if-function': if a completed engineering work is below the planned level described by 'initial project definition (E)' parameter, the construction 'work flow C' rate is zero; if the completed engineering work is equal to the planned level (established as 2,000 drawings in all models without counting additional risk impacts), the construction 'work flow (C)' rate becomes equal to planned level which is established as 8.9 units per month in all models without counting additional risk impacts. Actually, this is true only in the case of the risk-free models SD1. Occurred risks in the models SD2 and SD3 do add more drawings and construction units to rework to the initially planned engineering and construction works. This means that the construction rework cycle should be activated only when the total number of drawings (2,000 plus the reworked drawings) is produced.

It might be noticed that there is an apparent duplication as the two engineering rework cycles should be identical in the models SD1–SD3. That is correct in the case where the engineering and construction rework cycles are separate: drawings with errors and omissions cannot enter construction. The situation changes when two models SD4 and SD5 are considered in Chapter 16 and Appendix E. They have the engineering and construction rework cycles linked. This means that the construction rework cycle should not wait until the entire engineering work (including EC-rework) is done. The case of models SD4 and SD5 requires two separate engineering rework cycles (see appendix E): one should be connected with the construction rework cycle and gets EC-rework from construction and one to trigger the construction works (no EC-rework). Although the structure of all developed models SD1–SD5 regarding the construction rework cycle's activation is kept the same for consistency purposes.

C.2 Times to Detect Errors and Omissions

The models SD2 and SD3 use characteristic times to detect E-errors and C-errors. Until poor quality of works is discovered it constitutes part of 'Undiscovered E-rework' and 'Undiscovered C-rework' stocks. As soon as a rework is discovered it enters 'Work Remaining (E)' and 'Work Remaining (C)' stocks, correspondingly (see Figure 7.2).

Dynamics of models SD2 and SD3 directly depend on these discovery characteristic times. The higher the required time to detect the flaws, the longer the response times to addresses identified engineering and construction issues. Flaws occurred earlier in a project require more time to become detected. These discovery times are longer at the beginning of engineering and construction phases and shorter when engineering and construction phases approach completion. This is described by a fraction of engineering or construction works completed. Table C.1 defines error discovery times used in models SD2 and SD3 to define engineering and construction errors. Actually, the same times are used in the models SD4 and SD5 (see Chapter 16 and Appendix E).

Table C.1 Times to detect engineering or construction errors depending on fraction of engineering or construction work completion[1]

Fraction completed	0	0.25	0.5	0.75	0.8	0.9	1	2
Error detection time, month	6.0	5.0	4.0	2.5	2.0	1.0	0.5	0

C.3 Selected Timing of Risk Impacts in the Model SD3

Table C.2 clarifies timing of risk impacts used in the model SD3 for risks R01, R02, R05, and R06. (The model SD2 does not require this as these four risks are left out.) A key principle used in development of Table C.2 is that engineering impact may occur in engineering only and construction risks—in construction only, which depends on actual project phase durations. Table C.2 also shows when risk impacts should begin.

Table C.2 Timing of risk occurrences since the project start in the model SD3

Risk ID	Month of impact
R01:SD00	2.0
R01:PD00	10.0
R01:DE00	19.0
R02:SD00	0.5
R02:PD00	7.0
R02:DE00	16.0
R03:SD00	
R03:PD00	Impact during engineering
R03:DE00	
R04:CN00	Impact during construction
R04:CN00	
R05:CN00	36.0
R06:CS00	40.0

C.4 Recalculation of Risk Impacts: The General Approach

As a first step, mean values of schedule risk impacts on the mapped normal tasks after addressing are multiplied by probabilities of risk occurrences to estimate mean expected values of schedule risk

[1] Error detection time for fraction completed equal to one is established at 0.5 months. This is a purely technical requirement to keep the error detection time series 'smooth' and avoid numerical simulation issues called floating point errors.

impacts.[2] In case of the models SD2 and SD3 required information on the impact's distributions is collected in a L-SCRA model's risk register after addressing shown by Table A.5.

As a second step, these mean values are converted to corresponding numbers of drawings and construction units, correspondingly: 2,000 drawings (initially planned work) is divided by the overall duration of the engineering phase of 21.2 months yielding 94.5 drawings per month. In other words, every month of risk's expected value (multiplication of an impact by risk's probability) in engineering is evaluated as 94.5 extra drawings. Similarly, planned construction productivity is calculated by dividing the total number of planned construction units (200 units) by the overall duration of the construction phase including commissioning and startup (22.4 months) resulting in 8.9 units per month. This presumes that every month of a risk's expected value in construction corresponds to the rework of 8.9 construction units.

C.5 Recalculation of Risk Impacts for the Models SD2 and SD3

Table C.3 provides recalculation of the schedule part of the L-SCRA model discussed in Chapter 5 to parameters of the models SD2 and SD3. The model SD2 requires only recalculation for two quality-related risks R03 and R04. Besides information for risks R03 and R04 used in both models, the rest of the Table C.3 provides required information for the model SD3.

The expected value numbers for risks R03 and R04 are used to evaluate a share of engineering drawing and construction units that require rework. Namely, according to Table C.3, the entire impact of risk R03 on three engineering normal tasks SD00, PD00, and DE00 would be equal to 45 drawings, or 2.3 per cent of the total number of drawings. In case of construction, three units would require full rework which corresponds to 1.5 per cent of all units.

Table C.3 Recalculation of expected values of schedule risk impacts to numbers of drawings and construction units for the models SD2 and SD3

ID	Risk title	Mapped to	Phase	Mean expected value, month	Share of full-rework drawings or units	Drawing's or unit's # equivalent	% of total drawings or units
R01	Design Basis Change	SD00	E	0.36	100%	34.3	
	(Proponent-Driven)	PD00	E	0.36	100%	34.3	
		DE00	E	0.36	100%	34.3	
R02	Overlooked External	SD00	E	0.16	100%	14.8	
	Stakeholder's	PD00	E	0.16	100%	14.8	
	Requirements	DE00	E	0.16	100%	14.8	
R03	Errors and	SD00	E	0.16	100%	15.1	2.3%
	Omissions	PD00	E	0.16	100%	15.1	(E-quality)
	(Engineering)	DE00	E	0.16	100%	15.1	
R04	Poor Work Quality (Construction)	CN00	C	0.33	100%	3.0	1.5% (C-quality)
R05	Delays in Material Delivery	CN00	C	0.12	100%	1.0	
R06	Commissioning and Startup (CSU) Failure	CS00	C	0.17	100%	1.5	

C.6 The Workable Models SD1–SD3

The workable models SD1–SD3 are available on the book's companion website at www.oup.co.uk/companion/Raydugin.

[2] Triangular distributions for schedule impacts can be easily evaluated using @Risk, Crystal Ball, Safran Risk, Oracle Primavera Risk Analysis or any other suitable software package in order to directly derive the mean values.

APPENDIX D

The Non-linear Monte Carlo Schedule and Cost Risk Analysis Models N-SCRA-A, N-SCRA-B, and N-SCRA-C

This appendix provides details on required inputs to carry out N-SCRA modelling discussed in Chapter 15 for the three developed models N-SCRA-A, N-SCRA-B, and N-SCRA-C. Mathematical concepts and general requirements to build the N-SCRA models are overviewed in Chapters 13 and 14.

D.1 Factoring Risk Interactions to the SCRA Risk Register

According to a mathematical approach of Chapter 13, non-linear risk interactions should be represented through non-linearity parameters and non-linear multipliers. This should allow the developing a modified risk register for project Zemblanity.

Required schedule-related non-linearity parameters and multipliers (see Table D.1) are developed in accordance with Figure 14.1: Table D.1 converts the visually represented instances of cross-risk interactions in corresponding mathematical expressions. These are used for development of the modified N-SCRA risk register of the models N-SCRA-A, N-SCRA-B, and N-SCRA-C. The schedule

Table D.1 Schedule non-linearity parameters ρ_{ij} and non-linear multipliers for the models N-SCRA-A and N-SCRA-B[1]

Risk ID	Interaction type	Parameter ρ_{ij}	Non-linear interaction's multiplier M_{ij}	Total non-linear multiplier $\prod M_{ij}$
R01	Internal amplification: R01 by R01	ρ_{11}	$1+\rho_{11}\times R01$	$(1+\rho_{11}\times R01)\times[1+0.5\times\rho_{12}\times R02\times(1+\rho_{22}\times R02)]$
	Compounding: with R02	ρ_{12}	$1+0.5\times\rho_{12}\times R02\times(1+\rho_{22}\times R02)$	
R02	Internal amplification: R02 by R02	ρ_{22}	$1+\rho_{22}\times R02$	$(1+\rho_{22}\times R02)\times[1+0.5\times\rho_{21}\times R01\times(1+\rho_{11}\times R01)]$
	Compounding: with R01	ρ_{21}	$1+0.5\times\rho_{21}\times R01\times(1+\rho_{11}\times R01)$	
R03	Internal amplification: R03 by R03	ρ_{33}	$1+\rho_{33}\times R03$	$(1+\rho_{33}\times R03)\times[1+0.5\times\rho_{31}\times R01\times(1+\rho_{11}\times R01)]\times[1+0.5\times\rho_{32}\times R02\times(1+\rho_{22}\times R02)]$
	Knock-on: from R01	ρ_{31}	$1+0.5\times\rho_{31}\times R01\times(1+\rho_{11}\times R01)$	
	Knock-on: from R02	ρ_{32}	$1+0.5\times\rho_{32}\times R02\times(1+\rho_{22}\times R02)$	
R04	Internal amplification: R04 by R04	ρ_{44}	$1+\rho_{44}\times R04$	$(1+\rho_{44}\times R04)\times[1+\rho_{43}\times R03\times(1+\rho_{33}\times R03)]\times[1+0.5\times\rho_{45}\times R05\times(1+\rho_{55}\times R05)]$
	Knock-on: from R03	ρ_{43}	$1+\rho_{43}\times R03\times(1+\rho_{33}\times R03)$	
	Compounding: with R05	ρ_{45}	$1+0.5\times\rho_{45}\times R05\times(1+\rho_{55}\times R05)$	
R05	Internal amplification: R05 by R05	ρ_{55}	$1+\rho_{55}\times R05$	$(1+\rho_{55}\times R05)\times[1+0.5\times\rho_{54}\times R04\times(1+\rho_{44}\times R04)]$
	Compounding: with R04	ρ_{54}	$1+0.5\times\rho_{54}\times R04\times(1+\rho_{44}\times R04)$	
R06	Internal amplification: R06 by R06	ρ_{66}	$1+\rho_{66}\times R06$	$(1+\rho_{66}\times R06)\times[1+\rho_{64}\times R04\times(1+\rho_{44}\times R04)]$
	Knock-on: from R04	ρ_{64}	$1+\rho_{64}\times R04\times(1+\rho_{44}\times R04)$	

[1] The same table is applicable to cost risk interactions in accordance with Figure 14.2. This just requires the replacing of non-linearity parameters ρ_{ij} that represent schedule risk interactions by α_{ij} that represent cost risk interactions in Table D.1.

Table D.2 A modified non-linear schedule N-SCRA risk register template

ID	Risk title	Mapped to	Prob. to-be	Schedule impact to-be, d				Non-linearity parameter	Impact multiplier M_{ij}	Total impact with all multipliers
			P_i	Min.	ML	Max.	I_i	ρ_{ij}	$1+\rho_{ij}\times P_j\times I_j$	$I_i\times\prod(M_{ij})$
R01	Design Basis Change (Proponent-Driven) [Internally amplified]	SD00 PD00 DE00	25%	20	40	70	43	0.000	1.000	43
	Compounding: with R02							0.000	1.000	
R02	Overlooked External Stakeholder's Requirements [Internally amplified]	SD00 PD00 DE00	20%	10	20	40	23	0.000	1.000	23
	Compounding: with R01							0.000	1.000	
R03	Errors and Omissions (Engineering) [internally amplified]	SD00 PD00 DE00	15%	15	30	50	32	0.000	1.000	32
	Knock-on: from R01							0.000	1.000	
	Knock-on: from R02							0.000	1.000	
R04	Poor Work Quality (Construction) [internally amplified]	CN00	20%	30	50	70	50	0.000	1.000	50
	Knock-on: from R03							0.000	1.000	
	Compounding: with R05							0.000	1.000	
R05	Delays in Material Delivery [internally amplified]	CN00	15%	10	20	40	23	0.000	1.000	23
	Compounding: with R04							0.000	1.000	
R06	Commissioning & Startup (CSU) Failure [internally amplified]	CS00	15%	20	30	50	33	0.000	1.000	33
	Knock-on: from R04							0.000	1.000	

Table D.3 A modified non-linear cost N-SCRA risk register template

ID	Risk title	Prob. to-be	Cost impact to-be, $'000				Non-linearity parameter	Impact multiplier M_{ij}	Total impact with all multipliers
		P_i	Min.	ML	Max.	I_i	α_{ij}	$1+\alpha_{ij}\times P_j\times I_j$	$I_i\times\prod(M_{ij})$
R01	Design Basis Change (Proponent-Driven) [internally amplified]	25%	2,500	10,000	25,000	12,500	0.000	1.000	12,500
	Compounding: with R02						0.000	1.000	
R02	Overlooked External Stakeholder's Requirements [Internally amplified]	20%	3,500	7,500	12,500	7,833	0.000	1.000	7,833
	Compounding: with R01						0.000	1.000	
R03	Errors and Omissions (Engineering) [internally amplified]	15%	3,500	5,000	6,500	5,000	0.000	1.000	5,000
	Knock-on: from R01						0.000	1.000	
	Knock-on: from R02						0.000	1.000	
R04	Poor Work Quality (Construction) [internally amplified]	20%	20,000	30,000	50,000	33,333	0.000	1.000	33,333
	Knock-on: from R03						0.000	1.000	
	Compounding: with R05						0.000	1.000	
R05	Delays in Material Delivery [internally amplified]	15%	5,000	7,000	10,000	7,333	0.000	1.000	7,333
	Compounding: with R04						0.000	1.000	
R06	Commissioning and Startup (CSU) Failure [internally amplified]	15%	10,000	15,000	30,000	18,333	0.000	1.000	18,333
	Knock-on: from R04						0.000	1.000	

part of the modified risk register is shown by Table D.2, and the cost part by Table D.3. These tables are developed only for the after-addressing situation for simplicity purposes. Input cells for internal amplification, knock-on, and compounding non-linearity parameters and corresponding non-linear multipliers are established. All established non-linear multipliers attached to individual risk interactions for a given risk are combined to yield total impacts with all multipliers. All non-linearity parameters in the risk register of Tables D.2 and D.3 are established at zero level as a placeholder and correspondingly all non-linear multipliers are equal to one. This exactly corresponds to the L-SCRA risk register after addressing (see Table A.4). Required non-linearity parameters for the models N-SRCA-A and N-SCRA-B that come from project system assessments are introduced below in Tables D.5–D.8. Prior to this, calibration of the N-SCRA-A and SCRA-B models are required

(Table D.4). The model N-SCRA-C does not follow this path as it is established for demonstration purposes. It is further discussed in section D.6.

D.2 N-SCRA Model's Calibration

As discussed in Chapter 14, the N-SCRA models should be calibrated first to establish values of non-linearity parameters corresponding to 10 per cent and 25 per cent of schedule delay and overspending above P70 linear case duration and cost values.

Several runs of the modified risk register are undertaken. All schedule non-linearity parameters are considered equal in each calibration run although different in all runs. The same is applicable to cost non-linearity parameters. This allows values of non-linearity parameters to be pinned down that give rise to exactly 10 per cent and 25 per cent of duration and cost overruns, correspondingly. To achieve this the non-linear risk registers of Tables D.2 and D.3 become subject to simple Monte Carlo simulations.[2] The purpose of these simulations is to obtain modified schedule and cost risk impact distributions for all risks R01–R06 when all relevant risk interactions (see Figures 14.1 and 14.2 and Table D.1) are taken into account.

Figure D.1 shows a sample output for risk R01 after addressing when the same schedule non-linearity parameter $\rho = 0.083$ is used for all risk interactions in a given run. Three main distribution parameters shown by Figure D.1 (minimum = 32 d, mode (most likely) = 86 d, and maximum = 261 d) are used as approximating triangular distribution input to the L-SCRA model of Chapter 5, although after getting the modified non-linear inputs this model can be pronounced a non-linear N-SCRA model.

This exercise is repeated when using the cost risk register (see Table D.3). Corresponding modified cost risk distributions are used as inputs to the linear L-SCRA model (its cost part). These runs allow values for non-linearity parameters to be pinned down when cost overruns are equal to 10 per cent and 25 per cent above L-SCRA case.

It is important to keep in mind that when the calibration of the cost part is carried out, the schedule part is kept linear as in the model L-SCRA to keep the schedule-driven costs as in the linear case.

The calibration results are represented by Table D.4. Table D.4 allows the undertaking of evaluations of non-linearity parameters for models N-SCRA-A and N-SCRA-B as now it is clear which non-linearity parameter ranges correspond to a mature, ordinary, and immature project system characterized by weak, medium, and strong static and/or dynamic PSS–PDS mismatches, correspondingly.

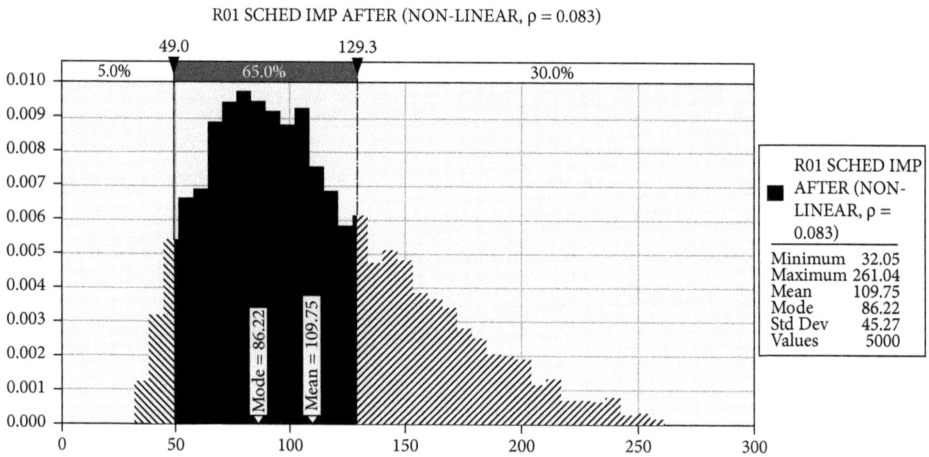

Figure D.1 A sample output for total schedule impact of risk R01 after addressing (non-linearity parameter $\rho = 0.083$)

[2] A version 7.6 of @Risk is used to run all simulations with 5,000 iterations.

Table D.4 Calibrated non-linearity parameters

Project system maturity	Schedule		Cost	
	Schedule slippage, %	Non-linearity parameter ρ	Overspending, %	Non-linearity parameter α
Mature	0–10%	0–0.055	0–10%	0–0.00035
Ordinary	10%–25%	0.055–0.12	10%–25%	0.00035–0.00052
Immature	>25%	>0.12	>25%	>0.00052

D.3 Assessment of Non-linearity Parameters: the Model N-SCRA-A

Tables D.5 and D.6 represent results of assessments for three types of identified schedule and cost risk interaction (see Table D.1). Non-linearity parameters are assessed for each instance of risk interactions keeping in mind calibration results of Table D.4. These assessments are done based on

Table D.5 Assessment of schedule risk non-linearity parameters for the N-SCRA-A case

N-SCRA-A	Schedule risk non-linearity parameters			
Risk interaction type	Non-linearity parameter ρ_{ij}	Agreed category	Agreed value	Rational: the current situation
Internal amplification: to R01 by R01	ρ_{11}	high	0.15	Stakeholder management and change management procedures are not implemented.
Compounding: to R01 by R02	ρ_{12}	medium	0.08	The same timing of risk R01 and R02 occurrence.
Internal amplification: to R02 by R02	ρ_{22}	high	0.14	Stakeholder management and change management procedures are not implemented.
Compounding: to R02 by R01	ρ_{21}	medium	0.08	The same timing of risk R01 and R02 occurrence.
Internal amplification: to R03 by R03	ρ_{33}	medium	0.07	Engineering quality could be lower than expected due to a lack of experience to design facilities in Russian Arctic.
Knock-on: to R03 by R01	ρ_{31}	medium	0.06	Stakeholder management and change management procedures are not implemented.
Knock-on: to R03 by R02	ρ_{32}	medium	0.06	Stakeholder management and change management procedures are not implemented.
Internal amplification: to R04 by R04	ρ_{44}	high	0.15	Construction quality could be lower than expected due to a lack of experience of building facilities in Russian Arctic.
Knock-on: R04 by R03	ρ_{43}	high	0.14	Lack of experience in designing of facilities for Russian Arctic: some drawings with errors and omissions may enter construction.
Compounding: to R04 by R05	ρ_{45}	medium	0.059	Delay in delivery of equipment and construction quality may be compounded.
Internal amplification: to R05 by R05	ρ_{55}	medium	0.065	Lack of experience in constructing facilities in Russian Arctic and difficulties with attracting experienced foremen.
Compounding: to R05 by R04	ρ_{54}	medium	0.059	Delay in delivery of equipment and construction quality may be compounded.
Internal amplification: to R06 by R06	ρ_{66}	high	0.16	Commissioning and strut up could be complicated due to lack of experience to operate in Russian Arctic.
Knock-on: to R06 by R04	ρ_{64}	high	0.16	Lack of experience to construct in Russian Arctic: low quality of construction may enforce risks in commissioning and startup.

Table D.6 Assessment of cost risk non-linearity parameters for the N-SCRA-A case

N-SCRA-A	Cost risk non-linearity parameters			
Risk interaction type	Non-linearity parameter α_{ij}	Agreed category	Agreed value	Rational: the current situation
Internal amplification: to R01 by R01	α_{11}	high	0.0007	Stakeholder management and change management procedures are not implemented.
Compounding: to R01 by R02	α_{12}	medium	0.00038	The same timing of risk R01 and R02 occurrence.
Internal amplification: to R02 by R02	α_{22}	medium	0.00049	Stakeholder management and change management procedures are not implemented.
Compounding: to R02 by R01	α_{21}	medium	0.00038	The same timing of risk R01 and R02 occurrence.
Internal amplification: to R03 by R03	α_{33}	medium	0.00038	Engineering quality could be lower than expected due to a lack of experience to design facilities in Russian Arctic.
Knock-on: to R03 by R01	α_{31}	medium	0.00038	Stakeholder management and change management procedures are not implemented.
Knock-on: to R03 by R02	α_{32}	medium	0.00038	Stakeholder management and change management procedures are not implemented.
Internal amplification: to R04 by R04	α_{44}	medium	0.00038	Construction quality could be lower than expected due to a lack of experience to build facilities in Russian Arctic.
Knock-on: R04 by R03	α_{43}	medium	0.00043	Lack of experience in designing of facilities for Russian Arctic: some drawings with errors and omissions may enter construction.
Compounding: to R04 by R05	α_{45}	medium	0.00039	Delay in delivery of equipment and construction quality may be compounded.
Internal amplification: to R05 by R05	α_{55}	medium	0.00047	Lack of experience in constructing facilities in Russian Arctic and difficulties with attracting experienced foremen.
Compounding: to R05 by R04	α_{54}	medium	0.00039	Delay in delivery of equipment and construction quality may be compounded.
Internal amplification: to R06 by R06	α_{66}	high	0.00073	Commissioning and strut up could be complicated due to lack of experience to operate in Russian Arctic.
Knock-on: to R06 by R04	α_{64}	high	0.00053	Lack of experience to construct in Russian Arctic: low quality of construction may enforce risks in commissioning and startup.

observations and opinions of project risk specialists and available empirical data from previous projects. In accordance with Chapter 13, schedule non-linearity parameters are denoted by ρ_{ij} and cost non-linearity parameters are denoted by α_{ij}, where i and j denote project risks R01–R06 (i,j = 1,2...6); $i = j$ in case of internal risk amplifications, and $i \neq j$ in case of cross-risk interactions.

D.4 Assessment of Non-linearity Parameters: the Model N-SCRA-B

Tables D.7 and D.8 summarize results of a workshop discussion on factoring in additional measures to improve PDS of project Zemblanity.

Table D.7 Assessment of schedule risks non-linearity parameters for the N-SCRA-B case

N-SCRA-B	**Schedule risk non-linearity parameters**			
Risk interaction type	Non-linearity parameter ρ_{ij}	Agreed category	Agreed value	Rational: system issue's addressing
Internal amplification: to R01 by R01	ρ_{11}	medium	0.1	Agreement by partners to implement stakeholder management and change management procedures.
Compounding: to R01 by R02	ρ_{12}	low	0.04	Planning and project control measures to avoid risk compounding.
Internal amplification: to R02 by R02	ρ_{22}	medium	0.09	Agreement by partners to implement stakeholder management and change management procedures.
Compounding: to R02 by R01	ρ_{21}	low	0.04	Planning and project control measures to avoid risk compounding.
Internal amplification: to R03 by R03	ρ_{33}	low	0.05	Agreement by partners to implement a QA/QC procedure.
Knock-on: to R03 by R01	ρ_{31}	low	0.01	Agreement by partners to implement stakeholder management, change management, and QA/QC procedures.
Knock-on: to R03 by R02	ρ_{32}	low	0.01	Agreement by partners to implement stakeholder management, change management, and QA/QC procedures.
Internal amplification: to R04 by R04	ρ_{44}	low	0.05	Agreement by partners to implement a QA/QC procedure.
Knock-on: to R04 by R03	ρ_{43}	medium	0.1	Agreement by partners to implement a QA/QC procedure.
Compounding: to R04 by R05	ρ_{45}	low	0.04	Planning and project control measures to avoid risk compounding.
Internal amplification: to R05 by R05	ρ_{55}	low	0.05	Prequalification of bidders, selection of an EPC contractor qualified to work in Russian Arctic.
Compounding: to R05 by R04	ρ_{54}	low	0.04	Planning and project control measures to avoid risk compounding.
Internal amplification: to R06 by R06	ρ_{66}	medium	0.11	Agreement by partners to implement a commissioning and startup procedure.
Knock-on: to R06 by R04	ρ_{64}	medium	0.1	Agreement by partners to implement a QA/QC procedure.

Table D.8 Assessment of cost risks non-linearity parameters for the N-SCRA-B case.

N-SCRA-B	**Schedule risk non-linearity parameters**			
Risk interaction type	Non-linearity parameter ρ_{ij}	Agreed category	Agreed value	Rational: system issue's addressing
Internal amplification: to R01 by R01	α_{11}	medium	0.0004	Agreement by partners to implement stakeholder management and change management procedures.
Compounding: to R01 by R02	α_{12}	low	0.0001	Planning and project control measures to avoid risk compounding.
Internal amplification: to R02 by R02	α_{22}	medium	0.00037	Agreement by partners to implement stakeholder management and change management procedures.
Compounding: to R02 by R01	α_{21}	low	0.0001	Planning and project control measures to avoid risk compounding.

Continued

Table D.8 Continued

N-SCRA-B	Schedule risk non-linearity parameters			
Internal amplification: to R03 by R03	a_{33}	low	0.0001	Agreement by partners to implement a QA/QC procedure.
Knock-on: to R03 by R01	a_{31}	medium	0.00036	Agreement by partners to implement stakeholder management, change management, and QA/QC procedures.
Knock-on: to R03 by R02	a_{32}	medium	0.00036	Agreement by partners to implement stakeholder management, change management, and QA/QC procedures.
Internal amplification: to R04 by R04	a_{44}	medium	0.00036	Agreement by partners to implement a QA/QC procedure.
Knock-on: to R04 by R03	a_{43}	medium	0.00036	Agreement by partners to implement a QA/QC procedure.
Compounding: to R04 by R05	a_{45}	low	0.0001	Planning and project control measures to avoid risk compounding.
Internal amplification: to R05 by R05	a_{55}	low	0.0001	Prequalification of bidders, selection of an EPC contractor qualified to work in Russian Arctic.
Compounding: to R05 by R04	a_{54}	low	0.0001	Planning and project control measures to avoid risk compounding.
Internal amplification: to R06 by R06	a_{66}	medium	0.00039	Agreement by partners to implement a commissioning and startup procedure.
Knock-on: to R06 by R04	a_{64}	medium	0.00036	Agreement by partners to implement a QA/QC procedure.

D.5 Modified Inputs to the Monte Carlo Models N-SCRA-A and N-SCRA-B

Specific values for schedule and cost non-linearity parameters collected in Tables D.5–D.8 serve as inputs to the modified non-linear schedule and cost risk registers (see Tables D.2 and D.3). It is done the same way as for calibration purposes in section D.2, although individual parameters for risk interactions (Tables D.5–D.8) are used this time.

Figure D.2 demonstrates how modified schedule impacts (min., most likely, max.) could be obtained using an example of risk R02 in the model N-SCRA-A. Essentially, this is the example similar to the one in figure D.1, although unique non-linearity parameters for risk R02 are used as defined for the model N-SCRA-A. (see Table D.9)

Any multiplier from the Table D.1 for a given risk has a following structure.

$$M = 1 + \Delta, \tag{D.1}$$

where Δ comprises all non-linearity parameters and their combinations with project risks depending of the nature and number of risk interactions. Apparently, when all risk interactions are ignored $(\Delta = 0)$, $M = 1$, which exactly corresponds to the linear case of the model L-SCRA.

As discussed in Chapters 13 and 14, non-linear multipliers of Table D.1 modify products of risk probabilities and impacts (expected values) $P \times I$. In a presence of risk interactions, the expected value $P \times I$ for a given risk should be modified as follows.

$$P \times I \times M = P \times I \times (1 + \Delta), \tag{D.2}$$

According to expression (D.2), it is not possible to tell apart portions of a non-linear multiplier that modify only a probability and only an impact. A multiplier modifies both. These inputs should be used in a linear model L-SCRA to replace initial linear inputs (Table A.5) to yield non-linear models N-SCRA-A and N-SCRA-B (see Tables D9 and D10).

It should be noticed that Tables D.9 and D.10 contain modified impacts although keep the same probabilities as Tables A.4 and A.5 for the L-SCRA case. The technical explanation to this is as follows. All non-linearity parameters should modify risk consequences $P \times I$ and not probabilities and

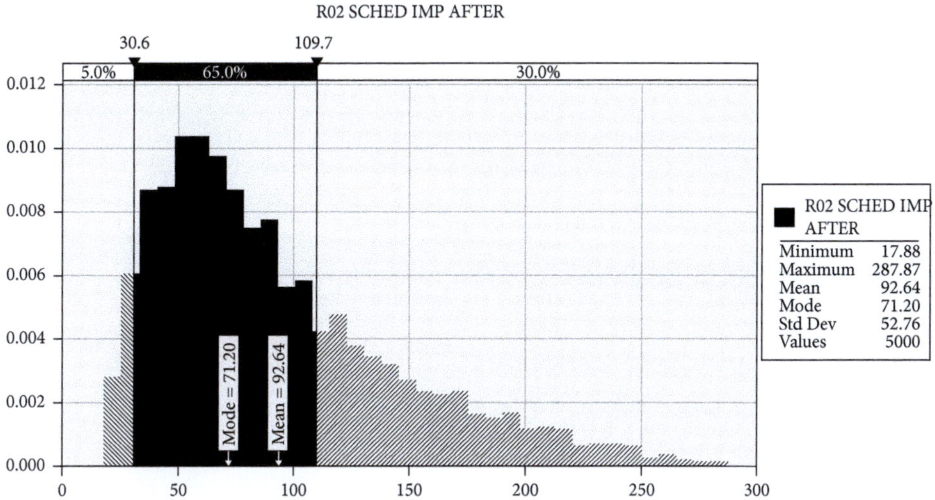

Figure D.2 A sample assessment of a modified schedule impact: an example for risk R02 in the model N-SCRA-A

Table D.9 Probabilities and modified risk impacts used as inputs to the model N-SCRA-A

N-SCRA-A	Probability	Modified Schedule Impact, d			Modified Cost Impact, $'000		
		Min.	Most likely	Max.	Min.	Most likely	Max.
R01	25%	40	142	413	4,706	28,446	273,104
R02	20%	18	71	288	5,653	26,121	202,451
R03	15%	32	109	702	8,059	30,619	372,887
R04	20%	80	235	623	78,618	192,986	549,700
R05	15%	15	43	124	20,443	46,898	171,086
R06	15%	85	216	849	140,504	425,717	3,293,752

Table D.10 Probabilities and modified risk impacts used as inputs to the model N-SCRA-B

N-SCRA-B	Probability	Modified Schedule Impact, d			Modified Cost Impact, $'000		
		Min.	Most likely	Max.	Min.	Most likely	Max.
R01	25%	33	88	242	3,378	19,387	104,282
R02	20%	14	40	134	4,561	14,912	45,608
R03	15%	19	41	114	7,689	32,509	239,800
R04	20%	53	106	273	70,811	133,665	442,830
R05	15%	13	35	76	14,712	30,877	103,998
R06	15%	50	97	316	106,780	318,009	2,215,349

impacts separately. Whereas Tables D.9 and D.10 actually contain impacts that consume entire non-linear multipliers keeping probabilities intact. When both the unmodified probabilities and the modified impacts are put in Monte Carlo models, the models do handle products $P \times I$ that are properly modified.

D.6 Inputs to the Monte Carlo Models N-SCRA-C

An additional model N-SCRA-C is considered in Chapter 17 to investigate project failure modes. Table D.11 contains inputs used to run the N-SCRA-C model. Cost impacts are not shown as the CRA part of the model is not used. General uncertainty inputs to the model N-SCRA-C are the same as in case of the models L-SCRA, N-SCRA-A, and N-SCRA-B (see Table A.2).

Table D.11 Inputs to the schedule part of the model N-SCRA-C

N-SCRA-C	Probability	Modified Schedule Impact, d		
		Min.	Most likely	Max.
R01	25%	150	200	250
R02	20%	150	200	250
R03	15%	70	120	150
R04	20%	70	120	150
R05	15%	20	30	40
R06	15%	100	150	200

D.7 The Workable Models N-SCRA-A, N-SCRA-B, and N-SCRA-C

The workable models N-SCRA-A, N-SCRA-B, and N-SCRA-C are available on the book's companion website at www.oup.co.uk/companion/Raydugin.

APPENDIX E

Inputs to the System Dynamics Models SD4 and SD5

Following Appendix C that provides input information to models SD2 and SD3 this appendix provides required inputs for the model SD4 and for its modification with risk compounding ('the model SD4 with compounding') and for the model SD5. As models SD2–SD5 have the same features (activation of construction rework cycle, time to detect errors and omissions, etc.), a reference to Appendix C is done to avoid repetition.

E.1 Activation of the Construction Rework Cycle in the Models SD4 and SD5

Features of activation of the construction rework cycle are explained in section C.1 of Appendix C.

E.2 Times to Detect Errors and Omissions in the Models SD and SD5

The same Table C.1 is applicable to the models SD4 and SD5 as is applicable to models SD2 and SD3.

E.3 Selected Timing of Risk Impacts

Table E.1 clarifies timing of risk impacts used in the models SD4, SD5, and in the model SD4 with compounding. As considered models yield different project durations, the key principle used in development of Table E.1 is that engineering risks may occur in engineering only and construction risks—in construction. For instance, risk R06 related to commissioning and startup may occur by the end of construction only, and not in the middle of the construction phase. So, risk impact timing

Table E.1 Timing of risk occurrences since the project start in the models SD4, SD5, and SD4 with compounding

Risk ID	Model SD4		Model SD5		Model SD4 with compounding	
	Start, month #	Finish, month #	Start, month #	Finish, month #	Start, month #	Finish, month #
R01:SD00	2.0	5.0	2.0	5.0	1.0	4.0
R01:PD00	10.0	13.0	10.0	13.0	9.0	12.0
R01:DE00	19.0	22.0	19.0	22.0	18.0	21.0
R02:SD00	0.5	3.5	0.5	3.5	1.0	4.0
R02:PD00	7.0	10.0	7.0	10.0	8.0	11.0
R02:DE00	16.0	19.0	16.0	19.0	17.0	20.0
R03:SD00						
R03:PD00	Impact during engineering					
R03:DE00						
R04:CN00	Impact during construction					
R05:CN00	43.0	48.0	40.0	45.0	45.0	50.0
R06:CS00	48.0	53.0	45.0	50.0	46.0	51.0

of Table E.1 takes into account actual project durations yielded by the models N-SCRA-A and SCRA-B (see Chapter 15).

Actually, timing of risk occurrences is selected to exclude risk compounding on purpose in the models SD4 and SD5. Namely, the difference in risk occurrence timing for the models SD4 without and with compounding is done to clearly demonstrate the impact of compounding.

Table E.1 also shows when risk impacts should smoothly vanish. It should happen in three months in the case of engineering risks and in five months in the case of construction risks. This information relating to temporary impacts of risks on work flow rates is discussed in section E.4.

E.4 Risk Multipliers in the Models SD4 and SD5

As soon as risks R01, R02, R05, and R06 occur, they permanently add a corresponding amount of work to 'initial project definition (E)' and 'initial project definition (C)' parameters (see Figure 16.1). However, there is another phenomenon—'work flow (E)' and 'work flow (C)' are tampered by occurred risks. This phenomenon relates mostly to emerged work constraints, extra work, and out-of-sequence work due to occurred risks. To model this phenomenon, so-called risk multipliers are introduced for engineering risks R01 and R02 ('E risk multiplier'), and for construction risks R05 and R06 ('C risk multiplier'). Presumably, a project team is to recover from the initial risk impacts to restore the normal 'work flow (E)' and 'work flow (C)' rates: 94.5 drawings per month and 8.9 construction units per month. Periods of temporary impacts on work flow rates are established as three months for engineering risks R01 and R02 and five months for construction risks R05 and R06 in all SD models (see Table E.1).

The temporary risk impact on workflows is set up as follows. An assumption is done that after an initial abrupt impact modelled by so-called 'step function', the impact soothes and eventually (in three months or five months, correspondingly) vanishes. The soothing part is formalized through using so-called 'delay function'. Corresponding risk multipliers in engineering and construction—'E risk multiplier' and 'C risk multiplier'—are shown in Figure 16.1. Their behaviour and impacts on the workflows are clearly shown in Figures 16.2, 16.3, 16.7, 16.8, and 16.9.

E.5 Recalculation of Risk Impacts for the Models SD4 and SD5: General Approach

Essentially the same general approach is used in case of the models SD4 and SD5 as in the case of models SD2 and SD3 (see Appendix C). Only a few distinctive features related to the linked engineering and construction rework cycles are discussed in this section.

One feature that is not used in the models SD2 and SD3 which have two separated rework cycles but is used in the models SD4 and SD5 is related to conversion of drawing to construction units. This is required when engineering drawings with errors and omissions enter construction rework cycle undetected as construction handles construction units, not drawings. Risks R05 and R06 impact construction activities. Upon discovery of errors and omissions an impacted number of construction units is recalculated back to drawings before they get back in the 'Work Remaining (E + EC)' stock. As the total number of initial drawings is established as 2,000, and a total number of initial construction units is set up as 200 in the discussed models SD4 and SD5, an 'exchange rate' 1:10 is used for these recalculations.

E.6 Recalculation of Risk Impacts for the Model SD4

The model SD4 differs from linear models SD2 and SD3 due to incorporation of internal risk amplifications and cross-risk interactions. First, risk impacts for internally amplified risks should be considered. Second, all relevant cross-risk interactions should be factored in. The key novelty due to the established link between the engineering and construction rework cycles is the knock-on impact of risk R03 (poor engineering quality) on risk R04 (poor construction quality). This corresponds to admitting poor-quality engineering drawings in construction. This is the loop (see Figure 16.1) that combines engineering and construction cycles.

The impact of risk R04 should be broken down: the internal amplification and compounding impacts[1] should be told apart from the knock-on impact by risk R03. This knock-on part should be defined by the 'poor EC-quality' variable. That is why the structure of Table E.2 for the model SD4 differs from the structure of Table C.3 for the models SD2 and SD3: an additional knock-on line is added for risk R04. According to Table D.1, risk R04 has three multipliers. Two of them describe internal risk amplification of risk R04 and the compounding with risk R05, correspondingly:

$$M_{44}^{int} = 1 + \rho_{44}^{int} \times R04, \tag{E.1}$$

$$M_{45}^{comp} = 1 + 0.5 \times \rho_{45}^{comp} \times R05 \times (1 + \rho_{55}^{int} \times R05) \tag{E.2}$$

The third multiplier shows the knock-on impact of risk R03 on risk R04 as follows.

$$M_{43}^{knock} = 1 + \rho_{43}^{knock} \times R03 \times (1 + \rho_{33}^{int} \times R03). \tag{E.3}$$

According to expressions (E.2) and (E.3), risks R03 and R05 are internally multiplied.

As Table E.2 shows, risk R03 corresponds to 9.2 per cent of engineering drawings that require full rework. This related to the internal rework in the engineering cycle. Risk R04 has two realizations. First, it corresponds to 3.4 per cent of construction units that require full rework. This is to be done internally in the construction cycle. Second, it accepts 2.1 per cent of drawings with errors and omissions that enter construction undetected. This does describe the knock-on impact of risk R03 on risk R04.

An assumption is used in Table E.2 that only 10 per cent of drawings require full rework (a full engineering rework equivalent) due to any risk and only 16 per cent of construction units require full rework (a full construction rework equivalent) due to poor quality (risks R03 and R04). It is also assumed that 40 per cent of construction units will require full rework due to risks R05 and R06.

Table E.2 Recalculation of expected values of schedule risk impacts to numbers of drawings and construction units for the model SD4

ID	Risk title	Mapped to	Phase	Mean expected value, month	Share of full-rework drawings or units	Drawing's or unit's # equivalent	% of total drawings or units
R01	Design Basis Change (Proponent-Driven)	SD00	E	5.33	10%	50.4	
		PD00	E	5.33	10%	50.4	
		DE00	E	5.33	10%	50.4	
R02	Overlooked External Stakeholder's Requirements	SD00	E	3.10	10%	29.3	
		PD00	E	3.10	10%	29.3	
		DE00	E	3.10	10%	29.3	
R03	Errors and Omissions (Engineering)	SD00	E	6.50	10%	61.4	9.2% (E-quality)
		PD00	E	6.50	10%	61.4	
		DE00	E	6.50	10%	61.4	
R04	Poor Work Quality (Construction)	CN00	C	4.80	16%	6.8	3.4% (C-quality)
			C	3.00	16%	4.3	2.1% (EC-quality)
R05	Delays in Material Delivery	CN00	C	1.7	40%	6.3	
R06	Commissioning & Startup (CSU) Failure	CS00	C	10.8	40%	38.5	

[1] Actually, the compounding part is left out in the models SD4 and SD5. It is considered in the model SD4 with compounding in section E.8.

Table E.3 Recalculation of expected values of schedule risk impacts to numbers of drawings and construction units for the model SD5

ID	Risk title	Mapped to	Phase	Mean expected value, month	Share of full-rework drawings or units	Drawing's or unit's # equivalent	% of total drawings or units
R01	Design Basis Change	SD00	E	3.57	10 %	33.7	
	(Proponent-Driven)	PD00	E	3.57	10 %	33.7	
		DE00	E	3.57	10 %	33.7	
R02	Overlooked External	SD00	E	1.73	10 %	16.4	
	Stakeholder's	PD00	E	1.73	10 %	16.4	
	Requirements	DE00	E	1.73	10 %	16.4	
R03	Errors and	SD00	E	1.77	10 %	16.7	2.5 %
	Omissions	PD00	E	1.77	10 %	16.7	(E-quality)
	(Engineering)	DE00	E	1.77	10 %	16.7	
R04	Poor Work Quality (Construction)	CN00	C	2.77	16%	3.9	2.0 % (C-quality)
			C	2.10	16%	3.0	1.5% (EC-quality)
R05	Delays in Material Delivery	CN00	C	1.23	40 %	4.4	
R06	Commissioning and Startup (CSU) Failure	CS00	C	4.70	40 %	16.8	

E.7 Recalculation of Risk Impacts for the Model SD5

Recalculation for the model SD5 is done the same way as for the model SD4, although information about probabilities and modified risk impacts should be taken from Table D.10.

Table E.3 evaluates values of risk impacts of the model N-SCRA-B recalculated for the model SD5. The same assumptions are used on the share of drawings and construction units that require full rework as in the case of the model SD4.

E.8 Recalculation of Risk Impacts for the Model SD4 with Compounding

The same risk impacts after interactions are applicable to the model SD4 with compounding as for the model SD4 (see Table E.2). The only difference in comparison with the model SD4 is timing of risk impacts (see Table E.1).

E.9 A Share of Engineering Errors and Omissions Resolved in Construction

When drawings with errors and omissions enter construction undetected, there are two scenarios related to handling of them upon detection. First, when detected errors and omissions could be resolved by the construction team at the construction site. Any amendments to constructed units should be reflected in 'as-built' drawings. This is usually the case when errors and omissions are not that significant. This option is shown as a link from 'Undiscovered Rework (EC)' to 'Work Remaining (C)' by Figure 16.1. Second, major errors and omissions do require engineering rework. This scenario is shown in Figure 16.1 by the link from 'Undiscovered Rework (EC)' to 'Work Remaining (E)'. It is established in the models SD4 and SD5 that 40 per cent of detected drawings may be resolved correctly in construction and 60 per cent require engineering rework.

E.10 The Workable Models SD4 and SD4

The workable models SD4 and SD5 as well as the model SD4 with compounding are available on the book's companion website at www.oup.co.uk/companion/Raydugin.

Glossary of Terms

Activity ranging: an early version of a Monte Carlo schedule risk analysis (SRA) that takes into account only general schedule uncertainties.

Adaptive system: a system that is capable of responding to changes in its internal or external environment.

Affinity of risks: interacting risks that form dynamic risk patterns supported by a project system.

Artificial neural network (ANN): an empirical inferential method of statistical data modelling with complex (non-linear) input–output relations that takes into account patterns in arrays of data by mimicking functions of the brain.

Bathtub model: a basic system dynamics model that describes a level of 'water' in a 'bathtub' (BT) depending on characteristics of the 'water' inflow and outflow.

Bias: a systematic error in a judgement, process, or a methodology.

Biased sampling: as opposed to random sampling, a sampling type when some members of an investigated population might be excluded from sampling.

Bowtie diagram (a standalone risk case): a risk identification and addressing framework that connects risk causes with a risk event and the risk event with impacted objectives.

Bowtie diagram (an interacting risks case): one of three risk identification and addressing frameworks that describes instances of three types of risk interactions—an internal risk amplification, a knock-on interaction, and a risk compounding.

Chaotic project: a project that results in abnormally high durations and/or costs due to overwhelming impact of the affinity of its interacting risks on schedule and/or cost objectives.

Conventional PRM methodology: a PRM methodology that follows the traditional PRM process steps to identify, assess, address, monitor, etc. risks.

Complete PSS–PDS match: a risk-free and dynamics-free state of a project system characterized by a full alignment of its two fundamental subsystems.

Complicated project: a project that consists of multiple parts that can be reduced to a number of simple non-interacting parts.

Complex projects: a project that consists of multiple parts that cannot be reduced to a number of simple parts due to intensive part's interactions.

Confidence interval: a range of parameter's values that should contain a statistical parameter of an investigated population with an assessed confidence or probability.

Confidence level: a parameter's value that should exceed a statistical parameter of an investigated population with an assessed confidence or probability.

Contingency: An amount of time or money added to project base duration or estimate, correspondingly, to allow for relevant general uncertainties and/or uncertain events.

Convenience sample: a type of biased sample that is easy to obtain and handle instead of obtaining and handling a random (unbiased) sample.

Cost risk analysis (CRA): a version of a Monte Carlo risk analysis that usually takes into account both general cost uncertainties and cost uncertain events although does not take into account schedule-driven costs.

Descriptive statistics: a branch of statistics that summarizes information collected about investigated samples although is not involved in predictive (inferential) activities to forecast statistical parameters of an investigated population.

Deterministic (scoring) PRM methodology: a conventional PRM methodology that uses scoring risk assessment matrices (RAM) in its risk assessment steps.

Dynamic (induced) project system issues: project system issues induced by standalone occurred risks or by interacting risks that is associated with corresponding dynamic PSS–PDS mismatches.

Dynamic PSS–PDS mismatch: a state of a project system characterized by dynamic misalignment of its two fundamental subsystems due to either occurrence of a standalone internally amplified risk or when occurred risks take part in cross-interactions.

Dynamic risk pattern: a pattern in an affinity of interacting project risks that is supported by a project system.

Feedback loop: a dynamic structure in a system that defines causal relations between system's stocks and rates.

Flow (or rate): a system dynamics parameter that measures and defines changes of a level (or a stock) over time.

General uncertainty: a representation of an inaccuracy of project durations or costs in a risk-free case through uncertainty distributions around deterministic duration or cost one-point values.

Hiding Hand principle: a concept that describes a special type of organizational and psychological bias stemming from risk ignorance or irrational risk acceptance in a hope that future circumstances and creativity would help negate negative risk consequences.

Inferential statistics: a branch of statistics that is aimed at predicting population's statistical parameters of interest through analysing corresponding parameters of available samples.

Internal risk (intra-risk) amplification: a non-linear amplification of consequences of a standalone risk due to presence of 'chronic' project system issues associated with static PSS–PDS mismatches.

Issue: a given or fact in a project system or its external environment that may serve as a cause of risks.

Joint confidence level (JCL): in decision-making, an additional condition applied to separately established schedule and cost confidence levels to ensure that both are obeyed simultaneously.

Judgement sample: a type of biased (non-random) preselected sample used to evaluate parameters of a population of interest.

Knock-on risk interaction: an impact of an up-the-line risk (usually a quality-related risk) on a down-the-line risk that can be realized with a delay.

Level (or stock): a system dynamics parameter representing an entity of interest that is accumulated or depleted over time due to corresponding inflows and outflows represented by flows or rates.

Line-item ranging: an early version of a Monte Carlo cost risk analysis (CRA) that takes into account only general cost uncertainties.

Management reserve: a confidential part of an overall project risk contingency that is not released to a project team.

Monte Carlo methodology (a linear case): an inferential simulation technique that is based on random sampling of parameters describing project general uncertainties and uncertain events.

Monte Carlo methodology (a non-linear case): an inferential simulation technique that is based on random sampling of parameters describing general uncertainties and interacting uncertain events.

Multiple failure mode: a mechanism to lead a complex project into chaos that is characterized by a one-peak probabilistic distribution when multiple interacting risks (an affinity of interacting risks) are responsible for a project failure.

Non-linearity parameters: parameters attached to individual instances of the risk interactions.

Non-linear multipliers: mathematical expressions that take into account non-linearity parameters to modify expected values of interacting risks.

Organizational bias: a systematic error in a process or a methodology stipulated by organizational procedures or stemming from factual practices.

Parametric method: an inferential multivariate-regression-analysis technique aimed at predicting population's parameters of interest through processing of corresponding parameters of available samples.

Political mathematics: a synonym to the project risk quantification referring to various realizations of subconscious and conscious bias in selecting of risk quantification methodologies, developing risk quantification results, and utilization of the results in decision-making.

Population: a set of alike tangible or intangible items of interest which statistical parameters are unknown.

Project (a simple projects case): a temporary endeavour undertaken to create a unique product or service.

Project (a complex projects case): a dynamic system accompanied by a phased process to deliver a new product, service or facility that is characterized by evolving phase-specific structure (the project structure subsystem (PSS)) matched by a required delivery organization (the project delivery subsystem (PDS)).

Project delivery subsystem (PDS): one of two fundamental interacting subsystems of a project system that defines project delivery through quality of applicable project development procedures and consistency of their implementation.

Project structure subsystem (PSS): one of two fundamental interacting subsystems of a project system that comprises project 'parts' associated with project deliverables and interactions among them.

Project risk management (PRM): one of the key project management disciplines that is aimed at identification, assessment, addressing, and monitoring of the project risk exposure.

PRM context: a top factor that shapes a PRM system in accordance with features and challenges of a given project based on organizational requirements.

PRM process: steps to identify, assess, address, monitor, etc. project risks and their interactions that shapes a PRM system in accordance with a PRM context.

Project rework cycle: a feedback loop in a project system that reflect on issues with work quality when a fraction of planned work is not accomplished correctly first time and should be redone.

PRM system: a system deployed by a project team to identify, assess, address, monitor, etc. project risk exposure that comprises a PRM context, a PRM process, and required PRM tools.

Project system: a system that comprises two fundamental interacting subsystems—a project structure subsystem (PSS) and a project delivery subsystem (PDS).

Project system issue: Any given or fact (permanent or induced) in a project system that results in either static or dynamic PSS–PDS mismatches associated with either project risks or their interactions.

Project system maturity: a project-level or discipline-level characterization that reflects on a degree of the PSS–PDS match, that is, a degree of a match between a level of project complexity and robustness of project delivery method.

Psychological bias: a systematic error in a judgement, process or methodology stemming from various realizations of subconscious (anchoring, lack of knowledge, etc.) and conscious (hidden agenda, the Hiding Hand, strategic misrepresentation, etc.) types of bias.

PSS–PDS mismatch: a static or dynamic misalignment between a project structure subsystem (PSS) and a project delivery subsystem (PDS) associated with static ('chronic') or dynamic (induced) project system issues.

Random error: unsystematic no-pattern variation of a statistical parameter that reflects on a precision of an inferential methodology and which is measured by a standard deviation of a resulting probabilistic distribution.

Random sampling: the unbiased sampling when any member of a population of interest can be randomly selected for inferential purposes.

Rate (or flow): a system dynamics parameter that measures and defines changes of a stock (or a level) over time.

Revealing Hand principle: as opposed to the Hiding Hand principle, an approach that allows various realizations of bias and systematic errors in risk quantification and decision-making to be suppressed.

Risk (a simple projects case): a possible impact on project objectives.

Risk (a complex projects case): a possible aggregated impact on project objectives by an affinity of interacting risks that form a dynamic risk pattern supported by a project system.

Risk addressing: actions and controls to avoid, mitigate, transfer, or accept identified risks and their interactions.

Risk assessment matrix (RAM): a framework that combines risk occurrence scores with scores of impacts on selected project objectives in order to assess and rank risk severities.

Risk compounding: a non-linear phenomenon in a project system when risks occur simultaneously resulting in unproportionally high risk consequences.

Risk interaction: an instance of a non-linear intra-risk amplification or a cross-risk action (knock-on or compounding) resulting in unproportionally high risk consequences.

Risk interaction addressing: actions and controls to reduce or disrupt risk interactions.

Risk register: a document that collects detailed information about identified risks, their interactions, and risk and risk-interaction addressing.

Sample: a population's subset which statistical parameters are known that is used to infer unknown population's parameters.

Simple projects: linear functional projects that have few non-interacting parts and do not support cross-risk interactions although may possess intra-risk interactions.

Schedule-driven cost: additional impact on the project cost objective resulting from schedule delays.

Schedule risk analysis (SRA): a version of Monte Carlo risk analysis that usually takes into account both general schedule uncertainties and schedule uncertain events.

Single failure mode: a mechanism to throw a complex project into chaos that is characterized by a two-peak probabilistic distribution when a single risk (or few risks) exposed to interactions with other risks becomes highly sensitive and responsible for a project failure.

Static PSS–PDS match: a state of a project system characterized by the absence of chronic issues although does not exclude the dynamic PSS–PDS mismatch when project risks occur and interact.

Static PSS–PDS mismatch: a state of a project system characterized by static (long-term) misalignment of its two fundamental subsystems due to possession of chronic issues that additionally obstruct project delivery through internal risk amplifications.

Statistical bias: a systematic error of an inferential methodology due to using of biased samples.

Stock (or level): a system dynamics parameter representing an entity of interest that is accumulated or depleted over time due to its inflows and outflows which are represented by rates or flows.

Stretched target: project duration or cost established as a target for a project team that has relatively low confidence level.

System dynamics: a concept and a methodology aimed at understanding and modelling the non-linear behaviour of complex systems.

Systematic error: inaccuracy inherent to an inferential methodology resulting in its inability to produce accurate evaluations of parameters describing a population of interest.

Schedule and cost risk analysis (SCRA): an advanced version of Monte Carlo risk analysis that integrates CRA and SRA models and allows schedule-driven costs to be taken into account consistently.

Static (chronic) project system issues: long-term issues either in a project structure subsystem (PSS) or in a project delivery subsystem (PDS) that give rise to static PSS–PDS mismatches and result in internal amplifications of corresponding risks.

Uncertain event: a future risk event that, if it occurs, can make an impact on project objectives.

Unconventional PRM methodology: a modelling methodology that is not based on steps of a standard PRM process to identify, address, monitor, etc. risks being aimed only at assessments of overall project durations or costs.

Work constraint: a project condition resulting from risk occurrence that is characterized by a lack of required information, labour, materials, equipment, or permits making fulfilment of planned works impossible.

References

AACE. 2008. *AACE International Recommended Practice 40R-08: Contingency Estimating–General Principles*. Morgantown, WV: AACE International.

AACE. 2011. *AACE International Recommended Practice 42R-08: Risk Analysis and Contingency Determination Using Parametric Estimating*. Morgantown, WV: AACE International.

AACE. 2012. *AACE International Recommended Practice 10S-90: Cost Engineering Terminology*. Morganton, WV: AACE International.

AACE. 2014. *AACE International Recommended Practice 34R-05: Basis of Estimate*. Morganton, WV: AACE International.

AACE. 2015a. *AACE International Recommended Practice 87R-14: Cost Estimate Classification System—As Applied for the Petroleum Explorations and Production Industry*. Morgantown, WV: AACE International.

AACE. 2015b. *AACE International Recommended Practice 35R-09: Development of Cost Estimate Plans—As Applied for the Building and General Construction Industries*. Morgantown, WV: AACE International.

Baccarini, D. 2005. *Estimating Project Cost Contingency—Beyond the 10% Syndrome*. Proceedings of National Conference of Australian Institute of Project Management. Victoria, Australia: Australian Institute of Project Management.

Boyd, W. 1998. *Armadillo*. Toronto, ON: Penguin Group.

Braunschweig, B., and Huot J. 1984. *A System Dynamic Project Management Model*. Transactions of the Annual 1984 AACE Conference (D.2). Morganton, VW: AACE International.

Chapman, R. 2017. 'System Dynamics for Project Risk Management', in Y. Raydugin (ed) *Handbook of Research on Leveraging Risk and Uncertainties for Effective Project Management*, Hershey, PA: IGI Global.

Chen, D., and Hartman, F. 2000. *A Neural Network Approach to Risk Assessment and Contingency Allocation*. Transactions of the 2000 Annual AACE Conference (Risk.07). Morganton, VW: AACE International.

CII. 2013. *Integrated Project Risk Assessment (Implementation Resource 181–2)*, Austin, TX: CII.

CII. 1996. *PDRI: Project Definition Rating Index—Industrial Projects (Implementation Resource 113-2)*, Austin, TX: CII.

Chao, D., and Skibniewski, M. 1994. 'Estimating Construction Productivity: Neural-network-based Approach', *Journal of Computing in Civil Engineering*, 8, 2: 234–51.

Cooper, K. 1980. 'Naval Ship Production: a Claim Settled and a Framework Built'. *INFORMS Journal of Applied Analytics*, 10, 6: 20–36.

Cooper, K. 2004. *Toward a Unifying Theory for Compounding and Cumulative Impacts of Project Risks and Changes*, Proceedings of the PMI Research Conference: Innovations, London, England, Newtown Square, PA: PMI.

Cooper. K., and Lee G. 2009. *Managing the Dynamics of Projects and Changes at Fluor*. Proceedings of the 27th International Conference of the Systems Dynamics Society, Albuquerque, New Mexico: Systems Dynamics Society.

Coyle, R. 1996. *System Dynamics Modeling: A Practical Approach*, London, UK: Chapman & Hall.

Curran, M. 1989. 'Range Estimating: Measuring Uncertainty and Reasoning with Risk', *Cost Engineering*, 31, 3.

Doane, D., Seward, L., Aneja, Y., and Miller, P. 2010. *Applied Statistics in Business and Economics*. Whitby, ON: McGraw-Hill Ryerson.

DoD (2011), *Technology Readiness Assessment Guidance*. Washington, DC: US Department of Defense.

DoE. 2008. *Technology Readiness Assessment Guide*. Washington, DC: US Department of Energy.

Ebeling, C. 1997. *An Introduction to Reliability and Maintainability Engineering*, Long Grove, IL: Waveland Press.

Eisberg, R., and Resnik, R. 1985. *Quantum Physics of Atoms, Molecules, Solids, Nuclei, and Particles*. New York, NY: John Wiley & Sons.

Ellenberg, J. 2014. *How Not to Be Wrong: The Power of Mathematical Thinking*. New York, NY: Penguin Press.

Feynman, R. 1997. *Surely You're Joking, Mr. Feynman! (Adventure of a Curious Character)*. New York, NY: W.W. Norton.

Ferrari, T. 1978. *Elements of System-dynamics Simulation: A Text Book with Exercises*. Wageningen, The Netherlands: Center for Agricultural Publishing and Documentation.

Flyvbjerg, B. 2011. 'Over Budget, Over Time, Over and Over Again: Managing Major Projects', in P. Morris, J. Pinto, and J. Soderlund (eds), *The Oxford Handbook of Project Management*, Oxford, UK: Oxford University Press.

Flyvbjerg, B. 2017. 'Introduction: The Iron Law of Megaproject Management', in B. Flyvbjerg (ed), *The Oxford Handbook Megaproject Management*, Oxford, UK: Oxford University Press.

Forrester, J. 1961. *Industrial Dynamics*. Cambridge, MA: MIT Press.

Forrester, T. J. 1968. *Principles of Systems*. Cambridge, MA: Wright-Allen Press.

Galway, L. 2004. *Quantitative Risk Analysis for Project Management. A Critical Review (Working Paper WR-112-RC)*. Santa Monica, CA: RAND Corporation.

GAO. 2015. *Schedule Assessment Guide GAO-16-89G. Best Practices for Project Schedules*. Washington, DC: U.S Government Accountability Office.

GAO. 2016. *Technology Readiness Assessment Guide GAO-16-410G. Best Practices for Evaluating the Readiness of Technology for Use in Acquisition Programs and Projects*. Washington, DC: U.S Government Accountability Office.

Garvey, P., Book, S., and Covert, R. 2016. *Probability Methods for Cost Uncertainty Analysis: A System Engineering Perspective*. Boca Raton, FL: Taylor & Francis Group.

Genidy, M. 2017. 'Multiple non-linear regression model for the maximum number of migratory bird types during migration years'. *Communications in Statistics Theory and Methods*, 46, 16: 7969–75.

George, E. 2015. *A Banquet of Consequences: A Lynley Novel*, New York, NY: Viking.

Goodchild, R. 2018. *Proven by Computer? System Dynamics and Disruption Claims. A Publication 212 by the Society of Construction Law*. London, UK: The Society of Construction Law.

Gray, M., and Shahidi, A. 2011. *Applying the Principles of System Dynamics to Project Risk Management or "the Domino" Effect, Proceeding of the PMI Global Congress, North America*. Dallas, Newtown Square, PA: PMI

Gull, S., and Williams, T. 2017. 'Project Complexity for the Uninitiated', in Y. Raydugin (ed) *Handbook of Research on Leveraging Risk and Uncertainties for Effective Project Management*, Hershey, PA: IGI Global.

Hass, K. 2011. *The Enterprise Business Analysis: Developing Creative Solutions to Complex Business Problems*. Oakland, CA: Berrett-Koehler Publishers.

Hass, K. 2017. 'Living on the Edge: Managing Project Complexity', in Y. Raydugin (ed) *Handbook of Research on Leveraging Risk and Uncertainties for Effective Project Management*, Hershey, PA: IGI Global.

Hackney, J. 1965. *Control and Management of Capital Projects*. Hoboken, NY: John Wiley & Sons.

Hines, J. 1996. *Molecules of Structure: Building Blocks for System Dynamics Models (Version 1.4)*. Cambridge, MA: Ventana Systems.

Hollmann, J. 2015. *Risk Analysis at the Edge of Chaos, Transactions of the 2015 Annual AACE Conference (Risk-1584)*. Morganton, VW: AACE International.

Hollmann, J. 2016. *Project Risk Quantification: A Practitioner's Guide to Realistic Cost and Schedule Risk Management*. Sugar Land, TX: Probabilistic Publishing.

Hollmann, J. 2017a. 'Systemic Risks and Parametric Modeling', in Y. Raydugin (ed) *Handbook of Research on Leveraging Risk and Uncertainties for Effective Project Management*, Hershey, PA: IGI Global.

Hollmann, J. 2017b 'Complexity Risk and Modeling Disorder', in Y. Raydugin (ed) *Handbook of Research on Leveraging Risk and Uncertainties for Effective Project Management*, Hershey, PA: IGI Global.

Hulett, D. 2009. *Practical Schedule Risk Analysis*, Aldershot, UK: Gower Publishing.

Hulett, D. 2011. *Integrated Cost-Schedule Risk Analysis*, Aldershot, UK: Gower Publishing.

Hulett, D. 2017. 'Monte Carlo Simulation for Integrated Cost-Schedule Risk Analysis: Concepts, Methods, and Tools for Risk Analysis and Mitigation', in Y. Raydugin (ed) *Handbook of Research on Leveraging Risk and Uncertainties for Effective Project Management*, Hershey, PA: IGI Global.

Hulett, D. 2019. 'Journey Map to a More Mature Schedule Risk Analysis (SRA) process', *Cost Engineering—The Journal of AACE International*, March/April issue: 8–17.

Hulett, D., and Whitehead, W. 2016, *The Monte Carlo Method for Modeling and Mitigating Systemic Risks. Transactions of the 2016 Annual AACE Conference (Risk-2142).* Morganton, VW: AACE International.

Ibbs., W., and McEniry, G. 2008. 'Evaluating the Cumulative Impact of Changes on Labor Productivity - an Evolving Discussion', *Cost Engineering*, 50, 12: 23–9.

ISO. 2018. *ISO 31000 International Standard: Risk Management—Guidelines*, Geneva, Switzerland: ISO.

Kreyszik, E. 1999. *Advanced Engineering Mathematics.* Hoboken, NY: John Wiley & Sons.

La Vega, J. 1957. *Confusion of Confusions*, Cambridge, MA: Harvard University Printing Office.

Lhee, S., Flood I., and Issa, R. 2014. 'Development of a Two-step Neural Network-based Model to Predict Construction Cost Contingency', *Journal of Informational Technology in Construction*, 19: 399–411.

Lorenz, E. 1963. 'Deterministic Nonperiodic Flow', *Journal of the Atmospheric Science*, 20, 2: 130–41.

Lyneis, J., and Ford, D. 2007. 'System Dynamics Applied to Project Management: a Survey, Assessment, and Directions for Future Research', *System Dynamics Review*, 23, 2/3: 157–89.

Mac Guinness, M. 2019. *Disruption Quantification in Small and Medium EPC Projects: A System Dynamics Model. Transactions of the 2019 Annual AACE Conference (CSC-3124).* Morganton, VW: AACE International.

Machiavelli, N. 1996. *Discourses on Livy (Book 3).* Chicago, IL: The University of Chicago Press.

Mankins, J. 1995. *Technology Readiness Level (a NASA white paper).* Washington, DC: NASA.

Merrow, E. 2011. *Industrial Megaprojects: Concepts, Strategies, and Practices for Success.* Hoboken, NJ: John Wiley & Sons, Inc.

Merrow, E., Phillips. K., and Myers, C. 1981. *Understanding Cost Growth and Performance Shortfalls in Pioneer Process Plants' (Working Paper R-2569-DOE).* Santa Monica, CA: RAND Corporation.

Murtaza, M., and Fisher, D. 1994. 'A Neural Network Model for Decision Making with Application in Construction Management', *Journal of International Information Management*, 3, 2: 27–38.

Nabokov, V. 1989. *Pale Fire*, New York, NY: Random House.

NASA. 2015. *NASA Cost Estimating Handbook (CEH) (Version 4.0).* Washington, DC: NASA.

National Post. 2018. '"Optimism Bias" Common Cause of Cost Overrun, Expert Says at Public Inquiry Over $12.7B Muskrat Falls Dam' (available at https://nationalpost.com/news/canada/optimism-bias-common-cause-of-cost-overrun-project-expert-on-muskrat-falls. Last visited 7-Aug-2019.)

Palisade Corporation. 2015. *Neural Tools: Neural Network Add-in for Microsoft Excel, Version 7. A NeuralTools User Manual.* Ithaca, NY: Palisade Corporation.

Patterson, D. 2018. 'How AI Is Changing Art & Science of CPM Scheduling', a presentation at InEight company web-site https://ineight.com/wp-content/uploads/2018/11/How-AI-is-Changing-the-Art-and-Science-of-CPM-Scheduling.pdf, last time accessed 5-Aug-2019.

PMI, 2009. *Practice Standard for Project Risk Management*, Newtown Square, PA: PMI

PMI, 2017. *A Guide to the Project Management Body of Knowledge (PMBOK Guide)*, Newtown Square, PA: PMI

Pushkin, A. 2017. *A Complete One-volume Collection.* Moscow, Russia: Alfa-Kniga.

Rassenfoss, S. 2017. 'The Future of Oil Sands Depends on One Thing: Totally Rethinking Everything', *Journal of Petroleum Technology*, 69, 4: 36–42.

Raydugin, Y. 2012. 'Consistent Application of Risk Management for Selection of Engineering Design Options in Mega-projects'. *International Journal of Risk and Contingency Management*, 1, 4: 44–55.

Raydugin, Y. 2013. *Project Risk Management: Essential Methods for Project Teams and Decision Makers.* Hoboken, NY: John Wiley & Sons.

Raydugin, Y. 2014. 'Holistic View on Unknown Unknown in project Risk management', in M. Tavana (ed) *Developing Business Strategies and Identifying Risk Factors in Modern Organizations*, Hershey, PA: IGI Global.

Raydugin, Y. 2017a. 'Project Risk Management: Popular Fallacies and Overlooked Best Practices', in Y. Raydugin (ed) *Handbook of Research on Leveraging Risk and Uncertainties for Effective Project Management*, Hershey, PA: IGI Global.

Raydugin, Y. 2017b. 'Integrated Risk-Based and Economic-Based (IRBEB) Methodology for Selection of Project Alternatives', in Y. Raydugin (ed) *Handbook of Research on Leveraging Risk and Uncertainties for Effective Project Management*, Hershey, PA: IGI Global.

Raydugin, Y. 2017c. *Role of Project Novelty, Execution and Bias in Risk Quantification. Transactions of the 2017 Annual AACE Conference (Risk-2512).* Morganton, VW: AACE International.

Raydugin, Y. 2018a. *Non-linear Probabilistic Modeling of Systemic Risks. Transactions of the 2018 Annual AACE Conference (Risk-2808).* Morganton, VW: AACE International.

Raydugin, Y. 2018b. 'Adequate Quantification of Project Cost Risks: Introduction of Non-linear Probabilistic (Monte Carlo) Technique', *International Journal of Risk and Contingency Management*, 7, 4: 1–20.

Sarcia, S., Cantone G., and Basili, V. 2007. 'A Statistical Neural Network Framework for Risk Management Process: From the Proposal to Its Preliminary Validation for Efficiency', *Proceedings of the 2nd International Conference on Software and Data Technologies (ICSOFT), Barcelona, Spain*: 168–77.

Savage, S. 2009. *The Flaw of Averages: Why We Underestimate Risk in The Face of Uncertainty*, Hoboken, NY: John Wiley & Sons.

Shenhar, A., Dvir, D., Lehler, T., and Poli, M. 2002. *One Size Does Not Fit All—True for Projects, True for Frameworks. Proceeding of the PMI Research Conference, Seattle, WA.* Newtown Square, PA: PMI.

Sommer, T. 2001. 'Suppression of Scientific Research: *Bahramdipity* and *Nulltiple Scientific Discoveries*', *Science and Engineering Ethics*, 7, 1: 77–104.

Spiegel, M., Stephens, L. 1999. *Theory and Problems of Statistics.* New York, NY: McGraw Hill.

Stacey, R., Griffin, D., and Shaw, P. 2000. *Complexity and Management: Fad or Radical Challenge to System Thinking?* London, UK: Routledge.

Stepanov, S., and Sheindlin, M. 2017. 'Statistical Analysis of Measurement Results in Multiwavelength Pyrometry', *High Temperature*, 55, 5: 802–7.

Sterman, J. 2000. *Business Dynamics: System Thinking and Modeling for a Complex World*, New York, NY: McGraw-Hill.

Strogatz, S. 2015. *Nonlinear Dynamics and Chaos with Applications to Physics, Biology, Chemistry, and Engineering.* Boulder, CO: Westview Press.

Teknobuild. 2019. http://www.teknobuild.com/?page_id=1120. Last visited 7 August 2019.

Telegram. 2017. 'N.L.'s NDP Backs Forensic Audit of Muskrat Falls', 10 May 2017 (available at https://www.thetelegram.com/news/local/nls-ndp-backs-forensic-audit-of-muskrat-falls-25122/). Last visited 7 August 2019.

Tolstoy, L. 2016. *Anna Karenina.* Oxford, UK: Oxford University Press.

Vanhoucke, M. 2013. *Project Management with Dynamic Scheduling: Baseline Scheduling, Risk Analysis and Project Controls.* Berlin, Germany: Springer-Verlag.

Vargas, R. 2015. *Applying Neural Networks and Analogous Estimating to Determine the Project Budget. Proceeding of the PMI Global Congress, North America, FL.* Newtown Square, PA: PMI.

Verhulst, P. 1845. 'Recherches mathèmatiques sur la loi d'accroissement de la population', *Nouv. M'em. Acad. R. Sci. B.-lett. Brux.* 18, 1–45.

Volden, G., and Samset, K. 2017. 'Quality Assurance in Megaproject Management: A Norwegian Way' in B. Flyvbjerg (ed), *The Oxford Handbook Megaproject Management*, Oxford, UK: Oxford University Press.

Waziri, B., Bala, K., and Bustani, S. 2017. 'Artificial Neural Networks in Construction Engineering and Management', *International Journal of Architecture, Engineering and Construction*, 6, 1: 50–60.

Westney, R., and Dahl, J. 2019. 'Enhanced Front-end Loading Aims for Assurance in Project Planning', *Oil & Gas*, 117, 6: 18–24.

Whetton, R., Zhao, Y., Shaddad, S., and Mouazen, A. 2017. 'Nonlinear parametric modeling study how soil properties affect crop yields and NDVI'. *Computers and Electronics in Agriculture*, 138: 127–36.

Williams, T. 2002. *Modeling Complex Projects*. Chichester, UK: John Wiley & Sons.

Williams, T. 2004. 'Why Monte-Carlo Simulations of Project Networks Can Mislead', *Project Management Journal*, 35, 3: 53–61.

Williams, T., Eden, C., Ackermann, F., and Tait, A. 1995. 'The Effect of Design Changes and Delays on Project Costs', *Journal of the Operational Research Society*, 46, 7: 809–18.

Index